STUDENT'S SOLUTIONS MAN

PREALGEBRA

SECOND EDITION

STUDENT'S SOLUTIONS MANUAL

PREALGEBRA

SECOND EDITION

Marvin L. Bittinger
Indiana University - Purdue University at Indianapolis

David J. Ellenbogen
St. Michael's College

Judith A. Penna

ADDISON-WESLEY PUBLISHING COMPANY
Reading, Massachusetts • Menlo Park, California • New York
Don Mills, Ontario • Wokingham, England • Amsterdam • Bonn
Sydney • Singapore • Tokyo • Madrid • San Juan • Milan • Paris

Reproduced by Addison-Wesley from camera ready copy supplied by the author.

ISBN 0-201-85525-9
1 2 3 4 5 6 7 8 9 10- ML-99989796

Table of Contents

Thanks are extended to
Patsy Hammond, Mike Penna, and Pam Smith
for their help in the preparation of this manual.

STUDENT'S SOLUTIONS MANUAL

PREALGEBRA

SECOND EDITION

Chapter 1

Operations on the Whole Numbers

Exercise Set 1.1

1. 5742 = 5 thousands + 7 hundreds + 4 tens + 2 ones

3. 27,342 = 2 ten thousands + 7 thousands + 3 hundreds + 4 tens + 2 ones

5. 5609 = 5 thousands + 6 hundreds + 0 tens + 9 ones, or 5 thousands + 6 hundreds + 9 ones

7. 2300 = 2 thousands + 3 hundreds + 0 tens + 0 ones, or 2 thousands + 3 hundreds

9. 2 thousands + 4 hundreds + 7 tens + 5 ones = 2475

11. 6 ten thousands + 8 thousands + 9 hundreds + 3 tens + 9 ones = 68,939

13. 7 thousands + 3 hundreds + 0 tens + 4 ones = 7304

15. 1 thousand + 9 ones = 1009

17. 85 = Eighty-five

19. 88,000

Eighty-eight thousand

21. 123,765

One hundred twenty-three thousand,
seven hundred sixty-five

23. 7, 754, 211, 577

Seven billion,
seven hundred fifty-four million,
two hundred eleven thousand,
five hundred seventy-seven

25. 6, 469, 952

Six million,
four hundred sixty-nine thousand,
nine hundred fifty-two

27. 1, 954, 116

One million,
nine hundred fifty-four thousand,
one hundred sixteen

29. Two million,
two hundred thirty-three thousand,
eight hundred twelve

Standard notation is 2, 233, 812.

31. Eight billion

Standard notation is 8,000,000,000.

33. One billion,
one hundred eighty-seven thousand,
five hundred forty-two

Standard notation is 1,000, 187,542.

35. Two hundred six million,
six hundred fifty-eight thousand

Standard notation is 206, 658,000.

37. 2 3 [5] , 8 8 8

The digit 5 means 5 thousands.

39. 4 8 8, [5] 2 6

The digit 5 means 5 hundreds.

41. 8 9, [3] 0 2

The digit 3 tells the number of hundreds.

43. 8 9, 3 [0] 2

The digit 0 tells the number of tens.

45.

47. All digits are 9's. Answers may vary. For an 8-digit readout, for example, it would be 99,999,999. This number has three periods.

49. Nine billion, six million, seven hundred forty-three thousand, fifty-eight

9, ___ 0 ___ ,7 4 ___ , ___ 5 8

There are no hundred millions, so the first missing digit is 0.

There are 6 millions, so the second missing digit is 6.

There are 3 thousands, so the third missing digit is 3.

There are no hundreds, so the last missing digit is 0.

We write 9,006,743,058.

Exercise Set 1.2

1.

They rent 3 videos one week.		They rent 6 videos the next week.		They rent 9 videos in all.
3 videos	+	6 videos	=	9 videos

3.

Earns \$23 one day.		Earns \$31 the next day.		The total earned is \$54.
\$23	+	\$31	=	\$54

5. 325 ft + 325 ft + 325 ft + 325 ft = 1300 ft

7. $\begin{array}{r} 364 \\ +\ 23 \\ \hline 387 \end{array}$ Add ones, add tens, then add hundreds.

9. $\begin{array}{r} 1 \\ 1716 \\ +3482 \\ \hline 5198 \end{array}$ Add ones: We get 8. Add tens: We get 9 tens. Add hundreds: We get 11 hundreds, or 1 thousand + 1 hundred. Write 1 in the hundreds column and 1 above the thousands. Add thousands: We get 5 thousands.

11. $\begin{array}{r} 1 \\ 86 \\ +78 \\ \hline 164 \end{array}$ Add ones: We get 14, or 1 ten + 4 ones. Write 4 in the ones column and 1 above the tens. Add tens: We get 16 tens.

13. $\begin{array}{r} 1 \\ 99 \\ +\ 1 \\ \hline 100 \end{array}$ Add ones: We get 10 ones, or 1 ten + 0 ones. Write 0 in the ones column and 1 above the tens. Add tens: We get 10 tens.

15. $\begin{array}{r} 11 \\ 789 \\ +111 \\ \hline 900 \end{array}$ Add ones: We get 10 ones, or 1 ten + 0 ones. Write 0 in the ones column and 1 above the tens. Add tens: We get 10 tens, or 1 hundred + 0 tens. Write 0 in the tens column and 1 above the hundreds. Add hundreds: We get 9 hundreds.

17. $\begin{array}{r} 1 \\ 909 \\ +101 \\ \hline 1010 \end{array}$ Add ones: We get 10. Write 0 in the ones column and 1 above the tens. Add tens: We get 1 ten. Add hundreds: We get 10 hundreds.

19. $\begin{array}{r} 1 \\ 811 \\ +390 \\ \hline 1201 \end{array}$ Add ones: We get 1. Add tens: We get 10 tens. Write 0 in the tens column and 1 above the hundreds. Add hundreds: We get 12 hundreds.

21. $\begin{array}{r} 1 \\ 356 \\ +491 \\ \hline 847 \end{array}$ Add ones: We get 7. Add tens: We get 14 tens. Write 4 in the tens column and 1 above the hundreds. Add hundreds: We get 8 hundreds.

23. $\begin{array}{r} 11 \\ 5093 \\ +3217 \\ \hline 8310 \end{array}$ Add ones: We get 10. Write 0 in the ones column and 1 above the tens. Add tens: We get 11. Write 1 in the tens column and 1 above the hundreds. Add hundreds: We get 3 hundreds. Add thousands: We get 8 thousands.

25. $\begin{array}{r} 11 \\ 4825 \\ +1783 \\ \hline 6608 \end{array}$ Add ones: We get 8. Add tens: We get 10 tens. Write 0 in the tens column and 1 above the hundreds. Add hundreds: We get 16 hundreds. Write 6 in the hundreds column and 1 above the thousands. Add thousands: We get 6 thousands.

27. $\begin{array}{r} 111 \\ 9999 \\ +6785 \\ \hline 16,784 \end{array}$ Add ones: We get 14. Write 4 in the ones column and 1 above the tens. Add tens: We get 18 tens. Write 8 in the tens column and 1 above the hundreds. Add hundreds: We get 17 hundreds. Write 7 in the hundreds column and 1 above the thousands. Add thousands: We get 16 thousands.

29. $\begin{array}{r} 1\ 11 \\ 23,443 \\ +10,989 \\ \hline 34,432 \end{array}$ Add ones: We get 12. Write 2 in the ones column and 1 above the tens. Add tens: We get 13 tens. Write 3 in the tens column and 1 above the hundreds. Add hundreds: We get 14 hundreds. Write 4 in the hundreds column and 1 above the thousands. Add thousands: We get 4 thousands. Add ten thousands: We get 3 ten thousands.

31. $\begin{array}{r} 1\ 11\ 1 \\ 77,543 \\ +23,767 \\ \hline 101,310 \end{array}$ Add ones: We get 10. Write 0 in the ones column and 1 above the tens. Add tens: We get 11 tens. Write 1 in the tens column and 1 above the hundreds. Add hundreds: We get 13 hundreds. Write 3 in the hundreds column and 1 above the thousands. Add thousands: We get 11 thousands. Write 1 in the thousands column and 1 above the ten thousands. Add ten thousands: We get 10 ten thousands.

33. $\begin{array}{r} 1\ 11\ 1 \\ 99,999 \\ +\quad 112 \\ \hline 100,111 \end{array}$ Add ones: We get 11. Write 1 in the ones column and 1 above the tens. Add tens: We get 11 tens. Write 1 in the tens column and 1 above the hundreds. Add hundreds: We get 11 hundreds. Write 1 in the hundreds column and 1 above the thousands. Add thousands: We get 10 thousands. Write 0 in the thousands column and 1 above the ten thousands. Add ten thousands: We get 10 ten thousands.

35. Add from the top.

We first add 7 and 9, getting 16; then 16 and 4, getting 20; then 20 and 8, getting 28.

$$\begin{array}{r} \boxed{\begin{array}{r} 7 \\ 9 \end{array}} \to \boxed{\begin{array}{r} 16 \\ 4 \end{array}} \to \boxed{\begin{array}{r} 20 \\ 8 \end{array}} \to 28 \\ 4 \\ +\ 8 \\ \hline 28 \end{array}$$

Check by adding from the bottom.

We first add 8 and 4, getting 12; then 12 and 9, getting 21; then 21 and 7, getting 28.

$$\begin{array}{r} 7 \\ 9 \\ 4 \\ +\ 8 \\ \hline 28 \end{array}$$

$\boxed{4 \atop 8} \to \boxed{9 \atop 12} \to \boxed{7 \atop 21} \to 28$

37. Add from the top.

$$\begin{array}{r} 4 \\ 3 \\ 9 \\ 1 \\ +\ 8 \\ \hline 28 \end{array}$$

$\boxed{4 \atop 3} \to \boxed{7 \atop 9} \to \boxed{16 \atop 1} \to \boxed{17 \atop 8} \to 25$

Check:

$$\begin{array}{r} 4 \\ 3 \\ 9 \\ 1 \\ +\ 8 \\ \hline 25 \end{array}$$

$\boxed{1 \atop 8} \to \boxed{9 \atop 9} \to \boxed{3 \atop 18} \to \boxed{4 \atop 21} \to 25$

39. Add from the top.

$$\begin{array}{r} 9 \\ 4 \\ 7 \\ 8 \\ +\ 7 \\ \hline 35 \end{array}$$

$\boxed{9 \atop 4} \to \boxed{13 \atop 7} \to \boxed{20 \atop 8} \to \boxed{28 \atop 7} \to 35$

Check:

$$\begin{array}{r} 9 \\ 4 \\ 7 \\ 8 \\ +\ 7 \\ \hline 35 \end{array}$$

$\boxed{8 \atop 7} \to \boxed{7 \atop 15} \to \boxed{4 \atop 22} \to \boxed{9 \atop 26} \to 35$

41. We look for pairs of numbers whose sums are 10, 20, 30, and so on.

23	→	23
16	→	16
11	→	30
18	→	18
+19	↗	
87		87

43. We look for pairs of numbers whose sums are 10, 20, 30, and so on.

45	→	70
25	↗	
36	→	80
44	↗	
+80	→	80
230		230

45.

$$\begin{array}{r} {}^{1} \\ 2\ 3 \\ 6\ 2 \\ +\ 4\ 5 \\ \hline 1\ 3\ 0 \end{array}$$

Add ones: We get 10. Write 0 in the ones column and 1 above the tens. Add tens: We get 13 tens.

47.

$$\begin{array}{r} 5\ 1 \\ 3\ 6 \\ +\ 6\ 2 \\ \hline 1\ 4\ 9 \end{array}$$

Add ones, then add tens.

49.

$$\begin{array}{r} 2\ 6 \\ 8\ 2 \\ +\ 6\ 1 \\ \hline 1\ 6\ 9 \end{array}$$

Add ones, then add tens.

51.

$$\begin{array}{r} 1\ 1\ \\ 2\ 0\ 7 \\ 2\ 9\ 5 \\ +\ 3\ 4\ 0 \\ \hline 8\ 4\ 2 \end{array}$$

Add ones: We get 12. Write 2 in the ones column and 1 above the tens. Add tens: We get 14 tens. Write 4 in the tens column and 1 above the hundreds. Add hundreds: We get 8 hundreds.

53.

$$\begin{array}{r} 2\ 4\ \\ 3\ 2\ 7 \\ 4\ 2\ 8 \\ 5\ 6\ 9 \\ 7\ 8\ 7 \\ +\ 2\ 0\ 9 \\ \hline 2\ 3\ 2\ 0 \end{array}$$

Add ones: We get 40. Write 0 in the ones column and 4 above the tens. Add tens: We get 22 tens. Write 2 in the tens column and 2 above the hundreds. Add hundreds: We get 23 hundreds.

55.

$$\begin{array}{r} 1\ 1\ 1\ \\ 2\ 0\ 3\ 7 \\ 4\ 9\ 2\ 3 \\ 3\ 4\ 7\ 1 \\ +\ 1\ 2\ 4\ 8 \\ \hline 1\ 1,6\ 7\ 9 \end{array}$$

Add ones: We get 19. Write 9 in the ones column and 1 above the tens. Add tens: We get 17 tens. Write 7 in the tens column and 1 above the hundreds. Add hundreds: We get 16 hundreds. Write 6 in the hundreds column and 1 above the thousands. Add thousands: We get 11 thousands.

57.

$$\begin{array}{r} 1\ \ 1\ \\ 3\ 4\ 2\ 0 \\ 8\ 7\ 1\ 9 \\ 4\ 3\ 1\ 2 \\ +\ 6\ 2\ 0\ 3 \\ \hline 2\ 2,6\ 5\ 4 \end{array}$$

Add ones: We get 14. Write 4 in the ones column and 1 above the tens. Add tens: We get 5 tens. Add hundreds: We get 16 hundreds. Write 6 in the hundreds column and 1 above the thousands. Add thousands: We get 22 thousands.

59.

$$\begin{array}{r} 2\ 2\ 3\ \ 3\ 1\ 1\ \\ 5,6\ 7\ 8,9\ 8\ 7 \\ 1,4\ 0\ 9,3\ 1\ 2 \\ 8\ 9\ 8,8\ 8\ 8 \\ +\ 4,7\ 7\ 7,9\ 1\ 0 \\ \hline 1\ 2,7\ 6\ 5,0\ 9\ 7 \end{array}$$

61. 7000 + 900 + 90 + 2 = 7 thousands + 9 hundreds + 9 tens + 2 ones = 7992

63. 4 $\boxed{8}$ 6, 2 0 5

The digit 8 tells the number of ten thousands.

65.

67. 5,987,943+328,959+49,738,765

Using a calculator to carry out the addition, we find that the sum is 56,055,667.

69. Pair 1 with 99, 2 with 98, and so on through 49 with 51. Then we have 49 pairs, each of which totals 100, and the numbers 50 and 100 which are not paired with other numbers. Now $49 \cdot 100 = 4900$, so the total is $4900 + 50 + 100 = 5050$.

Exercise Set 1.3

1. $\begin{pmatrix}\text{Number of}\\ \text{gallons to}\\ \text{start with}\end{pmatrix} \begin{pmatrix}\text{Number of}\\ \text{gallons sold}\end{pmatrix} \begin{pmatrix}\text{Number of}\\ \text{gallons left}\end{pmatrix}$

2400 gal − 800 gal = $\square$

3. $7 - 4 = 3$

This number gets added (after 3).

$7 = 3 + 4$

(By the commutative law of addition, $7 = 4 + 3$ is also correct.)

5. $13 - 8 = 5$

This number gets added (after 5).

$13 = 5 + 8$

(By the commutative law of addition, $13 = 8 + 5$ is also correct.)

7. $23 - 9 = 14$

This number gets added (after 14).

$23 = 14 + 9$

(By the commutative law of addition, $23 = 9 + 14$ is also correct.)

9. $43 - 16 = 27$

This number gets added (after 27).

$43 = 27 + 16$

(By the commutative law of addition, $43 = 16 + 27$ is also correct.)

11. $6 + 9 = 15$

This number gets subtracted (moved).

$6 = 15 - 9$

$6 + 9 = 15$

This number gets subtracted (moved).

$9 = 15 - 6$

13. $8 + 7 = 15$

This number gets subtracted (moved).

$8 = 15 - 7$

$8 + 7 = 15$

This number gets subtracted (moved).

$7 = 15 - 8$

15. $17 + 6 = 23$

This number gets subtracted (moved).

$17 = 23 - 6$

$17 + 6 = 23$

This number gets subtracted (moved).

$6 = 23 - 17$

17. $23 + 9 = 32$

This number gets subtracted (moved).

$23 = 32 - 9$

$23 + 9 = 32$

This number gets subtracted (moved).

$9 = 32 - 23$

19. We write an addition sentence first.

$\underbrace{\text{Number already sold}}\ \text{plus}\ \underbrace{\text{Number yet to be sold}}\ \text{is}\ \underbrace{\text{Sales goal}}$

$190 + \square = 220$

Now we write a related subtraction.

$190 + \square = 220$

$\square = 220 - 190$ 190 gets subtracted (moved)

21.
```
  1 6
-   4
-----
  1 2
```
Subtract ones, then subtract tens.

23.
```
  6 5
- 2 1
-----
  4 4
```
Subtract ones, then subtract tens.

25.
```
  8 6 6
- 3 3 3
-------
  5 3 3
```
Subtract ones, subtract tens, then subtract hundreds.

27.
```
  4 5 4 7
- 3 4 2 1
---------
  1 1 2 6
```
Subtract ones, subtract tens, subtract hundreds, then subtract thousands.

29.
```
  7 16
  8̸ 6̸
- 4  7
------
  3  9
```
We cannot subtract 7 ones from 6 ones. Borrow 1 ten to get 16 ones. Subtract ones, then subtract tens.

31.
```
    11
  5 1̸ 15
  6̸ 2̸ 5̸
- 3 2  7
--------
  2 9  8
```
We cannot subtract 7 ones from 5 ones. Borrow 1 ten to get 15 ones. Subtract ones. We cannot subtract 2 tens from 1 ten. Borrow 1 hundred to get 11 tens. Subtract tens, then subtract hundreds.

33.
```
    2 15
  8 3̸ 5̸
- 6 0  9
--------
  2 2  6
```
We cannot subtract 9 ones from 5 ones. Borrow 1 ten to get 15 ones. Subtract ones, subtract tens, then subtract hundreds.

35.
```
    7 11
  9 8̸ 1̸
- 7 4  7
--------
  2 3  4
```
We cannot subtract 7 ones from 1 one. Borrow 1 ten to get 11 ones. Subtract ones, subtract tens, then subtract hundreds.

37.
```
    6 16
  7 7̸ 6̸ 9
- 2 3  8 7
----------
  5 3  8 2
```
Subtract ones. We cannot subtract 8 tens from 6 tens. Borrow 1 hundred to get 16 tens. Subtract tens, subtract hundreds, then subtract thousands.

39.

$$\begin{array}{r} {}^{17} \\ {}^{8\ 7\ 12} \\ 3\,\not{9}\,\not{8}\,\not{2} \\ -\,2\,4\,8\,9 \\ \hline 1\,4\,9\,3 \end{array}$$

We cannot subtract 9 ones from 2 ones. Borrow 1 ten to get 12 ones. Subtract ones. We cannot subtract 8 tens from 7 tens. Borrow 1 hundred to get 17 tens. Subtract tens, subtract hundreds, then subtract thousands.

41.

$$\begin{array}{r} {}^{13} \\ {}^{4\ 9\ 3\ 16} \\ \not{5}\,\not{0}\,\not{4}\,\not{6} \\ -\,2\,8\,5\,9 \\ \hline 2\,1\,8\,7 \end{array}$$

We cannot subtract 9 ones from 6 ones. Borrow 1 ten to get 16 ones. Subtract ones. We cannot subtract 5 tens from 3 tens. We have 5 thousands or 50 hundreds. We borrow 1 hundred to get 13 tens. We have 49 hundreds. Subtract tens, subtract hundreds, then subtract thousands.

43.

$$\begin{array}{r} {}^{6\ 16\ 3\ 10} \\ \not{7}\,\not{6}\,\not{4}\,\not{0} \\ -\,3\,8\,0\,9 \\ \hline 3\,8\,3\,1 \end{array}$$

We cannot subtract 9 ones from 0 ones. Borrow 1 ten to get 10 ones. Subtract ones, then tens. We cannot subtract 8 hundreds from 6 hundreds. Borrow 1 thousand to get 16 hundreds. Subtract hundreds, then thousands.

45.

$$\begin{array}{r} {}^{11\ 15\ 13} \\ {}^{1\ 5\ 3\ 17} \\ \not{1}\not{2},\not{6}\,\not{4}\,\not{7} \\ -\quad 4\,8\,9\,9 \\ \hline 7\,7\,4\,8 \end{array}$$

47.

$$\begin{array}{r} {}^{16\ 16} \\ {}^{5\ 6\ 6\ 11} \\ 4\not{6},\not{7}\,\not{7}\,\not{1} \\ -\,1\,2,9\,7\,7 \\ \hline 3\,3,7\,9\,4 \end{array}$$

49.

$$\begin{array}{r} {}^{7\ 10} \\ \not{8}\,\not{0} \\ -\,2\,4 \\ \hline 5\,6 \end{array}$$

51.

$$\begin{array}{r} {}^{8\ 10} \\ \not{9}\,\not{0} \\ -\,5\,4 \\ \hline 3\,6 \end{array}$$

53.

$$\begin{array}{r} {}^{8\ 10} \\ 6\,\not{9}\,\not{0} \\ -\,2\,3\,6 \\ \hline 4\,5\,4 \end{array}$$

55.

$$\begin{array}{r} {}^{8\ 10} \\ \not{9}\,\not{0}\,3 \\ -\,1\,3\,2 \\ \hline 7\,7\,1 \end{array}$$

57.

$$\begin{array}{r} {}^{2\ 9\ 10} \\ 2\,\not{3}\,\not{0}\,\not{0} \\ -\quad 1\,0\,9 \\ \hline 2\,1\,9\,1 \end{array}$$

We have 3 hundreds or 30 tens. We borrow 1 ten to get 10 ones. We then have 29 tens. Subtract ones, then tens, then hundreds, then thousands.

59.

$$\begin{array}{r} {}^{7\ 9\ 18} \\ 6\,\not{8}\,\not{0}\,\not{8} \\ -\,3\,0\,5\,9 \\ \hline 3\,7\,4\,9 \end{array}$$

We have 8 hundreds or 80 tens. We borrow 1 ten to get 18 ones. We then have 79 tens. Subtract ones, then tens, then hundreds, then thousands.

61.

$$\begin{array}{r} {}^{6\ 9\ 9\ 10} \\ \not{7}\,\not{0}\,\not{0}\,\not{0} \\ -\,2\,7\,9\,4 \\ \hline 4\,2\,0\,6 \end{array}$$

We have 7 thousands or 700 tens. We borrow 1 ten to get 10 ones. We then have 699 tens. Subtract ones, then tens, then hundreds, then thousands.

63.

$$\begin{array}{r} {}^{7\ 9\ 9\ 10} \\ 4\,\not{8},\not{0}\,\not{0}\,\not{0} \\ -3\,7,6\,9\,5 \\ \hline 1\,0,3\,0\,5 \end{array}$$

We have 8 thousands or 800 tens. We borrow 1 ten to get 10 ones. We then have 799 tens. Subtract ones, then tens, then hundreds, then thousands, then ten thousands.

65. 6, 3 [7] 5, 6 0 2

The digit 7 means 7 ten thousands.

67.

69. $3,928,124 - 1,098,947$

Using a calculator to carry out the subtraction, we find that the difference is 2,829,177.

71.

$$\begin{array}{r} 9,\,\_\,4\,8,6\,2\,1 \\ -\,2,0\,9\,7,\,\_\,8\,1 \\ \hline 7,2\,5\,1,1\,4\,0 \end{array}$$

To subtract tens, we borrow 1 hundred to get 12 tens.

$$\begin{array}{r} {}^{5\ 12} \\ 9,\,\_\,4\,8,\not{6}\,\not{2}\,1 \\ -\,2,0\,9\,7,\,\_\,8\,1 \\ \hline 7,2\,5\,1,1\,4\,0 \end{array}$$

In order to have 1 hundred in the difference, the missing digit in the subtrahend must be 4 $(5 - 4 = 1)$.

$$\begin{array}{r} {}^{5\ 12} \\ 9,\,\_\,4\,8,\not{6}\,\not{2}\,1 \\ -\,2,0\,9\,7,4\,8\,1 \\ \hline 7,2\,5\,1,1\,4\,0 \end{array}$$

In order to subtract ten thousands, we must borrow 1 hundred thousand to get 14 ten thousands. The number of hundred thousands left must be 2 since the hundred thousands place in the difference is 2 $(2 - 0 = 2)$. Thus, the missing digit in the minuend must be $2 + 1$, or 3.

$$\begin{array}{r} {}^{2\ 14\quad 5\ 12} \\ 9,\not{3}\,\not{4}\,8,\not{6}\,\not{2}\,1 \\ -\,2,0\,9\,7,4\,8\,1 \\ \hline 7,2\,5\,1,1\,4\,0 \end{array}$$

Exercise Set 1.4

1. Round 48 to the nearest ten.

4 [8]
↑

The digit 4 is in the tens place. Consider the next digit to the right. Since the digit, 8, is 5 or higher, round 4 tens up to 5 tens. Then change the digit to the right of the tens digit to zero.

The answer is 50.

3. Round 67 to the nearest ten.

$6\boxed{7}$

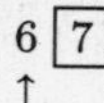

The digit 6 is in the tens place. Consider the next digit to the right. Since the digit, 7, is 5 or higher, round 6 tens up to 7 tens. Then change the digit to the right of the tens digit to zero.

The answer is 70.

5. Round 731 to the nearest ten.

$7\,3\boxed{1}$
↑

The digit 3 is in the tens place. Consider the next digit to the right. Since the digit, 1, is 4 or lower, round down, meaning that 3 tens stays as 3 tens. Then change the digit to the right of the tens digit to zero.

The answer is 730.

7. Round 895 to the nearest ten.

$8\,9\boxed{5}$
↑

The digit 9 is in the tens place. Consider the next digit to the right. Since the digit, 5, is 5 or higher, we round up. The 89 tens become 90 tens. Then change the digit to the right of the tens digit to zero.

The answer is 900.

9. Round 146 to the nearest hundred.

$1\boxed{4}6$
↑

The digit 1 is in the hundreds place. Consider the next digit to the right. Since the digit, 4, is 4 or lower, round down, meaning that 1 hundred stays as 1 hundred. Then change all digits to the right of the hundreds digit to zeros.

The answer is 100.

11. Round 957 to the nearest hundred.

$9\boxed{5}7$
↑

The digit 9 is in the hundreds place. Consider the next digit to the right. Since the digit, 5, is 5 or higher, round up. The 9 hundreds become 10 hundreds. Then change all digits to the right of the hundreds digit to zeros.

The answer is 1000.

13. Round 9079 to the nearest hundred.

$9\,0\boxed{7}9$
↑

The digit 0 is in the hundreds place. Consider the next digit to the right. Since the digit, 7, is 5 or higher, round 0 hundreds up to 1 hundred. Then change all digits to the right of the hundreds digit to zeros.

The answer is 9100.

15. Round 2850 to the nearest hundred.

$2\,8\boxed{5}0$
↑

The digit 8 is in the hundreds place. Consider the next digit to the right. Since the digit, 5, is 5 or higher, round 8 hundreds up to 9 hundreds. Then change all digits to the right of the hundreds digit to zeros.

The answer is 2900.

17. Round 5876 to the nearest thousand.

$5\boxed{8}7\,6$
↑

The digit 5 is in the thousands place. Consider the next digit to the right. Since the digit, 8, is 5 or higher, round 5 thousands up to 6 thousands. Then change all digits to the right of the thousands digit to zeros.

The answer is 6000.

19. Round 7500 to the nearest thousand.

$7\boxed{5}0\,0$
↑

The digit 7 is in the thousands place. Consider the next digit to the right. Since the digit, 5, is 5 or higher, round 7 thousands up to 8 thousands. Then change all the digits to the right of the thousands digit to zeros.

The answer is 8000.

21. Round 45,340 to the nearest thousand.

$4\,5,\boxed{3}4\,0$
↑

The digit 5 is in the thousands place. Consider the next digit to the right. Since the digit, 3, is 4 or lower, round down, meaning that 5 thousands stays as 5 thousands. Then change all the digits to the right of the thousands digit to zeros.

The answer is 45,000.

23. Round 373,405 to the nearest thousand.

$3\,7\,3,\boxed{4}0\,5$
↑

The digit 3 is in the thousands place. Consider the next digit to the right. Since the digit, 4, is 4 or lower, round down, meaning that 3 thousands stays as 3 thousands. Then change all the digits to the right of the thousands digit to zeros.

The answer is 373,000.

25.

	Rounded to the nearest ten
7828	7830
+ 9786	+ 9790
	17,620 ← Estimated answer

27.

	Rounded to the nearest ten
8074	8070
− 2347	− 2350
	5720 ← Estimated answer

29. Rounded to the nearest ten

$$\begin{array}{rr} 45 & 50 \\ 77 & 80 \\ 25 & 30 \\ +\,56 & +\,60 \\ \hline 343 & 220 \leftarrow \text{Estimated answer} \end{array}$$

The sum 343 seems to be incorrect since 220 is not close to 343.

31. Rounded to the nearest ten

$$\begin{array}{rr} 622 & 620 \\ 78 & 80 \\ 81 & 80 \\ +\,111 & +\,110 \\ \hline 932 & 890 \leftarrow \text{Estimated answer} \end{array}$$

The sum 932 seems to be incorrect since 890 is not close to 932.

33. Rounded to the nearest hundred

$$\begin{array}{rr} 7348 & 7300 \\ +\,9247 & +\,9200 \\ \hline & 16{,}500 \leftarrow \text{Estimated answer} \end{array}$$

35. Rounded to the nearest hundred

$$\begin{array}{rr} 6852 & 6900 \\ -\,1748 & -\,1700 \\ \hline & 5200 \leftarrow \text{Estimated answer} \end{array}$$

37. Rounded to the nearest hundred

$$\begin{array}{rr} 216 & 200 \\ 84 & 100 \\ 745 & 700 \\ +\,595 & +\,600 \\ \hline 1640 & 1600 \leftarrow \text{Estimated answer} \end{array}$$

The sum 1640 seems to be correct since 1600 is close to 1640.

39. Rounded to the nearest hundred

$$\begin{array}{rr} 750 & 800 \\ 428 & 400 \\ 63 & 100 \\ +\,205 & +\,200 \\ \hline 1446 & 1500 \leftarrow \text{Estimated answer} \end{array}$$

The sum 1446 seems to be correct since 1500 is close to 1446.

41. Rounded to the nearest thousand

$$\begin{array}{rr} 9643 & 10{,}000 \\ 4821 & 5000 \\ 8943 & 9000 \\ +\,7004 & +\,7000 \\ \hline & 31{,}000 \leftarrow \text{Estimated answer} \end{array}$$

43. Rounded to the nearest thousand

$$\begin{array}{rr} 92{,}149 & 92{,}000 \\ -\,22{,}555 & -\,23{,}000 \\ \hline & 69{,}000 \leftarrow \text{Estimated answer} \end{array}$$

45. 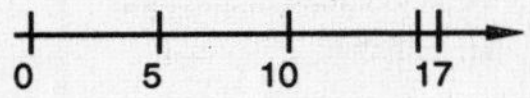

Since 0 is to the left of 17, $0 < 17$.

47. 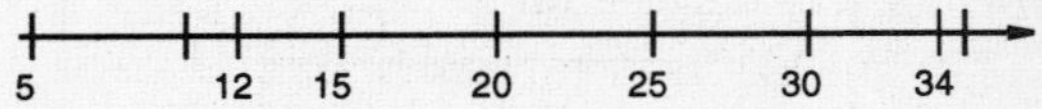

Since 34 is to the right of 12, $34 > 12$.

49. 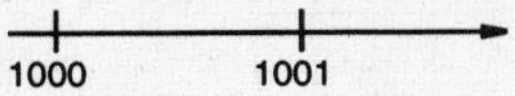

Since 1000 is to the left of 1001, $1000 < 1001$.

51. 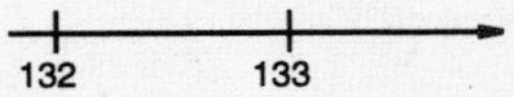

Since 133 is to the right of 132, $133 > 132$.

53.

$$\begin{array}{r} {}^{1\ 1\ 1\ 1} \\ 67{,}789 \\ +\,18{,}965 \\ \hline 86{,}754 \end{array}$$

Add ones. We get 14. Write 4 in the ones column and 1 above the tens. Add tens: We get 15 tens. Write 5 in the tens column and 1 above the hundreds. Add hundreds: We get 17 hundreds. Write 7 in the hundreds column and 1 above the thousands. Add thousands: We get 16 thousands. Write 6 in the thousands column and 1 above the ten thousands. Add ten thousands: We get 8 ten thousands.

55.

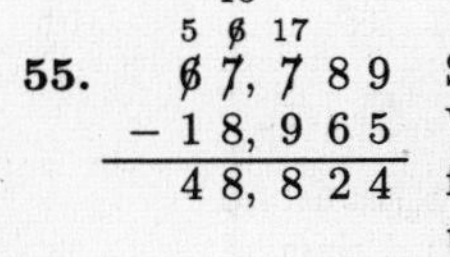

$$\begin{array}{r} {}^{\ \ \ 16} \\ {}^{5\ \ \not{7}\ \ 17} \\ \not{6}\not{7}{,}\not{7}89 \\ -\,18{,}965 \\ \hline 48{,}824 \end{array}$$

Subtract ones: We get 4. Subtract tens: We get 2. We cannot subtract 9 hundreds from 7 hundreds. We borrow 1 thousand to get 17 hundreds. Subtract hundreds. We cannot subtract 8 thousands from 6 thousands. We borrow 1 ten thousand to get 16 thousands. Subtract thousands, then ten thousands.

57.

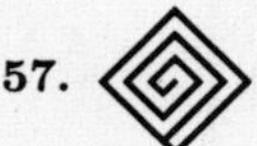

59. Using a calculator, we find that the sum is 30,411.

61. Using a calculator, we find that the difference is 69,594.

Exercise Set 1.5

1. Repeated addition fits best in this case.

$$\underbrace{\boxed{\$10} + \boxed{\$10} + \boxed{\$10} + \cdots + \boxed{\$10}}_{\text{32 addends}}$$

$$32 \cdot \$10 = \$320$$

3. We have a rectangular array.

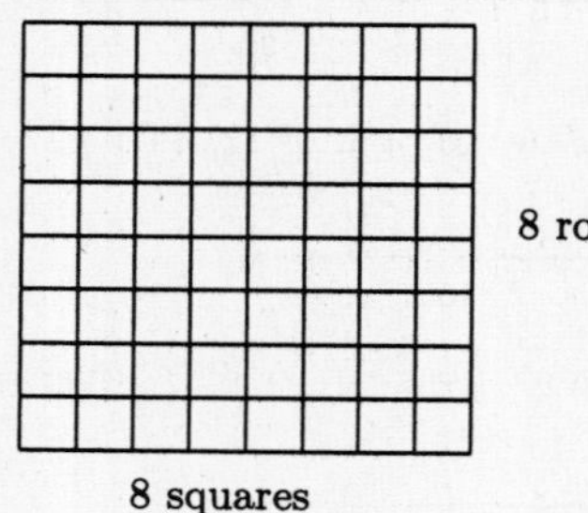

8 rows

8 squares

$$8 \cdot 8 = 64$$

5. If we think of filling the rectangle with square feet, we have a rectangular array.

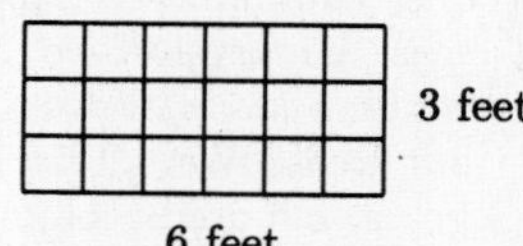

3 feet

6 feet

$$A = l \times w = 6 \times 3 = 18 \text{ ft}^2$$

7. $\begin{array}{r} 87 \\ \times\ 10 \\ \hline 870 \end{array}$ Multiplying by 1 ten (We write 0 and then multiply 87 by 1.)

9. $\begin{array}{r} 2340 \\ \times\ 1000 \\ \hline 2,340,000 \end{array}$ Multiplying by 1 thousand (We write 000 and then multiply 2340 by 1.)

11. $\begin{array}{r} {\scriptstyle 4} \\ 65 \\ \times\ 8 \\ \hline 520 \end{array}$ Multiplying by 8

13. $\begin{array}{r} {\scriptstyle 2} \\ 94 \\ \times\ 6 \\ \hline 564 \end{array}$ Multiplying by 6

15. $\begin{array}{r} {\scriptstyle 2} \\ 509 \\ \times\ 3 \\ \hline 1527 \end{array}$ Multiplying by 3

17. $\begin{array}{r} {\scriptstyle 1\ 2\ 6} \\ 9229 \\ \times\ 7 \\ \hline 64,603 \end{array}$ Multiplying by 7

19. $\begin{array}{r} {\scriptstyle 2} \\ 53 \\ \times\ 90 \\ \hline 4770 \end{array}$ Multiplying by 9 tens (We write 0 and then multiply 53 by 9.)

21.

$$\begin{array}{rl} {\scriptstyle 2} & \\ {\scriptstyle 3} & \\ 85 & \\ \times\ 47 & \\ \hline 595 & \text{Multiplying by 7} \\ 3400 & \text{Multiplying by 40} \\ \hline 3995 & \text{Adding} \end{array}$$

23.

$$\begin{array}{rl} {\scriptstyle 2} & \\ 640 & \\ \times\ 72 & \\ \hline 1280 & \text{Multiplying by 2} \\ 44800 & \text{Multiplying by 70} \\ \hline 46,080 & \text{Adding} \end{array}$$

25.

$$\begin{array}{rl} {\scriptstyle 1\ 1} & \\ {\scriptstyle 1\ 1} & \\ 444 & \\ \times\ 33 & \\ \hline 1332 & \text{Multiplying by 3} \\ 13320 & \text{Multiplying by 30} \\ \hline 14,652 & \text{Adding} \end{array}$$

27.

$$\begin{array}{rl} {\scriptstyle 3} & \\ {\scriptstyle 7} & \\ 509 & \\ \times\ 408 & \\ \hline 4072 & \text{Multiplying by 8} \\ 203600 & \text{Multiplying by 4 hundreds (We write 00} \\ \hline 207,672 & \text{and then multiply 509 by 4.)} \end{array}$$

29.

$$\begin{array}{rl} {\scriptstyle 4\ 2} & \\ {\scriptstyle 1} & \\ {\scriptstyle 3\ 1} & \\ 853 & \\ \times\ 936 & \\ \hline 5118 & \text{Multiplying by 6} \\ 25590 & \text{Multiplying by 30} \\ 767700 & \text{Multiplying by 900} \\ \hline 798,408 & \text{Adding} \end{array}$$

31.

$$\begin{array}{rl} {\scriptstyle 2\ 2} & \\ {\scriptstyle 3\ 3} & \\ 489 & \\ \times\ 340 & \text{Multiplying by 4 tens (We write 0 and} \\ \hline 19560 & \leftarrow \text{then multiply 489 by 4.)} \\ 146700 & \leftarrow \text{Multiplying by 3 hundreds (We write 00} \\ \hline 166,260 & \text{and then multiply 489 by 3.)} \end{array}$$

33.

$$\begin{array}{rl} {\scriptstyle 1\ 1} & \\ {\scriptstyle 2\ 4\ 4} & \\ {\scriptstyle 3\ 7\ 7} & \\ {\scriptstyle 1\ 3\ 3} & \\ 4378 & \\ \times\ 2694 & \\ \hline 17512 & \text{Multiplying by 4} \\ 394020 & \text{Multiplying by 90} \\ 2626800 & \text{Multiplying by 600} \\ 8756000 & \text{Multiplying by 2000} \\ \hline 11,794,332 & \text{Adding} \end{array}$$

35.
```
    1 3
    3 4 8 2
  ×   1 0 4
  1 3 9 2 8   Multiplying by 4
3 4 8 2 0 0   Multiplying by 1 hundred (We write 00
3 6 2,1 2 8   and then multiply 3482 by 1.)
```

37.
```
            2   3
            3   4
            5 6 0 8
        × 4 5 0 0   Multiplying by 5 hundreds (We write
  2 8 0 4 0 0 0   00 and then multiply 5608 by 5.)
2 2 4 3 2 0 0 0   Multiplying by 4000
2 5,2 3 6,0 0 0   Adding
```

39.
```
      2 1
      3 2
      3 3
      8 7 6
    × 3 4 5
    4 3 8 0   Multiplying by 5
  3 5 0 4 0   Multiplying by 40
2 6 2 8 0 0   Multiplying by 300
3 0 2,2 2 0   Adding
```

41.
```
          5 5 5
          1 1 1
          1 1 1
          3 3 3
        7 8 8 9
      × 6 2 2 4
      3 1 5 5 6   Multiplying by 4
    1 5 7 7 8 0   Multiplying by 20
  1 5 7 7 8 0 0   Multiplying by 200
4 7 3 3 4 0 0 0   Multiplying by 6000
4 9,1 0 1,1 3 6   Adding
```

43.
```
    2 2
    2 2
    5 5 5
  ×   5 5
  2 7 7 5   Multiplying by 5
2 7 7 5 0   Multiplying by 50
3 0,5 2 5   Adding
```

45.
```
      1 1
      2 2
      7 3 4
    × 4 0 7
    5 1 3 8   Multiplying by 7
2 9 3 6 0 0   Multiplying by 4 hundreds (We write 00
2 9 8,7 3 8   and then multiply 734 by 4.)
```

47.
```
             Rounded to
           the nearest ten
  4 5            5 0
× 6 7          × 7 0
             3 5 0 0 ← Estimated answer
```

49.
```
             Rounded to
           the nearest ten
  3 4            3 0
× 2 9          × 3 0
                 9 0 0 ← Estimated answer
```

51.
```
               Rounded to
           the nearest hundred
  8 7 6            9 0 0
× 3 4 5          × 3 0 0
             2 7 0,0 0 0 ← Estimated answer
```

53.
```
               Rounded to
           the nearest hundred
  4 3 2            4 0 0
× 1 9 9          × 2 0 0
               8 0,0 0 0 ← Estimated answer
```

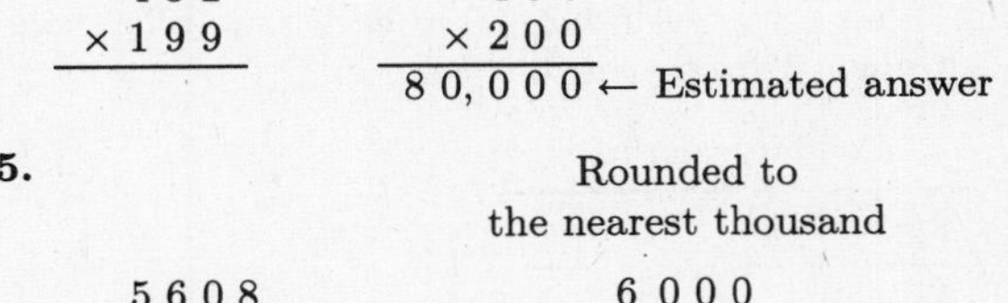

55.
```
                     Rounded to
                 the nearest thousand
  5 6 0 8              6 0 0 0
× 4 5 7 6            × 5 0 0 0
               3 0,0 0 0,0 0 0 ← Estimated
                                 answer
```

57.
```
                     Rounded to
                 the nearest thousand
  7 8 8 8              8 0 0 0
× 6 2 2 4            × 6 0 0 0
               4 8,0 0 0,0 0 0 ← Estimated
                                 answer
```

59.
```
  1
        2 0
      8 5 0
  + 3 5 0 0
    4 3 7 0
```
Add ones: We get 0. Add tens: We get 7 tens. Add hundreds: We get 13 hundreds. Write 3 in the hundreds column and 1 above the thousands. Add thousands: We get 4 thousands.

61. 

63. Use a calculator to perform the computations in this exercise.

First find the total area of each floor:

$A = l \times w = 172 \times 84 = 14,448$ square feet

Find the area lost to the elevator and the stairwell:

$A = l \times w = 35 \times 20 = 700$ square feet

Subtract to find the area available as office space on each floor:

$14,448 - 700 = 13,748$ square feet

Finally, multiply by the number of floors, 18, to find the total area available as office space:

$18 \times 13,748 = 247,464$ square feet

Exercise Set 1.6

1. Think of an array with 4 lb in each row. The cheese in each row will go in a box.

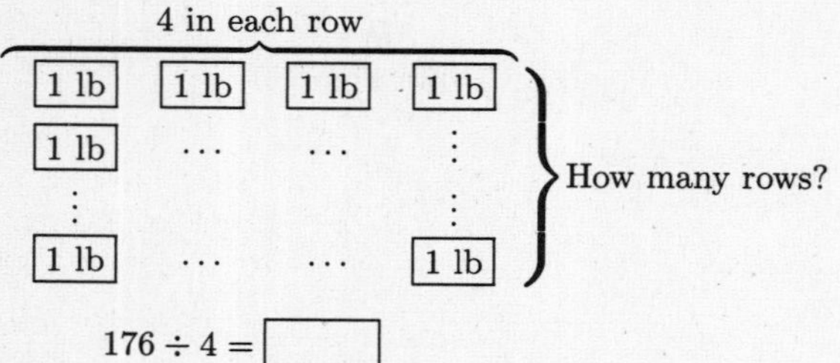

$176 \div 4 = \square$

3. Think of an array with \$23,000 in each row. The money in each row will go to a child.

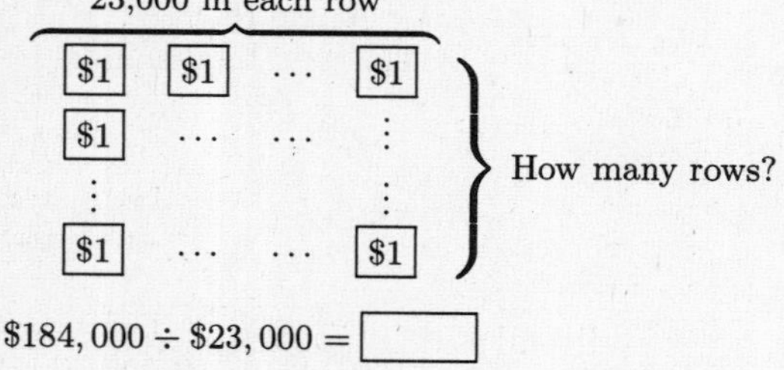

$\$184,000 \div \$23,000 = \square$

5. $18 \div 3 = 6$ The 3 moves to the right. A related multiplication sentence is $18 = 6 \cdot 3$. (By the commutative law of multiplication, there is also another multiplication sentence: $18 = 3 \cdot 6$.)

7. $22 \div 22 = 1$ The 22 on the right of the $\div$ symbol moves to the right. A related multiplication sentence is $22 = 1 \cdot 22$. (By the commutative law of multiplication, there is also another multiplication sentence: $22 = 22 \cdot 1$.)

9. $54 \div 6 = 9$ The 6 moves to the right. A related multiplication sentence is $54 = 9 \cdot 6$. (By the commutative law of multiplication, there is also another multiplication sentence: $54 = 6 \cdot 9$.)

11. $37 \div 1 = 37$ The 1 moves to the right. A related multiplication sentence is $37 = 37 \cdot 1$. (By the commutative law of multiplication, there is also another multiplication sentence: $37 = 1 \cdot 37$.)

13. $9 \times 5 = 45$

Move a factor to the other side and then write a division.

$9 \times 5 = 45$ $\quad$ $9 \times 5 = 45$

$9 = 45 \div 5$ $\quad$ $5 = 45 \div 9$

15. Two related division sentences for $37 \cdot 1 = 37$ are:

$37 = 37 \div 1$ $\quad$ $(37 \cdot 1 = 37)$

and

$1 = 37 \div 37$ $\quad$ $(37 \cdot 1 = 37)$

17. $8 \times 8 = 64$

Since the factors are both 8, moving either one to the other side gives the related division sentence $8 = 64 \div 8$.

19. Two related division sentences for $11 \cdot 6 = 66$ are:

$11 = 66 \div 6$ $\quad$ $(11 \cdot 6 = 66)$

and

$6 = 66 \div 11$ $\quad$ $(11 \cdot 6 = 66)$

21.
$$\begin{array}{r} 55 \\ 5\overline{)277} \\ \underline{250} \\ 27 \\ \underline{25} \\ 2 \end{array}$$

Think: 2 hundreds $\div$ 5. There are no hundreds in the quotient.
Think: 27 tens $\div$ 5. Estimate 5 tens.
Think: 27 ones $\div$ 5. Estimate 5 ones.

The answer is 55 R 2.

23.
$$\begin{array}{r} 108 \\ 8\overline{)864} \\ \underline{800} \\ 64 \\ \underline{64} \\ 0 \end{array}$$

Think: 8 hundreds $\div$ 8. Estimate 1 hundred.
Think: 6 tens $\div$ 8. There are no tens in the quotient (other than the tens in 100). Write a 0 to show this.
Think: 64 ones $\div$ 8. Estimate 8 ones.

The answer is 108.

25.
$$\begin{array}{r} 307 \\ 4\overline{)1228} \\ \underline{1200} \\ 28 \\ \underline{28} \\ 0 \end{array}$$

Think: 12 hundreds $\div$ 4. Estimate 3 hundreds.
Think: 2 tens $\div$ 4. There are no tens in the quotient (other than the tens in 300). Write a 0 to show this.
Think: 28 ones $\div$ 4. Estimate 7 ones.

The answer is 307.

27.
$$\begin{array}{r} 753 \\ 6\overline{)4521} \\ \underline{4200} \\ 321 \\ \underline{300} \\ 21 \\ \underline{18} \\ 3 \end{array}$$

Think: 45 hundreds $\div$ 6. Estimate 7 hundreds.
Think: 32 tens $\div$ 6. Estimate 5 tens.
Think: 21 ones $\div$ 6. Estimate 3 ones.

The answer is 753 R 3.

29.
$$\begin{array}{r} 74 \\ 4\overline{)297} \\ \underline{280} \\ 17 \\ \underline{16} \\ 1 \end{array}$$

Think: 29 tens $\div$ 4. Estimate 7 tens.
Think: 17 ones $\div$ 4. Estimate 4 ones.

The answer is 74 R 1.

31.
$$\begin{array}{r} 92 \\ 8\overline{)738} \\ \underline{720} \\ 18 \\ \underline{16} \\ 2 \end{array}$$

Think: 73 tens $\div$ 8. Estimate 9 tens.
Think: 18 ones $\div$ 8. Estimate 2 ones.

The answer is 92 R 2.

33.

```
    1 7 0 3
5 ) 8 5 1 5
    5 0 0 0
    3 5 1 5
    3 5 0 0
        1 5
        1 5
          0
```

Think: 8 thousands ÷ 5. Estimate 1 thousand.

Think: 35 hundreds ÷ 5. Estimate 7 hundreds.

Think: 1 ten ÷ 5. There are no tens in the quotient (other than the tens in 1700). Write a 0 to show this.

Think: 15 ones ÷ 5. Estimate 3 ones.

The answer is 1703.

35.

```
      9 8 7
9 ) 8 8 8 8
    8 1 0 0
      7 8 8
      7 2 0
        6 8
        6 3
          5
```

Think: 88 hundreds ÷ 9. Estimate 9 hundreds.

Think: 78 ten ÷ 9. Estimate 8 tens.

Think: 68 ones ÷ 9. Estimate 7 ones.

The answer is 987 R 5.

37.

```
          5 2
7 0 ) 3 6 9 2
      3 5 0 0
        1 9 2
        1 4 0
          5 2
```

Think: 369 tens ÷ 70. Estimate 5 tens.

Think: 192 ones ÷ 70. Estimate 2 ones.

The answer is 52 R 52.

39.

```
          2 9
3 0 ) 8 7 5
      6 0 0
      2 7 5
      2 7 0
          5
```

Think: 87 tens ÷ 30. Estimate 2 tens.

Think: 275 ones ÷ 30. Estimate 9 ones.

The answer is 29 R 5.

41.

```
        4 0
2 1 ) 8 5 2
      8 4 0
        1 2
```

Round 21 to 20.

Think: 85 tens ÷ 20. Estimate 4 tens.

Think: 12 ones ÷ 20. There are no ones in the quotient (other than the ones in 40). Write a 0 to show this.

The answer is 40 R 12.

43.

```
          8
8 5 ) 7 6 7 2
      6 8 0 0
      [8 7] 2
```

Round 85 to 90.

Think: 767 tens ÷ 90. Estimate 8 tens.

Since 87 is larger than the divisor, the estimate is too low.

```
          9 0
8 5 ) 7 6 7 2
      7 6 5 0
          2 2
```

Think: 767 tens ÷ 90. Estimate 9 tens.

Think: 22 ones ÷ 90. There are no ones in the quotient (other than the ones in 90). Write a 0 to show this.

The answer is 90 R 22.

45.

```
              3
1 1 1 ) 3 2 1 9
        3 3 3 0
```

Round 111 to 100.

Think: 321 tens ÷ 100. Estimate 3 tens.

Since we cannot subtract 3330 from 3219, the estimate is too high.

```
            2 9
1 1 1 ) 3 2 1 9
        2 2 2 0
          9 9 9
          9 9 9
              0
```

Think: 321 tens ÷ 100. Estimate 2 tens.

Think: 999 ones ÷ 100. Estimate 9 ones.

The answer is 29.

47.

```
    1 0 5
8 ) 8 4 3
    8 0 0
      4 3
      4 0
        3
```

Think: 8 hundreds ÷ 8. Estimate 1 hundred.

Think: 4 tens ÷ 8. There are no tens in the quotient (other than the tens in 100). Write a 0 to show this.

Think: 43 ones ÷ 8. Estimate 5 ones.

The answer is 105 R 3.

49.

```
    1 6 0 9
5 ) 8 0 4 7
    5 0 0 0
    3 0 4 7
    3 0 0 0
        4 7
        4 5
          2
```

Think: 8 thousands ÷ 5. Estimate 1 thousand.

Think: 30 hundreds ÷ 5. Estimate 6 hundreds.

Think: 4 tens ÷ 5. There are no tens in the quotient (other than the tens in 1600). Write a 0 to show this.

Think: 47 ones ÷ 5. Estimate 9 ones.

The answer is 1609 R 2.

51.

```
    1 0 0 7
5 ) 5 0 3 6
    5 0 0 0
        3 6
        3 5
          1
```

Think: 5 thousands ÷ 5. Estimate 1 thousand.

Think: 0 hundreds ÷ 5. There are no hundreds in the quotient (other than the hundreds in 1000). Write a 0 to show this.

Think: 3 tens ÷ 5. There are no tens in the quotient (other than the tens in 1000). Write a 0 to show this.

Think: 36 ones ÷ 5. Estimate 7 ones.

The answer is 1007 R 1.

53.

```
          2 2
4 6 ) 1 0 5 8
        9 2 0
        1 3 8
          9 2
          4 6
```

Round 46 to 50.

Think: 105 tens ÷ 50. Estimate 2 tens.

Think: 138 ones ÷ 50. Estimate 2 ones.

Since 46 is not smaller than the divisor, 46, the estimate is too low.

```
          2 3
4 6 ) 1 0 5 8
        9 2 0
        1 3 8
        1 3 8
            0
```

Think: 138 ones ÷ 50. Estimate 3 ones.

The answer is 23.

55.

```
      1 0 7
3 2 ) 3 4 2 5
      3 2 0 0
        2 2 5
        2 2 4
            1
```

Round 32 to 30.
Think: 34 hundreds ÷ 30. Estimate 1 hundred.
Think: 22 tens ÷ 30. There are no tens in the quotient (other than the tens in 100). Write 0 to show this.
Think: 225 ones ÷ 30. Estimate 7 ones.

The answer is 107 R 1.

57.

```
      4
2 4 ) 8 8 8 0
      9 6 0 0
```

Round 24 to 20.
Think: 88 hundreds ÷ 20. Estimate 4 hundreds.

Since we cannot subtract 9600 from 8880, the estimate is too high.

```
      3 8
2 4 ) 8 8 8 0
      7 2 0 0
      1 6 8 0
      1 9 2 0
```

Think: 88 hundreds ÷ 20. Estimate 3 hundreds.
Think: 168 tens ÷ 20. Estimate 8 tens.

Since we cannot subtract 1920 from 1680, the estimate is too high.

```
      3 7 0
2 4 ) 8 8 8 0
      7 2 0 0
      1 6 8 0
      1 6 8 0
            0
```

Think: 168 tens ÷ 20. Estimate 7 tens.
Think: 0 ones ÷ 20. There are no ones in the quotient (other than the ones in 370). Write a 0 to show this.

The answer is 370.

59.

```
        5
2 8 ) 1 7, 0 6 7
      1 4  0 0 0
        [3 0] 6 7
```

Round 28 to 30.
Think: 170 hundreds ÷ 30. Estimate 5 hundreds.

Since 30 is larger than the divisor, 28, the estimate is too low.

```
        6 0 8
2 8 ) 1 7, 0 6 7
      1 6  8 0 0
           2 6 7
           2 2 4
           [4 3]
```

Think: 170 hundreds ÷ 30. Estimate 6 hundreds.
Think: 26 tens ÷ 30. There are no tens in the quotient (other than the tens in 600.) Write a zero to show this.
Think: 267 ones ÷ 30. Estimate 8 ones.

Since 43 is larger than the divisor, 28, the estimate is too low.

```
        6 0 9
2 8 ) 1 7, 0 6 7
      1 6  8 0 0
           2 6 7
           2 5 2
             1 5
```

Think: 267 ones ÷ 30. Estimate 9 ones.

The answer is 609 R 15.

61.

```
          3 0 4
8 0 ) 2 4, 3 2 0
      2 4  0 0 0
           3 2 0
           3 2 0
               0
```

Think: 243 hundreds ÷ 80. Estimate 3 hundreds.
Think: 32 tens ÷ 80. There are no tens in the quotient (other than the tens in 300). Write a 0 to show this.
Think: 320 ones ÷ 80. Estimate 4 ones.

The answer is 304.

63.

```
             3 5 0 8
2 8 5 ) 9 9 9, 9 9 9
        8 5 5  0 0 0
        1 4 4  9 9 9
        1 4 2  5 0 0
             2 4 9 9
             2 2 8 0
               2 1 9
```

The answer is 3508 R 219.

65. 7882 = 7 thousands + 8 hundreds + 8 tens + 2 ones

67.

69. We divide 1231 by 42:

```
        2 9
4 2 ) 1 2 3 1
        8 4 0
        3 9 1
        3 7 8
          1 3
```

The answer is 29 R 13. Since 13 students will be left after 29 buses are filled, then 30 buses are needed.

Exercise Set 1.7

1. $x + 0 = 14$

We replace x by different numbers until we get a true equation. If we replace x by 14, we get a true equation: $14 + 0 = 14$. No other replacement makes the equation true, so the solution is 14.

3. $y \cdot 17 = 0$

We replace y by different numbers until we get a true equation. If we replace y by 0, we get a true equation: $0 \cdot 17 = 0$. No other replacement makes the equation true, so the solution is 0.

5.

$$\begin{aligned} 13 + x &= 42 \\ 13 + x - 13 &= 42 - 13 && \text{Subtracting 13 on both sides} \\ 0 + x &= 29 && \text{13 plus } x \text{ minus 13 is } 0 + x. \\ x &= 29 \end{aligned}$$

The solution is 29.

7.

$$\begin{aligned} 12 &= 12 + m \\ 12 - 12 &= 12 + m - 12 && \text{Subtracting 12 on both sides} \\ 0 &= 0 + m && \text{12 plus } m \text{ minus 12 is } 0 + m. \\ 0 &= m \end{aligned}$$

The solution is 0.

9. $3 \cdot x = 24$

$\frac{3 \cdot x}{3} = \frac{24}{3}$ Dividing by 3 on both sides

$x = 8$ 3 times x divided by 3 is x.

The solution is 8.

11. $112 = n \cdot 8$

$\frac{112}{8} = \frac{n \cdot 8}{8}$ Dividing by 8 on both sides

$14 = n$

The solution is 14.

13. $45 \times 23 = x$

To solve the equation we carry out the calculation.

$$\begin{array}{r} 45 \\ \times\ 23 \\ \hline 135 \\ 900 \\ \hline 1035 \end{array}$$

The solution is 1035.

15. $t = 125 \div 5$

To solve the equation we carry out the calculation.

$$\begin{array}{r} 25 \\ 5\,\overline{)\,125} \\ 100 \\ \hline 25 \\ 25 \\ \hline 0 \end{array}$$

The solution is 25.

17. $p = 908 - 458$

To solve the equation we carry out the calculation.

$$\begin{array}{r} 908 \\ -\ 458 \\ \hline 450 \end{array}$$

The solution is 450.

19. $x = 12,345 + 78,555$

To solve the equation we carry out the calculation.

$$\begin{array}{r} 12,345 \\ +\ 78,555 \\ \hline 90,900 \end{array}$$

The solution is 90,900.

21. $3 \cdot m = 96$

$\frac{3 \cdot m}{3} = \frac{96}{3}$ Dividing by 3 on both sides

$m = 32$

The solution is 32.

23. $715 = 5 \cdot z$

$\frac{715}{5} = \frac{5 \cdot z}{5}$ Dividing by 5 on both sides

$143 = z$

The solution is 143.

25. $10 + x = 89$

$10 + x - 10 = 89 - 10$

$x = 79$

The solution is 79.

27. $61 = 16 + y$

$61 - 16 = 16 + y - 16$

$45 = y$

The solution is 45.

29. $6 \cdot p = 1944$

$\frac{6 \cdot p}{6} = \frac{1944}{6}$

$p = 324$

The solution is 324.

31. $5 \cdot x = 3715$

$\frac{5 \cdot x}{5} = \frac{3715}{5}$

$x = 743$

The solution is 743.

33. $47 + n = 84$

$47 + n - 47 = 84 - 47$

$n = 37$

The solution is 37.

35. $x + 78 = 144$

$x + 78 - 78 = 144 - 78$

$x = 66$

The solution is 66.

37. $165 = 11 \cdot n$

$\frac{165}{11} = \frac{11 \cdot n}{11}$

$15 = n$

The solution is 15.

39. $624 = t \cdot 13$

$\frac{624}{13} = \frac{t \cdot 13}{13}$

$48 = t$

The solution is 48.

41. $x + 214 = 389$

$x + 214 - 214 = 389 - 214$

$x = 175$

The solution is 175.

43. $567 + x = 902$

$567 + x - 567 = 902 - 567$

$x = 335$

The solution is 335.

45. $18 \cdot x = 1872$

$\frac{18 \cdot x}{18} = \frac{1872}{18}$

$x = 104$

The solution is 104.

47. $x \cdot 233 = 22,135$

$$\frac{x \cdot 233}{233} = \frac{22,135}{233}$$

$$x = 95$$

The solution is 95.

49.
$$2344 + y = 6400$$
$$2344 + y - 2344 = 6400 - 2344$$
$$y = 4056$$

The solution is 4056.

51. $8322 + 9281 = x$

$17,603 = x$ Doing the addition

The solution is 17,603.

53. $7 + 8 = 15$

↑

This number gets subtracted (moved).

↓

$7 = 15 - 8$

$7 + 8 = 15$

↑

This number gets subtracted (moved).

↓

$8 = 15 - 7$

55. Since 123 is to the left of 789 on a number line, $123 < 789$.

57.

59. $23,465 \cdot x = 8,142,355$

We replace x by different numbers until we get a true equation. Rounding and estimating can be used to choose the first trial solution. Use a calculator to perform the computations. We find that the solution is 347.

Exercise Set 1.8

1. a) First we find how many hits Ty Cobb and Pete Rose got together.

Familiarize. We visualize the situation. We let n = the number of hits they got together.

Ty Cobb's hits	+	Pete Rose's hits	=	Number of hits together
↓	↓	↓	↓	↓
4191	+	4256	=	n

Translate. We write the number sentence that corresponds to the situation:

$4191 + 4256 = n$

Solve. We carry out the addition:

$$\begin{array}{r} \scriptstyle 1 \\ 4\,1\,9\,1 \\ +\,4\,2\,5\,6 \\ \hline 8\,4\,4\,7 \end{array}$$

Check. We can repeat the calculation. We note that the answer seems reasonable since it is larger than either of the separate hit totals. We can also find an estimated answer by rounding: $4191 + 4256 \approx 4200 + 4300 = 8500 \approx 8447$. The answer checks.

State. Ty Cobb and Pete Rose got 8447 hits together.

Next we find how many more hits Rose got than Cobb.

Familiarize. We visualize the situation. We let h = the number by which Rose's hits exceed Cobb's.

Translate. We see that this is a "take-away" situation. We write the equation:

Pete Rose's hits	−	Ty Cobb's hits	=	Number by which Rose's hits exceed Cobb's
↓	↓	↓	↓	↓
4256	−	4191	=	h

Solve. We carry out the subtraction:

$$\begin{array}{r} \scriptstyle 1\ 15 \\ 4\,\not{2}\,\not{5}\,6 \\ -\,4\,1\,9\,1 \\ \hline 6\,5 \end{array}$$

Thus, $65 = h$, or $h = 65$.

Check. We can repeat the calculation. We note that the answer seems reasonable since it is less than the number of Pete Rose's hits. We can add the answer to 4191, the number being subtracted: $4191 + 65 = 4256$. We can also estimate: $4256 - 4191 \approx 4300 - 4200 = 100 \approx 65$. The answer checks.

State. Pete Rose got 65 more hits than Ty Cobb.

b) First we find how many hits Hank Aaron and Stan Musial got together.

Familiarize. We visualize the situation. We let n = the number of hits they got together.

Hank Aaron's hits	+	Stan Musial's hits	=	Number of hits together
↓	↓	↓	↓	↓
3771	+	3630	=	n

Translate. We write the number sentence that corresponds to the situation:

$3771 + 3630 = n$

Solve. We carry out the addition:

$$\begin{array}{r} \scriptstyle 1\,1 \\ 3\,7\,7\,1 \\ +\,3\,6\,3\,0 \\ \hline 7\,4\,0\,1 \end{array}$$

Check. We can repeat the calculation. We note that the answer seems reasonable since it is larger than either of the separate hit totals. We can also find an estimated answer by rounding: $3771 + 3630 \approx 3800 + 3700 = 7500 \approx 7401$. The answer checks.

State. Hank Aaron and Stan Musial got 7401 hits together.

Next we find how many more hits Aaron got than Musial.

Familiarize. We visualize the situation. We let h = the number by which Aaron's hits exceed Musial's.

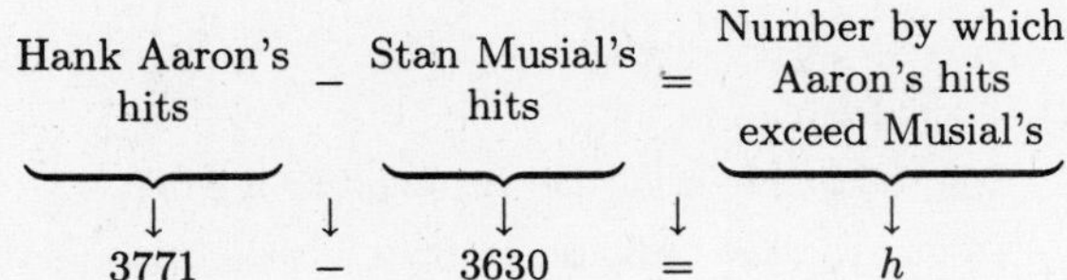

Translate. We see that this is a "take-away" situation. We write the equation:

Hank Aaron's hits − Stan Musial's hits = Number by which Aaron's hits exceed Musial's

$$3771 - 3630 = h$$

Solve. We carry out the subtraction:

$$\begin{array}{r} 3\,7\,7\,1 \\ -\ 3\,6\,3\,0 \\ \hline 1\,4\,1 \end{array}$$

Thus, $141 = h$, or $h = 141$.

Check. We can repeat the calculation. We note that the answer seems reasonable since it is less than the number of Hank Aaron's hits. We can also add the answer to 3630, the number being subtracted: $3630 + 141 = 3771$. We can also estimate: $3771 - 3630 \approx 3800 - 3600 = 200 \approx 141$. The answer checks.

State. Hank Aaron got 141 more hits than Stan Musial.

3. ***Familiarize***. We first make a drawing.

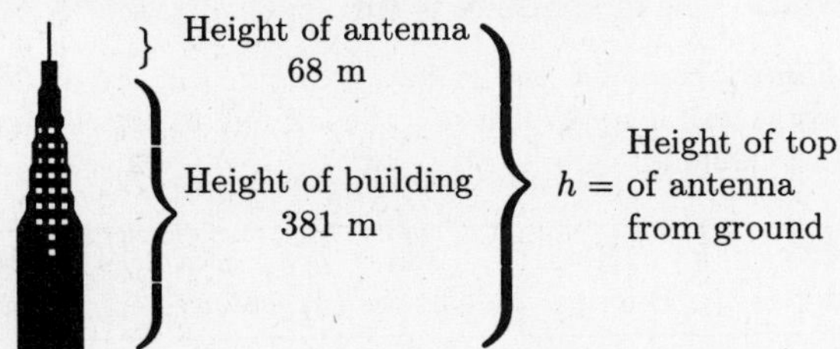

Since we are combining lengths, addition can be used. We let h = the height of the top of the antenna from the ground.

Translate. We translate to the following addition sentence:

$$381 + 68 = h$$

Solve. To solve we carry out the addition.

$$\begin{array}{r} 3\,8\,1 \\ +\ \ \ 6\,8 \\ \hline 4\,4\,9 \end{array}$$

Check. We can repeat the calculation. We can also find an estimated answer by rounding: $381 + 68 \approx 380 + 70 = 450 \approx 449$. The answer checks.

State. The top of the antenna is 449 m from the ground.

5. ***Familiarize***. We first make a drawing.

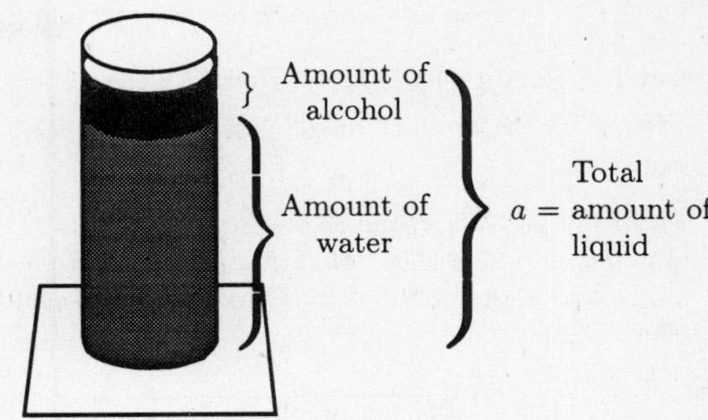

Since we are combining amounts, addition can be used. We let a = the total amount of liquid.

Translate. We translate to the following addition sentence:

$$2340 + 655 = a$$

Solve. To solve we carry out the addition.

$$\begin{array}{r} 2\,3\,4\,0 \\ +\ \ \ 6\,5\,5 \\ \hline 2\,9\,9\,5 \end{array}$$

Check. We can repeat the calculation. We can also find an estimated answer by rounding: $2340 + 655 \approx 2300 + 700 = 3000 \approx 2995$. The answer checks.

State. A total of 2995 cubic centimeters of liquid was poured.

7. ***Familiarize***. This is a multistep problem. We visualize the situation. We let p = the number of megabytes of memory used by the three programs that have been installed and m = the number of megabytes of memory left over.

Word processor: 15 megabytes	Database: 6 megabytes	Spreadsheet: 12 megabytes	Remaining memory
p			m
170 megabytes			

Translate. First we write a number sentence that corresponds to the number of megabytes of memory that have been used:

$$15 + 6 + 12 = p$$

Solve. We carry out the addition.

$$\begin{array}{r} 1\,5 \\ 6 \\ +\ 1\,2 \\ \hline 3\,3 \end{array}$$

Thus $33 = p$, or $p = 33$.

Now we must find the remaining memory. We have a "take-away" situation. We write the corresponding number sentence.

$$170 - 33 = m$$
$$137 = m \quad \text{Carrying out the subtraction}$$

Check. Since $15+6+12+137 = 170$, the answer checks.

State. There are 137 megabytes of memory left on the computer.

9. ***Familiarize.*** We visualize the situation. We let $p =$ the number by which the population of the Tokyo-Yokohama area exceeds the population of the New York-northeastern New Jersey area.

NY – NJ population 18,087,521	Excess p
Tokyo – Yokohama population 29,971,000	

Translate. This is a "how-much-more" situation. Translate to an equation.

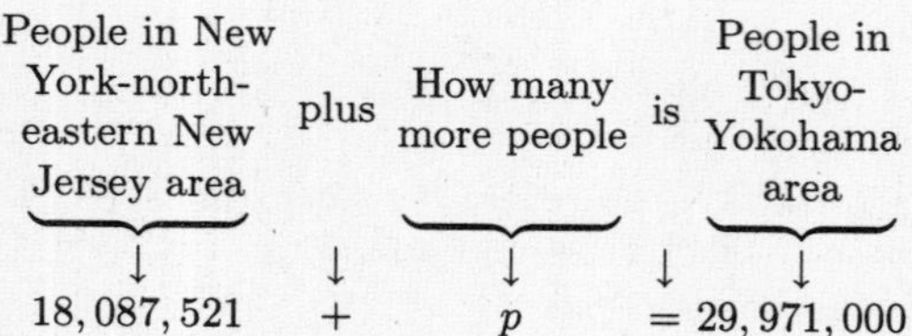

Solve. We solve the equation.

$$18,087,521 + p = 29,971,000$$
$$18,087,521 + p - 18,087,521 = 29,971,000 - 18,087,521$$
Subtracting 18,087,521 on both sides
$$p = 11,883,479$$

Check. We can add the result to $18,087,521$:

$18,087,521 + 11,883,479 = 29,971,000.$

We can also estimate:

$18,087,521 + 11,883,479 \approx 18,000,000 + 12,000,000 = 30,000,000 \approx 29,971,000$

The answer checks.

State. There are $11,883,479$ more people in the Tokyo-Yokohama area.

11. ***Familiarize.*** This is a multistep problem.

To find the new balance, find the total amount of the three checks. Then take that amount away from the original balance. We visualize the situation. We let $t =$ the total amount of the three checks and $n =$the new balance.

$246			
$45	$78	$32	n
t			n

Translate. To find the total amount of the three checks, write an addition sentence.

Amount of first check	+	Amount of second check	+	Amount of third check	=	Total amount
↓	↓	↓	↓	↓	↓	↓
$45	+	$78	+	$32	=	t

Solve. To solve the equation we carry out the addition.

$$\begin{array}{r} \$\ 4\ 5 \\ 7\ 8 \\ +\ 3\ 2 \\ \hline \$1\ 5\ 5 \end{array}$$

The total amount of the three checks is $155.

To find the new balance, we have a "take-away" situation.

Original balance	–	Amount of checks	=	New balance
↓	↓	↓	↓	↓
$246	–	$155	=	n

To solve we carry out the subtraction.

$$\begin{array}{r} \$\ 2\ 4\ 6 \\ -\ 1\ 5\ 5 \\ \hline \$\ \ \ 9\ 1 \end{array}$$

Check. We add the amounts of the three checks and the new balance. The sum should be the original balance.

$$\$45 + \$78 + \$32 + \$91 = \$246$$

The answer checks

State. The new balance is $91.

13. ***Familiarize.*** We first make a drawing. Repeated addition works well here. We let $n =$ the number of calories burned in 5 hours.

133 calories	133 calories	133 calories	133 calories	133 calories
1	2	3	4	5

Translate. Translate to a number sentence.

Number of hours	times	Number of calories	is	Total calories burned
↓	↓	↓	↓	↓
5	$\cdot$	133	=	n

Solve. To solve the equation, we carry out the multiplication.

$$5 \cdot 133 = n$$
$$665 = n \quad \text{Doing the multiplication}$$

Check. We can repeat the calculation. We can also estimate: $5 \cdot 133 \approx 5 \cdot 130 = 650 \approx 665$. The answer checks.

State. In 5 hours, 665 calories would be burned.

15. ***Familiarize.*** We first make a drawing. Repeated addition works well here. We let $x =$ the number of seconds in an hour.

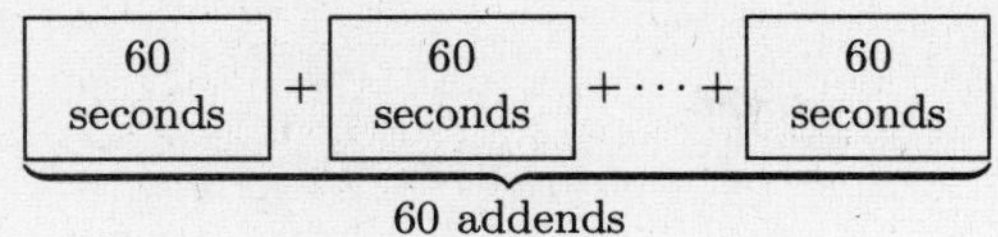

Translate. We translate to an equation.

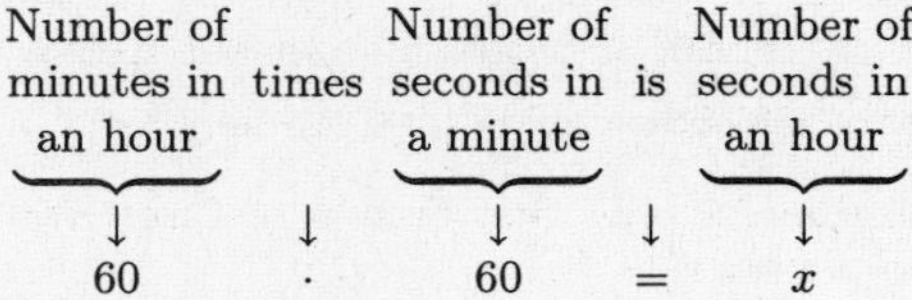

Solve. We carry out the multiplication.

$60 \cdot 60 = x$

$3600 = x$ Doing the multiplication

Check. We can check by repeating the calculation. The answer checks.

State. There are 3600 seconds in an hour.

17. ***Familiarize.*** We first make a drawing.

36 ft

78 ft

Translate. Using the formula for area, we have $A = l \cdot w = 78 \cdot 36$.

Solve. Carry out the multiplication.

$$\begin{array}{r} 78 \\ \times\ 36 \\ \hline 468 \\ 2340 \\ \hline 2808 \end{array}$$

Thus $A = 2808$.

Check. We repeat the calculation. The answer checks.

State. The area is 2808 ft^2.

19. ***Familiarize.*** We visualize the situation. We let $d =$ the diameter of the earth.

d	*d*	*d*	···	*d*
85,965 mi				

Translate. Repeated addition applies here. The following multiplication corresponds to the situation.

The diameter of Jupiter is 11 times the diameter of the earth

$$85,965 = 11 \cdot d$$

Solve. To solve the equation, we divide by 11 on both sides.

$$85,965 = 11 \cdot d$$

$$\frac{85,965}{11} = \frac{11 \cdot d}{11}$$

$$7815 = d$$

Check. To check, we multiply 7815 by 11:

$11 \cdot 7815 = 85,965$

This checks.

State. The diameter of the earth is 7815 mi.

21. ***Familiarize.*** This is a multistep problem.

We must find the total cost of the 8 suits and the total cost of the 3 shirts. The amount spent is the sum of these two totals.

Repeated addition works well in finding the total cost of the 8 suits and the total cost of the 3 shirts. We let $x =$ the total cost of the suits, and $y =$ the total cost of the shirts.

\$195 + \$195 + ··· + \$195

8 addends

\$46 + \$46 + \$46

3 addends

Translate. We translate to two equations.

Number of suits times Cost per suit is Total cost of suits

$$8 \cdot 195 = x$$

Number of shirts times Cost per shirt is Total cost of shirts

$$3 \cdot 46 = y$$

Solve. To solve these equations, we carry out the multiplications.

$$\begin{array}{r} 195 \\ \times\ 8 \\ \hline 1560 \end{array}$$ Thus, $x = \$1560$.

$$\begin{array}{r} 46 \\ \times\ 3 \\ \hline 138 \end{array}$$ Thus $y = \$138$.

We let $a =$ the total amount spent.

Total cost of suits plus Total cost of shirts is Amount spent

$$\$1560 + \$138 = a$$

To solve the equation, carry out the addition.

$$\begin{array}{r} \$\,1560 \\ +\ \ 138 \\ \hline \$\,1698 \end{array}$$

Check. We repeat the calculation. The answer checks.

State. The amount spent is \$1698.

23. ***Familiarize.*** This is a multistep problem.

We must find the area of the lot and the area of the garden. Then we take the area of the garden away from the area of the lot. We let $A =$ the area of the lot and $G =$ the area of the garden.

First, make a drawing of the lot with the garden. The area left over is shaded.

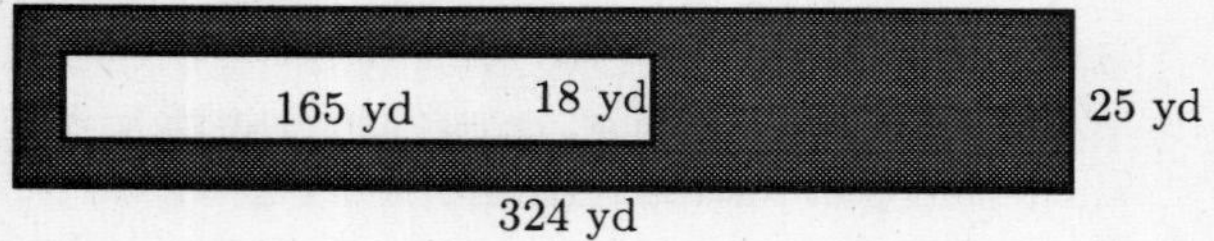

Translate. We use the formula for area twice.

$A = l \cdot w = 324 \cdot 25$

$G = l \cdot w = 165 \cdot 18$

Solve. We carry out the multiplications.

```
    3 2 4
  ×   2 5
  -------
  1 6 2 0
  6 4 8 0
  -------
  8 1 0 0
```

$A = 8100\text{ yd}^2$

```
    1 6 5
  ×   1 8
  -------
  1 3 2 0
  1 6 5 0
  -------
  2 9 7 0
```

$G = 2970\text{ yd}^2$

To find the area left over we have a "take-away" situation. We let $a =$ the area left over.

Area of lot	minus	Area of garden	is	Area left over
↓	↓	↓	↓	↓
8100	−	2970	=	a

To solve we carry out the subtraction .

```
    8 1 0 0
  − 2 9 7 0
  ---------
    5 1 3 0
```

Check. We repeat the calculations. The answer checks.

State. The area left over is 5130 yd^2.

25. ***Familiarize.*** We first draw a picture. We let $n =$ the number of bottles to be filled.

16 oz in each row.

How many rows?

Translate and Solve. We translate to an equation and solve as follows:

$608 \div 16 = n$

```
        3 8
  1 6 ) 6 0 8
        4 8 0
        -----
        1 2 8
        1 2 8
        -----
            0
```

Check. We can check by multiplying the number of bottles by 16: $16 \cdot 38 = 608$. The answer checks.

State. 38 sixteen-oz bottles can be filled.

27. ***Familiarize.*** We first draw a picture. We let $y =$ the number rows.

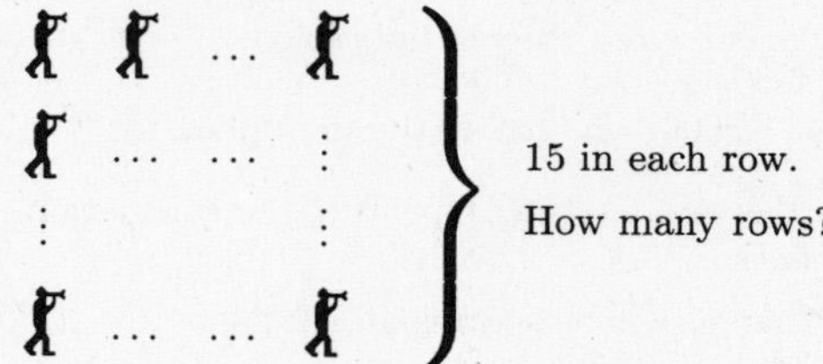

Translate and Solve. We translate to an equation and solve as follows:

$225 \div 15 = y$

```
        1 5
  1 5 ) 2 2 5
        1 5 0
        -----
          7 5
          7 5
        -----
            0
```

Check. We can check by multiplying the number of rows by 15: $15 \cdot 15 = 225$. The answer checks.

State. There are 15 rows.

29. ***Familiarize.*** We first draw a picture. We let $x =$ the amount of each payment.

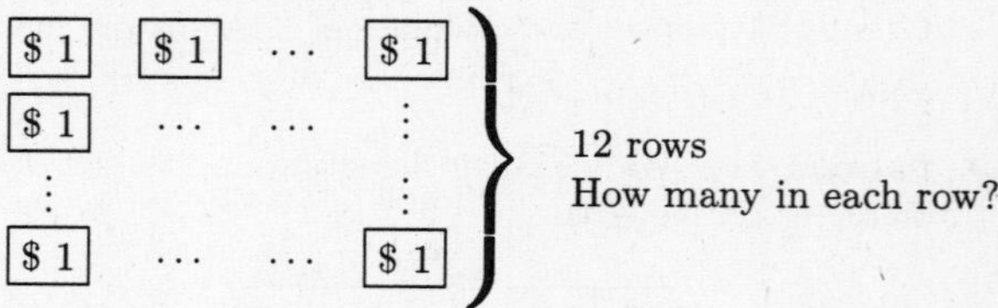

Translate and Solve. We translate to an equation and solve as follows:

$324 \div 12 = x$

```
        2 7
  1 2 ) 3 2 4
        2 4 0
        -----
          8 4
          8 4
        -----
            0
```

Check. We can check by multiplying 27 by 12: $12 \cdot 27 = 324$. The answer checks.

State. Each payment is $27.

31. ***Familiarize.*** We draw a picture. We let $n =$ the number of bags to be filled.

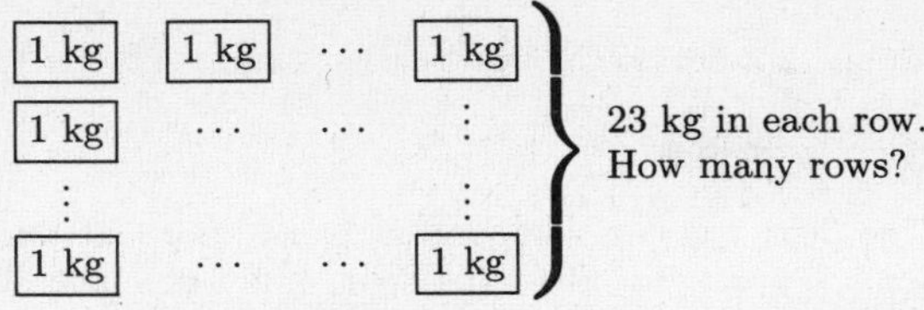

Translate and Solve. We translate to an equation and solve as follows:

$885 \div 23 = n$

$$\begin{array}{r} 38 \\ 23\overline{)885} \\ \underline{690} \\ 195 \\ \underline{184} \\ 11 \end{array}$$

Check. We can check by multiplying the number of bags by 23 and adding the remainder of 11:

$$23 \cdot 38 = 874$$
$$874 + 11 = 885$$

The answer checks.

State. 38 twenty-three-kg bags can be filled. There will be 11 kg of sand left over.

33. ***Familiarize.*** This is a multistep problem.

We must find the total price of the 5 coats. Then we must find how many 20's there are in the total price. Let $p =$ the total price of the coats.

To find the total price of the 5 coats we can use repeated addition.

$$\underbrace{\boxed{\$64} + \boxed{\$64} + \boxed{\$64} + \boxed{\$64} + \boxed{\$64}}_{\text{5 addends}}$$

Translate.

$$\underbrace{\text{Price per coat}}_{\downarrow\ 64} \quad \text{times}\ \overset{\downarrow}{\cdot} \quad \underbrace{\text{Number of coats}}_{\downarrow\ 5} \quad \text{is}\ \overset{\downarrow}{=} \quad \underbrace{\text{Total price of coats}}_{\downarrow\ p}$$

Solve. First we carry out the multiplication.

$$64 \cdot 5 = p$$
$$320 = p$$

The total price of the 5 coats is \$320. Repeated addition can be used again to find how many 20's there are in \$320. We let $x =$ the number of \$20 bills required.

\$320			
\$20	\$20	···	\$20

Translate to an equation and solve.

$$20 \cdot x = 320$$
$$\frac{20 \cdot x}{20} = \frac{320}{20}$$
$$x = 16$$

Check. We repeat the calculation. The answer checks.

State. It took 16 twenty dollar bills.

35. ***Familiarize.*** To find how far apart on the map the cities are we must find how many 55's there are in 605. Repeated addition applies here. We let $d =$ the distance between the cities on the map.

605 mi			
55 mi	55 mi	···	55 mi

Translate.

$$\underbrace{\text{55 mi}}_{\downarrow\ 55} \quad \text{times}\ \overset{\downarrow}{\cdot} \quad \underbrace{\text{How many 55's}}_{\downarrow\ d} \quad \text{is}\ \overset{\downarrow}{=} \quad \underbrace{\text{605 mi?}}_{\downarrow\ 605}$$

Solve. We divide on both sides by 55.

$$55 \cdot d = 605$$
$$\frac{55 \cdot d}{55} = \frac{605}{55}$$
$$d = 11$$

Check. We multiply the number of inches by 55: $55 \cdot 11 = 605$. The answer checks.

State. The cities are 11 inches apart on the map.

Familiarize. When two cities are 14 in. apart on the map, we can use repeated addition to find how far apart they are in reality. We let $n =$ the distance between the cities in reality.

Since 1 in. represents 55 mi, we can draw the following picture:

$$\underbrace{\boxed{\text{55 mi}} + \boxed{\text{55 mi}} + \cdots + \boxed{\text{55 mi}}}_{\text{14 addends}}$$

Translate.

$$\underbrace{\text{Number of inches}}_{\downarrow\ 14} \quad \text{times}\ \overset{\downarrow}{\cdot} \quad \underbrace{\text{Number of miles per inch}}_{\downarrow\ 55} \quad \text{is}\ \overset{\downarrow}{=} \quad \underbrace{\text{Real distance between cities}}_{\downarrow\ n}$$

Solve. We carry out the multiplication.

$14 \times 55 = n$

$770 = n$ Doing the multiplication

Check. We repeat the calculation. The answer checks.

State. The cities are 770 mi apart in reality.

37. ***Familiarize.*** This is a multistep problem.

We must first find the total number of ounces of iced tea. Then we must find how many 16's there are in that total. We let $x =$ the total number of ounces of iced tea.

To find the total number of ounces of iced tea we can use repeated addition.

$$\underbrace{▯ + ▯ + ▯ + ▯ + ▯ + ▯ + \cdots + ▯}_{\text{640 addends}}$$

Translate.

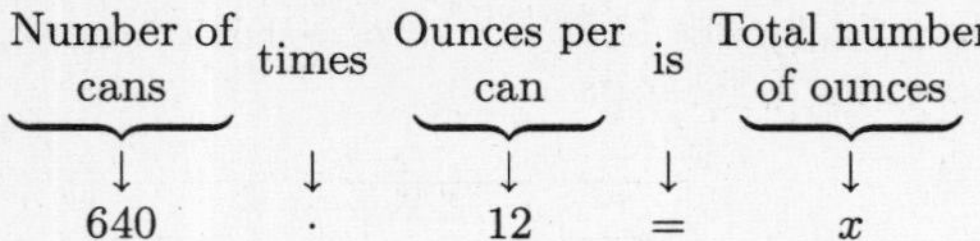

Solve. We carry out the multiplication.

$$640 \cdot 12 = x$$
$$7680 = x \qquad \text{Doing the multiplication}$$

There are 7680 oz of iced tea. We can also use repeated addition to find how many 16's there are in 7680. Let $y =$ this number.

7680 oz			
16 oz	16 oz	···	16 oz

Translate to an equation and solve.

$$16 \cdot y = 7680$$
$$\frac{16 \cdot y}{16} = \frac{7680}{16} \qquad \text{Dividing by 16 on both sides}$$
$$y = 480$$

Check. If we multiply 480 by 16 we get 7680, and $7680 \div 12 = 640$. The answer checks.

State. The iced tea will fill 480 sixteen-oz can.

39. ***Familiarize.*** This is a multistep problem.

We must find how many 100's there are in 3500. Then we must find that number times 15.

First we draw a picture

One pound			
3500 calories			
100 cal	100 cal	···	100 cal
15 min	15 min	···	15 min

In Example 11 it was determined that there are 35 100's in 3500. We let $t =$ the time you have to bicycle to lose a pound.

Translate and Solve. We know that bicycling at 9 mph for 15 min burns off 100 calories, so we need to bicycle for 35 times 15 min in order to burn off one pound. Translate to an equation and solve. We let $t =$ the time required to lose one pound by bicycling.

$$35 \times 15 = t$$
$$525 = t$$

Check. Suppose you bicycle for 525 minutes. If we divide 525 by 15, we get 35, and 35 times 100 is 3500, the number of calories that must be burned off to lose one pound. The answer checks.

State. You must bicycle at 9 mph for 525 min, or 8 hr 45 min, to lose one pound.

41. Round 234,562 to the nearest thousand.

2 3 4, [5] 6 2
↑

The digit 4 is in the thousands place. Consider the next digit to the right. Since the digit, 5, is 5 or higher, round 4 thousands up to 5 thousands. Then change all digits to the right of the thousands place to zeros.

The answer is 235,000.

43. 

45. ***Familiarize.*** We visualize the situation. Let $d =$ the distance light travels in 1 sec.

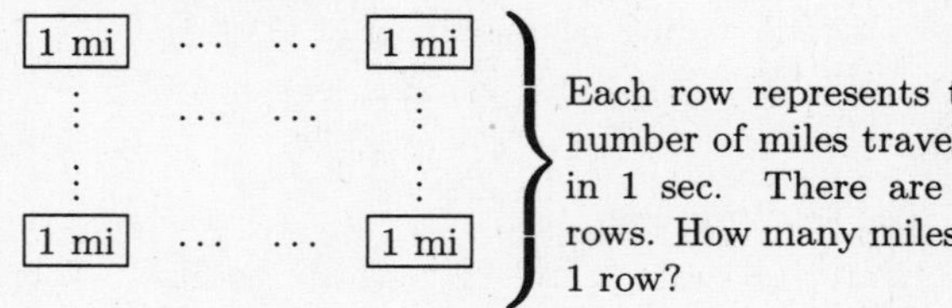

Translate and Solve. We translate to an equation and solve as follows:

$$8,370,000 \div 45 = d$$

```
        1 8 6 0 0 0
4 5 ) 8, 3 7 0, 0 0 0
      4 5 0 0 0 0 0
      3 8 7 0 0 0 0
      3 6 0 0 0 0 0
        2 7 0 0 0 0
        2 7 0 0 0 0
                  0
```

Check. We can check by multiplying the number of miles traveled in 1 sec by 45 sec:

$$45 \cdot 186,000 = 8,370,000$$

The answer checks.

State. Light travels 186,000 mi in 1 sec.

Exercise Set 1.9

1. Exponential notation for $3 \cdot 3 \cdot 3 \cdot 3$ is 3^4.

3. Exponential notation for $5 \cdot 5$ is 5^2.

5. Exponential notation for $7 \cdot 7 \cdot 7 \cdot 7 \cdot 7$ is 7^5.

7. Exponential notation for $10 \cdot 10 \cdot 10$ is 10^3.

9. $7^2 = 7 \cdot 7 = 49$

11. $9^3 = 9 \cdot 9 \cdot 9 = 729$

13. $12^4 = 12 \cdot 12 \cdot 12 \cdot 12 = 20,736$

15. $11^2 = 11 \cdot 11 = 121$

17. $12 + (6 + 4) = 12 + 10$ Doing the calculation inside the parentheses
$= 22$ Adding

19. $52-(40-8)=52-32$ Doing the calculation inside the parentheses
$=20$ Subtracting

21. $1000\div(100\div10)$
$=1000\div10$ Doing the calculation inside the parentheses
$=100$ Dividing

23. $(256\div64)\div4=4\div4$ Doing the calculation inside the parentheses
$=1$ Dividing

25. $(2+5)^2=7^2$ Doing the calculation inside the parentheses
$=49$ Evaluating the exponential expression

27. $15-3+7=12+7$ Doing all additions and subtractions in order from left to right
$=19$

29. $16\cdot24+50=384+50$ Doing all multiplications and divisions in order from left to right
$=434$ Doing all additions and subtractions in order from left to right

31. $83-7\cdot6=83-42$ Doing all multiplications and divisions in order from left to right
$=41$ Doing all additions and subtractions in order from left to right

33. $4+5^2=4+25$ Evaluating the exponential expression
$=29$ Doing all additions and subtractions in order from left to right

35. $40\div5+3\cdot7$
$=8+3\cdot7$ Doing all multiplications and divisions in order from left to right
$=8+21$
$=29$ Doing all additions and subtractions in order from left to right

37. $17\cdot20-(17+20)$
$=17\cdot20-37$ Carrying out the operation inside parentheses
$=340-37$ Doing all multiplications and divisions in order from left to right
$=303$ Doing all additions and subtractions in order from left to right

39. $6\cdot10-4\cdot10$
$=60-40$ Doing all multiplications and divisions in order from left to right
$=20$ Doing all additions and subtractions in order from left to right

41. $300\div5+10$
$=60+10$ Doing all multiplications and divisions in order from left to right
$=70$ Doing all additions and subtractions in order from left to right

43. $3\cdot(2+8)^2-5\cdot(4-3)^2$
$=3\cdot10^2-5\cdot1^2$ Carrying out operations inside parentheses
$=3\cdot100-5\cdot1$ Evaluating the exponential expressions
$=300-5$ Doing all multiplications and divisions in order from left to right
$=295$ Doing all additions and subtractions in order from left to right

45.
$$\begin{aligned}4^2+8^2\div2^2&=16+64\div4\\&=16+16\\&=32\end{aligned}$$

47.
$$\begin{aligned}10^3-10\cdot6-(4+5\cdot6)&=10^3-10\cdot6-(4+30)\\&=10^3-10\cdot6-34\\&=1000-10\cdot6-34\\&=1000-60-34\\&=940-34\\&=906\end{aligned}$$

49.
$$\begin{aligned}6\times11-(7+3)\div5-(6-4)&=6\times11-10\div5-2\\&=66-2-2\\&=64-2\\&=62\end{aligned}$$

51.
$$\begin{aligned}120-3^3\cdot4\div(30-24)&=120-3^3\cdot4\div6\\&=120-27\cdot4\div6\\&=120-108\div6\\&=120-18\\&=102\end{aligned}$$

53. We add the numbers and then divide by the number of addends.

$$\begin{aligned}(\$64+\$97+\$121)\div3&=\$282\div3\\&=\$94\end{aligned}$$

55. $8\times13+\{42\div[18-(6+5)]\}$
$=8\times13+\{42\div[18-11]\}$
$=8\times13+\{42\div7\}$
$=8\times13+6$
$=104+6$
$=110$

57. $[14-(3+5)\div2]-[18\div(8-2)]$
$=[14-8\div2]-[18\div6]$
$=[14-4]-3$
$=10-3$
$=7$

59. $(82-14)\times[(10+45\div5)-(6\cdot6-5\cdot5)]$
$=(82-14)\times[(10+9)-(36-25)]$
$=(82-14)\times[19-11]$
$=68\times8$
$=544$

61.

$$\begin{aligned} x+341 &= 793 \\ x+341-341 &= 793-341 \\ x &= 452 \end{aligned}$$

The solution is 452.

63.

65. $15(23-4\cdot 2)^3 \div (3\cdot 25)$

$= 15(23-8)^3 \div 75$	Multiplying inside parentheses
$= 15\cdot 15^3 \div 75$	Subtracting inside parentheses
$= 15\cdot 3375 \div 75$	Evaluating the exponential expression
$= 50,625 \div 75$	Doing all multiplications and divisions in order from left to right
$= 675$	

67. $1+5\cdot 4+3 = 1+20+3$
$= 24$ Correct answer

To make the incorrect answer correct we add parentheses:

$1+5\cdot(4+3) = 36$

69. $12\div 4+2\cdot 3-2 = 3+6-2$
$= 7$ Correct answer

To make the incorrect answer correct we add parentheses:

$12\div(4+2)\cdot 3-2 = 4$

Chapter 2

Introduction to Integers and Algebraic Expressions

Exercise Set 2.1

1. The integer -70 corresponds to a temperature of 70° below zero.

3. The integer -3 corresponds to 3 sec before liftoff; the integer 128 corresponds to 128 sec after liftoff.

5. Since 6 is to the right of 0, we have $6 > 0$.

7. Since -9 is to the left of 5, we have $-9 < 5$.

9. Since -6 is to the left of 6, we have $-6 < 6$.

11. Since -8 is to the left of -5, we have $-8 < -5$.

13. Since -5 is to the right of -11, we have $-5 > -11$.

15. Since -6 is to the left of -5, we have $-6 < -5$.

17. The distance from -3 to 0 is 3, so $|-3| = 3$.

19. The distance from 10 to 0 is 10, so $|10| = 10$.

21. The distance from 0 to 0 is 0, so $|0| = 0$.

23. The distance from -24 to 0 is 24, so $|-24| = 24$.

25. The distance from 53 to 0 is 53, so $|53| = 53$.

27. This distance from -8 to 0 is 8, so $|-8| = 8$.

29. To find the opposite of x when x is -6, we reflect -6 to the other side of 0.

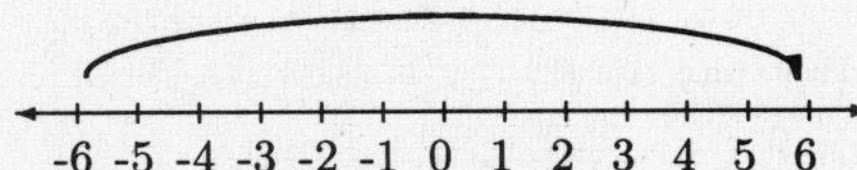

We have $-(-6) = 6$. The opposite of -6 is 6.

31. To find the opposite of x when x is 6, we reflect 6 to the other side of 0.

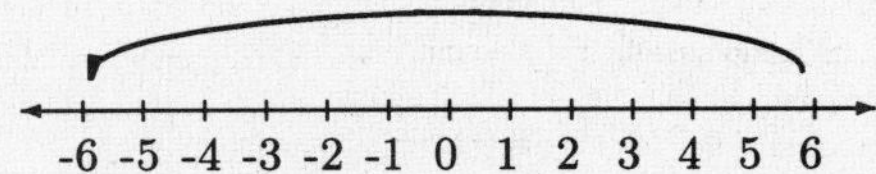

We have $-(6) = -6$. The opposite of 6 is -6.

33. To find the opposite of x when x is -12, we reflect -12 to the other side of 0. We have $-(-12) = 12$.

35. To find the opposite of x when x is 70, we reflect 70 to the other side of 0. We have $-(70) = -70$.

37. The opposite of -1 is 1. $\quad -(-1) = 1$

39. The opposite of 7 is -7 $\quad -(7) = -7$

41. The opposite of -14 is 14. $\quad -(-14) = 14$

43. The opposite of 0 is 0. $\quad -(0) = 0$

45. We replace x by -7. We wish to find $-(-(-7))$. Reflecting -7 to the other side of 0 gives us 7 and then reflecting back gives us -7. Thus, $-(-x) = -7$ when x is -7.

47. We replace x by 1. We wish to find $-(-1)$. Reflecting 1 to the other side of 0 gives us -1 and then reflecting back gives us 1. Thus, $-(-x) = 1$ when x is 1.

49. We replace x by 0. We wish to find $-(-0)$. When we try to reflect 0 "to the other side of 0" we go nowhere. The same thing happens when we try to reflect back. Thus, $-(-x) = 0$ when x is 0.

51. We replace x by -34. We wish to find $-(-(-34))$. Reflecting -34 to the other side of 0 gives us 34 and then reflecting back gives us -34. Thus, $-(-x) = -34$ when x is -34.

53.
$$\begin{array}{r} {\scriptstyle 1\,1\;\;} \\ 3\,2\,7 \\ +\,4\,9\,8 \\ \hline 8\,2\,5 \end{array}$$

55.
$$\begin{array}{r} {\scriptstyle 2\;\;} \\ {\scriptstyle 3\;\;} \\ 2\,0\,9 \\ \times\;\;3\,4 \\ \hline 8\,3\,6 \\ 6\,2\,7\,0 \\ \hline 7\,1\,0\,6 \end{array}$$

57. $9^2 = 9 \cdot 9 = 81$

59.

61. Answers may vary. On many scientific calculators you would first find the sum and then use the +/− key.

9	7	2	+	5	8	9	=	+/−

63. The integers whose distance from 0 is less than 2 are -1, 0, and 1. These are the solutions.

65. First note that $2^{10} = 1024$, $|-6| = 6$, $|3| = 3$, $2^7 = 128$, $7^2 = 49$, and $10^2 = 100$. Listing the entire set of integers in order from least to greatest, we have -100, -5, 0, $|3|$, 4, $|-6|$, 7^2, 10^2, 2^7, 2^{10}.

Exercise Set 2.2

1. Add: $-8+3$

Start at -8. Move 3 units to the right.

-9 -8 -7 -6 -5 -4 -3 -2 -1 0 1 2 3 4 5 6 7 8 9

$-8+3=-5$

3. Add: $-9+5$

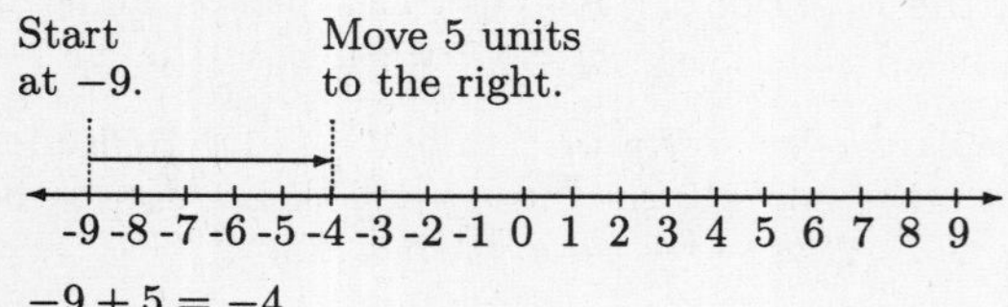

$-9+5=-4$

5. Add: $5+(-5)$

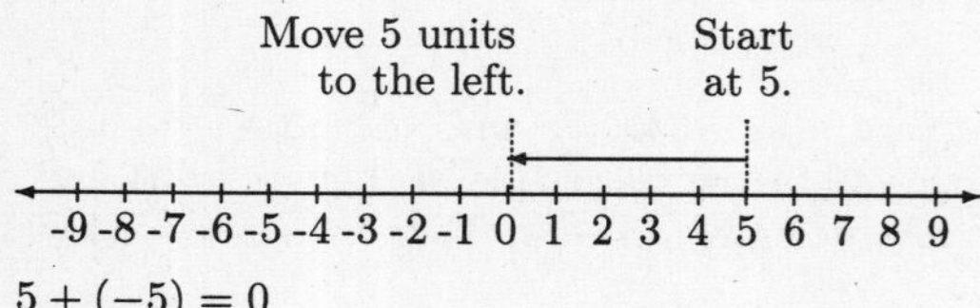

$5+(-5)=0$

7. Add: $-8+(-5)$

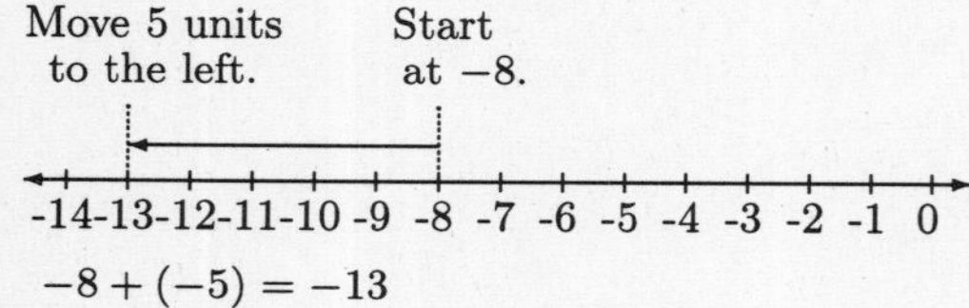

$-8+(-5)=-13$

9. $-5+(-11)$ Two negative integers

Add the absolute values: $5+11=16$
Make the answer negative: $-5+(-11)=-16$

11. $-6+(-5)$ Two negative integers

Add the absolute values: $6+5=11$
Make the answer negative: $-6+(-5)=-11$

13. $9+(-9)=0$

For any integer a, $a+(-a)=0$.

15. $-2+2=0$

For any integer a, $-a+a=0$.

17. $0+8=8$

For any integer a, $0+a=a$.

19. $0+(-8)=-8$

For any integer a, $0+a=a$.

21. $-25+0=-25$

For any integer a, $a+0=a$.

23. $0+(-27)=-27$

For any integer a, $0+a=a$.

25. $17+(-17)=0$

For any integer a, $a+(-a)=0$.

27. $-18+18=0$

For any integer a, $-a+a=0$.

29. $8+(-5)$ The absolute values are 8 and 5. The difference is 3. The positive number has the larger absolute value, so the answer is positive. $8+(-5)=3$

31. $-4+(-5)$ Two negative integers

Add the absolute values: $4+5=9$
Make the answer negative: $-4+(-5)=-9$

33. $0+(-5)=-5$

For any integer a, $0+a=a$.

35. $14+(-5)$ The absolute values are 14 and 5. The difference is 9. The positive number has the larger absolute value, so the answer is positive. $14+(-5)=9$

37. $-11+8$ The absolute values are 11 and 8. The difference is 3. Since the negative number has the larger absolute value, the answer is negative. $-11+8=-3$

39. $-19+19=0$

For any integer a, $-a+a=0$.

41. $-17+7$ The absolute values are 17 and 7. The difference is 10. Since the negative number has the larger absolute value, the answer is negative. $-17+7=-10$

43. $-17+(-7)$ Two negative integers

Add the absolute values: $17+7=24$
Make the answer negative: $-17+(-7)=-24$

45. $11+(-16)$ The absolute values are 11 and 16. The difference is 5. Since the negative number has the larger absolute value, the answer is negative. $11+(-16)=-5$

47. $-15+(-6)$ Two negative integers

Add the absolute values: $15+6=21$
Make the answer negative: $-15+(-6)=-21$

49. $11+(-9)$ The absolute values are 11 and 9. The difference is 2. The positive number has the larger absolute value, so the answer is positive. $11+(-9)=2$

51. $-20+(-6)$ Two negative integers

Add the absolute values: $20+6=26$
Make the answer negative: $-20+(-6)=-26$

53. We will add from left to right.

$$\begin{aligned}-15+(-7)+1 &= -22+1\\ &= -21\end{aligned}$$

55. We will add from left to right.

$$\begin{aligned}30+(-10)+5 &= 20+5\\ &= 25\end{aligned}$$

57. We will add from left to right.

$$-23+(-9)+15 = -32+15$$
$$= -17$$

59. We will add from left to right.

$$40+(-40)+6 = 0+6$$
$$= 6$$

61. We will add from left to right.

$$85+(-65)+(-12) = 20+(-12)$$
$$= 8$$

63. We will add from left to right.

$$-24+(-37)+(-19)+(-45)+(-35)$$
$$= -61+(-19)+(-45)+(-35)$$
$$= -80+(-45)+(-35)$$
$$= -125+(-35)$$
$$= -160$$

65. $28+(-44)+17+31+(-94)$

a) $28+17+31=76$ Adding the positive numbers

b) $-44+(-94)=-138$ Adding the negative numbers

c) $76+(-138)=-62$ Adding the results

67. 3 ten thousands + 9 thousands + 4 hundreds + 1 ten + 7 ones

69. a) Locate the digit in the thousands place.

3 2, [8] 3 1
↑

b) Then consider the next digit to the right.

c) Since the digit is 5 or higher, round 2 thousands up to 3 thousands.

d) Change all digits to the right of thousands to zeros.

The answer is 33,000.

71.

$$\begin{array}{r} 32 \\ 9\overline{)288} \\ \underline{270} \\ 18 \\ \underline{18} \\ 0 \end{array}$$

The answer is 32.

73. 

75. We use a calculator.

$-3496+(-2987)=-6483$

77. Think of starting at x on a number line and moving 7 units to the left. When x is 7 we move to 0. When x is less than 7, we move to a negative number, and when x is greater than 7 we move to a positive number. Thus, for all numbers x greater than 7, $x+(-7)$ is positive.

79. If n is positive, $-n$ is negative. Then $-n+m$, the sum of two negative numbers, is negative.

Exercise Set 2.3

1. $2-9=2+(-9)=-7$

3. $0-4=0+(-4)=-4$

5. $-8-(-2)=-8+2=-6$

7. $-11-(-11)=-11+11=0$

9. $12-16=12+(-16)=-4$

11. $20-27=20+(-27)=-7$

13. $-9-(-3)=-9+3=-6$

15. $-40-(-40)=-40+40=0$

17. $7-7=7+(-7)=0$

19. $5-(-5)=5+5=10$

21. $8-(-3)=8+3=11$

23. $-6-8=-6+(-8)=-14$

25. $-4-(-9)=-4+9=5$

27. $1-8=1+(-8)=-7$

29. $-6-(-5)=-6+5=-1$

31. $8-(-10)=8+10=18$

33. $0-10=0+(-10)=-10$

35. $-5-(-2)=-5+2=-3$

37. $-7-14=-7+(-14)=-21$

39. $0-(-5)=0+5=5$

41. $-8-0=-8+0=-8$

43. $7-(-5)=7+5=12$

45. $2-25=2+(-25)=-23$

47. $-42-26=-42+(-26)=-68$

49. $-71-2=-71+(-2)=-73$

51. $24-(-92)=24+92=116$

53. $-50-(-50)=-50+50=0$

55. $9-(-4)+7-(-2)=9+4+7+2=22$

57. $-31+(-28)-(-14)-17$

$= -31+(-28)+14+(-17)$

$= -31+(-28)+(-17)+14$ Using a commutative law

$= -76+14$ Adding the negative numbers

$= -62$

59. $-34-28+(-33)-44=(-34)+(-28)+(-33)+(-44)=-139$

61. $-93-(-84)-41-(-56)$
$=-93+84+(-41)+56$
$=-93+(-41)+84+56$ Using a commutative law
$=-134+140$ Adding negatives and adding positives
$=6$

63. $-5-(-30)+30+40-(-12)$
$=-5+30+30+40+12$
$=-5+112$ Adding the positive numbers
$=107$

65. $132-(-21)+45-(-21)=132+21+45+21=219$

67. The integer 8 corresponds to 8 lb above the ideal weight, and -9 corresponds to 9 lb below it. We subtract the lower weight from the higher weight:

$$8-(-9)=8+9=17$$

Rosa lost 17 lb.

69. To find the elevation that is 360 ft deeper than -2860 ft, we subtract the additional depth from the current depth:

$$-2860-360=-2860+(-360)=-3220$$

In 1998 the elevation of the world's deepest offshore oil well will be -3220 ft.

71. We draw a picture of the situation.

Sea level
$-28,538$ ft
Puerto Rico Trench
$-34,370$ ft
To find
Marianas Trench

We subtract the lower altitude from the higher altitude: $-28,538-(-34,370)=-28,538+34,370=5832$. The Puerto Rico Trench is 5832 ft higher than the Marianas Trench.

73. From the submarine's original depth of -30 m we subtract 50 m and add 18 m:

$$-30-50+18=-30+(-50)+18=-62$$

The submarine's new depth is -62 m, or 62 m below the surface.

75. $4^3=4\cdot 4\cdot 4=64$

77. ***Familiarize***. Let $n=$ the number of 12-oz cans that can be filled. We think of an array consisting of 96 oz with 12 oz in each row.

The number n corresponds to the number of rows in the array.

Translate and Solve. We translate to an equation and solve it.

$96\div 12=n$

$$\begin{array}{r} 8 \\ 12\overline{)96} \\ \underline{96} \\ 0 \end{array}$$

Check. We multiply the number of cans by 12: $8\cdot 12=96$. The result checks.

State. Eight 12-oz cans can be filled.

79. 

81. Use a calculator to do this exercise.

$123,907-433,789=-309,882$

83. True

85. True

87. True by the definition of opposites.

89. True

91. The changes during weeks 1 to 5 are represented by the integers -13, -16, 36, -11, and 19, respectively. We add to find the total rise or fall:

$$-13+(-16)+36+(-11)+19=15$$

The market rose 15 points during the 5 week period.

Exercise Set 2.4

1. $-2\cdot 5=-10$

3. $-9\cdot 2=-18$

5. $9\cdot(-5)=-45$

7. $-10\cdot 3=-30$

9. $-2\cdot(-5)=10$

11. $-9\cdot(-2)=18$

13. $-9\cdot(-5)=45$

15. $-10(-3)=30$

17. $12(-10)=-120$

19. $-6(-50)=300$

21. $(-35)(-1)=35$

23. $(-20)17=-340$

25. $-23\cdot 0=0$

27. $0(-14)=0$

29. $2\cdot(-7)\cdot 5$
$=-14\cdot 5$ Multiplying the first two numbers
$=-70$

31. $7(-4)(-3)5$
$=7\cdot 12\cdot 5$ Multiplying the negative numbers
$=84\cdot 5$
$=420$

33. $-2(-5)(-7)$
$=10\cdot(-7)$ Multiplying the first two numbers
$=-70$

35. $(-3)(-5)(-2)(-1)$
$= 15 \cdot (-2) \cdot (-1)$ Multiplying the first two numbers
$= 15 \cdot 2$ Multiplying the last two numbers
$= 30$

37. $-14(34)12(-1)$
$= -476(-12)$ Multiplying the first two numbers and the last two numbers
$= 5712$

39. $(-17)(-29)0 \cdot 3$
$= 493 \cdot 0$ Multiplying the first two numbers and the last two numbers
$= 0$

(We might have noted at the outset that the product would be 0 since one of the numbers in the product is 0.)

41. $(-5)^2 = (-5)(-5) = 25$

43. $(-10)^3 = (-10)(-10)(-10)$
$= 100(-10)$
$= -1000$

45. $(-5)^4 = (-5)(-5)(-5)(-5)$
$= 25 \cdot 25$
$= 625$

47. $(-2)^7 = (-2)(-2)(-2)(-2)(-2)(-2)(-2)$
$= 4 \cdot 4 \cdot 4 \cdot (-2)$
$= 16(-8)$
$= -128$

49. $(-3)^5 = (-3)(-3)(-3)(-3)(-3)$
$= 9 \cdot 9 \cdot (-3)$
$= 81(-3)$
$= -243$

51. $(-1)^{12}$
$= (-1) \cdot (-1) \cdot (-1) \cdot (-1) \cdot (-1) \cdot (-1) \cdot (-1) \cdot (-1) \cdot$
$(-1) \cdot (-1) \cdot (-1) \cdot (-1)$
$= 1 \cdot 1 \cdot 1 \cdot 1 \cdot 1 \cdot 1$
$= 1 \cdot 1 \cdot 1$
$= 1 \cdot 1$
$= 1$

53. a) Locate the digit in the hundreds place.

5 3 2, 4 [5] 1
↑

b) Then consider the next digit to the right.

c) Since that digit is 5 or higher, round 4 hundreds up to 5 hundreds.

d) Change all digits to the right of hundreds to zeros.

The answer is 532,500.

55.
```
      8 0
3 6 ) 2 8 8 0
      2 8 8 0
            0
            0
            0
```
The answer is 80.

57. ***Familiarize***. We first make a drawing. We let A = the area.

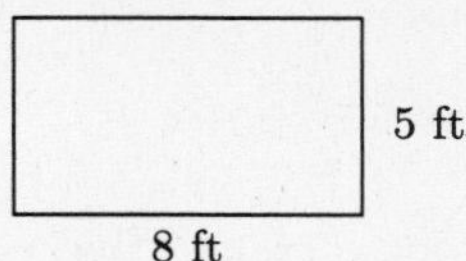

Translate. Using the formulas for area, we have $A = l \cdot w = 8 \cdot 5$.

Solve. We carry out the multiplication.

$A = 8 \cdot 5 = 40$

Check. We repeat our calculation.

State. The area of the rug is 40 ft^2.

59. 

61. We use a calculator.
$-935(238 - 243)^3 = -935(-5)^3$
$= -935(-125)$
$= 116,875$

63. The new balance will be
$\$68 - 7(\$13) = \$68 - \$91 = -\$23$

65. $(-5)^3(-1)^{421} = -125(-1) = 125$

67. $|(-2)^3 + 4^2| - (2-7)^2 = |(-2)^3 + 4^2| - (-5)^2 =$
$|-8+16| - 25 = |8| - 25 = 8 - 25 = -17$

69. a) If $[(-5)^m]^n$ is to be negative, first m must be an odd number so that $(-5)^m$ is negative. Similarly, n must also be odd in order for $[(-5)^m]^n$ to be negative. Thus, both m and n must be odd numbers.

b) If $[(-5)^m]^n$ is to be positive, at least one of m and n must be an even number. For example, if m is even then $(-5)^m$ is positive and so is $[(-5)^m]^n$ regardless of whether n is even or odd. If m is odd, then $(-5)^m$ is negative and n must be even in order for $[(-5)^m]^n$ to be positive.

Exercise Set 2.5

1. $42 \div (-6) = -7$ Check: $-7(-6) = 42$

3. $\dfrac{28}{-2} = -14$ Check: $-14(-2) = 28$

5. $\dfrac{-16}{8} = -2$ Check: $-2 \cdot 8 = -16$

7. $\dfrac{-48}{-12} = 4$ Check: $4(-12) = -48$

9. $\dfrac{-72}{8} = -9$ Check: $-9 \cdot 8 = -72$

11. $-100 \div (-50) = 2$ Check: $2(-50) = -100$

13. $-344 \div 8 = -43$ Check: $-43 \cdot 8 = -344$

15. $\dfrac{200}{-25} = -8$ Check: $-8(-25) = 200$

17. Undefined

19. $\dfrac{88}{-11} = -8$ Check: $-8(-11) = 88$

21. $-\dfrac{276}{12} = \dfrac{-276}{12} = -23$ Check: $-23 \cdot 12 = -276$

23. $\dfrac{0}{-9} = 0$ Check: $0 \cdot (-9) = 0$

25. $\dfrac{19}{-1} = -19$ Check: $-19(-1) = 19$

27. $-41 \div 1 = -41$ Check: $-41 \cdot 1 = -41$

29. $8 - 2 \cdot 3 - 9 = 8 - 6 - 9$ Multiplying
$= 2 - 9$ Doing all additions and subtractions in order
$= -7$ from left to right

31. $8 - 2(3 - 9) = 8 - 2(-6)$ Subtracting inside parentheses
$= 8 + 12$ Multiplying
$= 20$ Adding

33. $16 \cdot (-24) + 50 = -384 + 50$ Multiplying
$= -334$ Adding

35. $40 - 3^2 - 2^3$
$= 40 - 9 - 8$ Evaluating the exponential expressions
$= 31 - 8$ Doing all additions and subtractions
$= 23$ in order from left to right

37. $4 \cdot (6 + 8)/(4 + 3)$
$= 4 \cdot 14/7$ Adding inside parentheses
$= 56/7$ Doing all multiplications and divisions
$= 8$ in order from left to right

39. $4 \cdot 5 - 2 \cdot 6 + 4 = 20 - 12 + 4$ Multiplying
$= 8 + 4$
$= 12$

41. $\dfrac{9^2 - 1}{1 - 3^2}$
$= \dfrac{81 - 1}{1 - 9}$ Evaluating the exponential expressions
$= \dfrac{80}{-8}$ Subtracting in the numerator and in the denominator
$= -10$

43. $8(-7) + 6(-5) = -56 - 30$ Multiplying
$= -86$

45. $20 \div 5(-3) + 3 = 4(-3) + 3$ Dividing
$= -12 + 3$ Multiplying
$= -9$ Adding

47. $9 \div (-3) \cdot 16 \div 8$
$= -3 \cdot 16 \div 8$ Doing all multiplications
$= -48 \div 8$ and divisions in order
$= -6$ from left to right

49. $2 \cdot 3^2 \div 6 = 2 \cdot 9 \div 6$ Evaluating the exponential expression
$= 18 \div 6$ Multiplying
$= 3$ Dividing

51. $(3 - 8)^2 \div (-1)$
$= (-5)^2 \div (-1)$ Subtracting inside parentheses
$= 25 \div (-1)$ Evaluating the exponential expression
$= -25$

53. $12 - 20^3 = 12 - 8000$
$= -7988$

55. $2 \times 10^3 - 5000 = 2 \times 1000 - 5000$
$= 2000 - 5000$
$= -3000$

57. $6[9 - (3 - 4)] = 6[9 - (-1)]$ Subtracting inside the innermost parentheses
$= 6[9 + 1]$
$= 6[10]$
$= 60$

59. $-1000 \div (-100) \div 10 = 10 \div 10$ Doing the divisions in order
$= 1$ from left to right

61. $8 - |7 - 9| \cdot 3 = 8 - |-2| \cdot 3$
$= 8 - 2 \cdot 3$
$= 8 - 6$
$= 2$

63. $9 - |7 - 3^2| = 9 - |7 - 9|$
$= 9 - |-2|$
$= 9 - 2$
$= 7$

65. $\dfrac{(-5)^3 + 17}{10(2 - 6) - 2(5 + 2)}$
$= \dfrac{-125 + 17}{10(2 - 6) - 2(5 + 2)}$ Evaluating the exponential expression
$= \dfrac{-125 + 17}{10(-4) - 2 \cdot 7}$ Doing the calculations within parentheses
$= \dfrac{-125 + 17}{-40 - 14}$ Multiplying
$= \dfrac{-108}{-54}$ Adding and subtracting
$= 2$

67. ***Familiarize***. We first make a drawing. We let c = the number of chairs in the classroom. Repeated addition works well here.

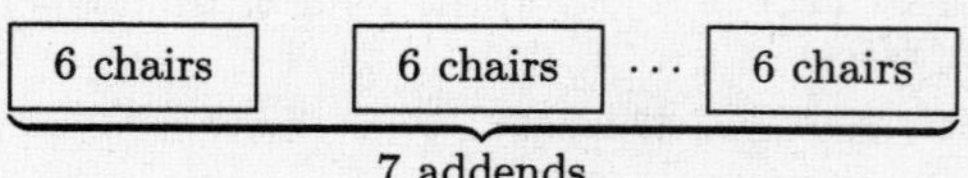

Translate. We translate to an equation.

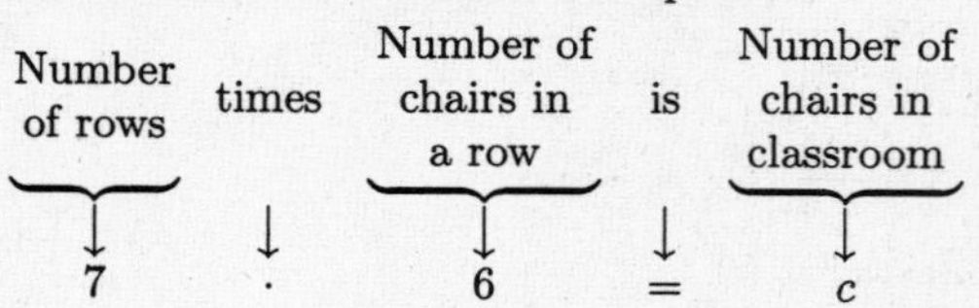

Solve. We carry out the multiplication.

$7 \cdot 6 = 42$

Check. We can check by repeating our calculation or by repeated addition. The result checks.

State. There are 42 chairs in the classroom.

69. *Familiarize*. We let $c =$ the number of calories in a 1-oz serving. Think of a rectangular array consisting of 1050 calories arranged in 7 rows. The number c is the number of calories in each row.

Translate and Solve. We translate to an equation and solve it.

$1050 \div 7 = c$

$$\begin{array}{r} 150 \\ 7\overline{)1050} \\ \underline{700} \\ 350 \\ \underline{350} \\ 0 \\ \underline{0} \\ 0 \end{array}$$

Check. We multiply the number of calories in 1 oz by 7:

$150 \cdot 7 = 1050$

The result checks

State. There are 150 calories in a 1-oz serving.

71. 

73. We use a calculator.

$$\frac{19-17^2}{13^2-34} = \frac{19-289}{169-34} = \frac{-270}{135} = -2$$

75. $-n$ and m are both negative, so $\frac{-n}{m}$ is the quotient of two negative numbers and, thus, is positive.

77. $\frac{-n}{m}$ is positive (see Exercise 75), so $-\left(\frac{-n}{m}\right)$ is the opposite of a positive number and, thus, is negative.

79. $-n$ is negative and $-m$ is positive, so $\frac{-n}{-m}$ is the quotient of a negative and a positive number and, thus, is negative. Then $-\left(\frac{-n}{-m}\right)$ is the opposite of a negative number and, thus, is positive.

Exercise Set 2.6

1. $6x = 6 \cdot 7 = 42$

3. $\frac{x}{y} = \frac{9}{-3} = -3$

5. $\frac{3p}{q} = \frac{3 \cdot 2}{6} = \frac{6}{6} = 1$

7. $\frac{x+y}{5} = \frac{-10+20}{5} = \frac{10}{5} = 2$

9. $3 + 5 \cdot x = 3 + 5 \cdot 2 = 3 + 10 = 13$

11. $2l + 2w = 2 \cdot 3 + 2 \cdot 4 = 6 + 8 = 14$

13. $2(l + w) = 2(3 + 4) = 2 \cdot 7 = 14$

15. $7a - 7b = 7 \cdot 5 - 7 \cdot 2 = 35 - 14 = 21$

17. $7(a - b) = 7(5 - 2) = 7 \cdot 3 = 21$

19. $16t^2 = 16 \cdot 3^2 = 16 \cdot 9 = 144$

21. $$\begin{aligned} 9m - m^2 &= 9(-4) - (-4)^2 \\ &= 9(-4) - 16 \\ &= -36 - 16 \\ &= -52 \end{aligned}$$

23. $$\begin{aligned} a + (b - a)^2 &= 6 + (4 - 6)^2 \\ &= 6 + (-2)^2 \\ &= 6 + 4 \\ &= 10 \end{aligned}$$

25. $$\begin{aligned} a + b - a^2 &= 6 + 4 - 6^2 \\ &= 6 + 4 - 36 \\ &= 10 - 36 \\ &= -26 \end{aligned}$$

27. $$\begin{aligned} \frac{5(F-32)}{9} &= \frac{5(68-32)}{9} \\ &= \frac{5 \cdot 36}{9} \\ &= \frac{180}{9} \\ &= 20 \end{aligned}$$

29. $-\frac{a}{b}$, $\frac{-a}{b}$, and $\frac{a}{-b}$ all represent the same number. Thus we can also write $-\frac{3}{a}$ as $\frac{-3}{a}$ and $\frac{3}{-a}$.

31. $-\frac{a}{b}$, $\frac{-a}{b}$, and $\frac{a}{-b}$ all represent the same number. Thus we can also write $\frac{-n}{b}$ as $-\frac{n}{b}$ and $\frac{n}{-b}$.

33. $-\frac{a}{b}$, $\frac{-a}{b}$, and $\frac{a}{-b}$ all represent the same number. Thus we can also write $\frac{9}{-p}$ as $-\frac{9}{p}$ and $\frac{-9}{p}$.

35. $-\frac{a}{b}$, $\frac{-a}{b}$, and $\frac{a}{-b}$ all represent the same number. Thus we can also write $\frac{-14}{w}$ as $-\frac{14}{w}$ and $\frac{14}{-w}$.

37. $$\frac{-a}{b} = \frac{-35}{7} = -5$$
$$\frac{a}{-b} = \frac{35}{-7} = -5$$
$$-\frac{a}{b} = -\frac{35}{7} = -5$$

39. $$\frac{-a}{b} = \frac{-81}{3} = -27$$
$$\frac{a}{-b} = \frac{81}{-3} = -27$$
$$-\frac{a}{b} = -\frac{81}{3} = -27$$

41. $(-3x)^2 = (-3 \cdot 7)^2 = (-21)^2 = 441$
$-3x^2 = -3(7)^2 = -3 \cdot 49 = -147$

43. $5x^2 = 5(2)^2 = 5 \cdot 4 = 20$
$5x^2 = 5(-2)^2 = 5 \cdot 4 = 20$

45. $x^3 = 6^3 = 6 \cdot 6 \cdot 6 = 216$
$x^3 = (-6)^3 = (-6) \cdot (-6) \cdot (-6) = -216$

47. $x^6 = 1^6 = 1 \cdot 1 \cdot 1 \cdot 1 \cdot 1 \cdot 1 = 1$
$x^6 = (-1)^6 = (-1) \cdot (-1) \cdot (-1) \cdot (-1) \cdot (-1) \cdot (-1) = 1$

49. $a^7 = 2^7 = 2 \cdot 2 \cdot 2 \cdot 2 \cdot 2 \cdot 2 \cdot 2 = 128$
$a^7 = (-2)^7 = (-2)(-2)(-2)(-2)(-2)(-2)(-2) = -128$

51.
$$\begin{aligned} m^4 - 5m^2 &= 3^4 - 5 \cdot 3^2 \\ &= 81 - 5 \cdot 9 \\ &= 81 - 45 \\ &= 36 \end{aligned}$$
$$\begin{aligned} m^4 - 5m^2 &= (-3)^4 - 5(-3)^2 \\ &= 81 - 5 \cdot 9 \\ &= 81 - 45 \\ &= 36 \end{aligned}$$

53.
$$\begin{aligned} a + 4a^3 &= 5 + 4 \cdot 5^3 \\ &= 5 + 4 \cdot 125 \\ &= 5 + 500 \\ &= 505 \end{aligned}$$
$$\begin{aligned} a + 4a^3 &= -5 + 4(-5)^3 \\ &= -5 + 4(-125) \\ &= -5 - 500 \\ &= -505 \end{aligned}$$

55.
$$\begin{aligned} x^2 + 5x \div 2 &= 6^2 + 5 \cdot 6 \div 2 \\ &= 36 + 5 \cdot 6 \div 2 \\ &= 36 + 30 \div 2 \\ &= 36 + 15 \\ &= 51 \end{aligned}$$
$$\begin{aligned} x^2 + 5x \div 2 &= (-6)^2 + 5(-6) \div 2 \\ &= 36 + 5(-6) \div 2 \\ &= 36 - 30 \div 2 \\ &= 36 - 15 \\ &= 21 \end{aligned}$$

57.
$$\begin{aligned} m^3 - m^2 &= 5^3 - 5^2 \\ &= 125 - 25 \\ &= 100 \end{aligned}$$
$$\begin{aligned} m^3 - m^2 &= (-5)^3 - (-5)^2 \\ &= -125 - 25 \\ &= -150 \end{aligned}$$

59. Twenty-three million, forty-three thousand, nine hundred twenty-one

61.
$$\begin{array}{r} 5\,2\,8\,3 \\ -\,2\,4\,7\,5 \\ \hline \end{array} \qquad \begin{array}{r} 5\,2\,8\,0 \\ -\,2\,4\,8\,0 \\ \hline 2\,8\,0\,0 \end{array}$$

63. ***Familiarize.*** Since we are combining snowfall amounts, addition can be used. We let $s =$ the total snowfall.

Translate. We translate to an equation.

$5 + 8 = s$

Solve. We carry out the addition.

$5 + 8 = 13$

Thus, $13 = s$, or $s = 13$.

Check. We repeat the calculation.

State. It snowed 13 in. altogether.

65. 

67. We use a calculator.
$$\begin{aligned} a - b^3 + 17a &= 19 - (-16)^3 + 17 \cdot 19 \\ &= 19 - (-4096) + 17 \cdot 19 \\ &= 19 - (-4096) + 323 \\ &= 19 + 4096 + 323 \\ &= 4115 + 323 \\ &= 4438 \end{aligned}$$

69.
$$\begin{aligned} a^{1996} - a^{1997} &= (-1)^{1996} - (-1)^{1997} \\ &= 1 - (-1) \\ &= 1 + 1 \\ &= 2 \end{aligned}$$

71.
$$\begin{aligned} (m^3 - mn)^m &= (4^3 - 4 \cdot 6)^4 \\ &= (64 - 4 \cdot 6)^4 \\ &= (64 - 24)^4 \\ &= 40^4 \\ &= 2,560,000 \end{aligned}$$

73. True

75. True

Exercise Set 2.7

1. $3(a + b) = 3 \cdot a + 3 \cdot b = 3a + 3b$

3. $4(x + 1) = 4 \cdot x + 4 \cdot 1 = 4x + 4$

5. $2(b + 5) = 2 \cdot b + 2 \cdot 5 = 2b + 10$

7. $7(1 - t) = 7 \cdot 1 - 7 \cdot t = 7 - 7t$

9. $6(5x + 2) = 6 \cdot 5x + 6 \cdot 2 = 30x + 12$

11. $7(x + 4 + 6y) = 7 \cdot x + 7 \cdot 4 + 7 \cdot 6y = 7x + 28 + 42y$

13. $-7(y - 2) = -7 \cdot y - (-7) \cdot 2 = -7y - (-14) = -7y + 14$

15. $-9(-5x - 6y + 8) = -9(-5x) - (-9)6y + (-9)8 =$
$45x - (-54y) + (-72) = 45x + 54y - 72$

17. $-4(x - 3y - 2z) = -4 \cdot x - (-4)3y - (-4)2z =$
$-4x - (-12y) - (-8z) = -4x + 12y + 8z$

19. $8(a - 3b + c) = 8 \cdot a - 8 \cdot 3b + 8 \cdot c =$
$8a - 24b + 8c$

21. $3(x - 2y - 9z) = 3 \cdot x - 3 \cdot 2y - 3 \cdot 9z =$
$3x - 6y - 27z$

23. $5(4a - 5b + c - 2d) = 5 \cdot 4a - 5 \cdot 5b + 5 \cdot c - 5 \cdot 2d =$
$20a - 25b + 5c - 10d$

25. $9a + 10a = (9 + 10)a = 19a$

27. $10a - a = 10a - 1 \cdot a = (10 - 1)a = 9a$

29. $2x + 9z + 6x = 2x + 6x + 9z$
$= (2 + 6)x + 9z$
$= 8x + 9z$

31. $41a + 90 - 60a - 2 = 41a - 60a + 90 - 2$
$= (41 - 60)a + (90 - 2)$
$= -19a + 88$

33. $23 + 5t + 7y - t - y - 27$
$= 23 - 27 + 5t - 1 \cdot t + 7y - 1 \cdot y$
$= (23 - 27) + (5 - 1)t + (7 - 1)y$
$= -4 + 4t + 6y$, or $4t + 6y - 4$

35. $3x - 11x = (3 - 11)x = -8x$

37. $y - 17y = (1 - 17)y = -16y$

39. $-8 + 11a - 5b + 6a - 7b + 7$
$= 11a + 6a - 5b - 7b - 8 + 7$
$= (11 + 6)a + (-5 - 7)b + (-8 + 7)$
$= 17a - 12b - 1$

41. $9x + 2y - 5x = 9x - 5x + 2y$
$= (9 - 5)x + 2y$
$= 4x + 2y$

43. $11x + 2y - 4x - y = 11x - 4x + 2y - y$
$= (11 - 4)x + (2 - 1)y$
$= 7x + y$

45. $a + 3b + 5a - 2 + b$
$= a + 5a + 3b + b - 2$
$= (1 + 5)a + (3 + 1)b - 2$
$= 6a + 4b - 2$

47. $6x^3 + 2x - 5x^3 + 7x = 6x^3 - 5x^3 + 2x + 7x$
$= (6 - 5)x^3 + (2 + 7)x$
$= x^3 + 9x$

49. $3a^2 + 7a^3 - a^2 + 5 + a^3$
$= 7a^3 + a^3 + 3a^2 - a^2 + 5$
$= (7 + 1)a^3 + (3 - 1)a^2 + 5$
$= 8a^3 + 2a^2 + 5$

51. $9xy + 4y^2 - 2xy + 2y^2 - 1$
$= 9xy - 2xy + 4y^2 + 2y^2 - 1$
$= (9 - 2)xy + (4 + 2)y^2 - 1$
$= 7xy + 6y^2 - 1$

53. $8a^2b - 3ab^2 - 4a^2b + 2ab$
$= 8a^2b - 4a^2b - 3ab^2 + 2ab$
$= (8 - 4)a^2b - 3ab^2 + 2ab$
$= 4a^2b - 3ab^2 + 2ab$

55. $3x^4 - 2y^4 + 8x^4y^4 - 7x^4 + 8y^4$
$= 3x^4 - 7x^4 - 2y^4 + 8y^4 + 8x^4y^4$
$= (3 - 7)x^4 + (-2 + 8)y^4 + 8x^4y^4$
$= -4x^4 + 6y^4 + 8x^4y^4$

57. Perimeter = 4 mm + 6 mm + 7 mm
= (4 + 6 + 7) mm
= 17 mm

59. Perimeter = 4 m + 4 m + 4 m + 5 m + 1 m
= (4 + 4 + 4 + 5 + 1) m
= 18 m

61. $P = 4s$ Perimeter of a square
$P = 4 \cdot 5$ in.
$P = 20$ in.

63. $P = 2l + 2w = 2 \cdot 11$ in. $+ 2 \cdot 14$ in.
$= 22$ in. $+ 28$ in. $= 50$ in.

65. $P = 2(l + w) = 2(10$ ft $+ 12$ ft$)$
$= 2 \cdot 22$ ft $= 44$ ft

67. $P = 4s$
$= 4 \cdot 14$ in. $= 56$ in.

69. $P = 4s$
$= 4 \cdot 65$ cm $= 260$ cm

71. $P = 4s$
$= 4 \cdot 85$ cm $= 340$ cm

73. $P = 2l + 2w = 2 \cdot 3$ ft $+ 2 \cdot 5$ ft
$= 6$ ft $+ 10$ ft $= 16$ ft

75. ***Familiarize***. We first draw a picture. We express one hour as 60 minutes.

|4 min|4 min|4 min| . . . |4 min|
←—— 60 minutes ——→

Let $c =$ the number of contractions in one hour.

Translate. Repeated addition applies here. We translate to a multiplication sentence.

Time between onset of contractions	times	Number of contractions	is	60 minutes
↓	↓	↓	↓	↓
4	·	c	=	60

Solve. To solve the equation we divide on both sides.

$$4 \cdot c = 60$$
$$\frac{4 \cdot c}{4} = \frac{60}{4}$$
$$c = 15$$

Check. To check, we multiply 15 by 4: $4 \cdot 15 = 60$

State. The woman will experience 15 contractions in one hour.

77.
$$\begin{array}{r} {}^{1\,5} \\ 5\,2\,9 \\ \times\quad 6 \\ \hline 3\,1\,7\,4 \end{array}$$

79.
$$\begin{array}{r} 7\,0\,9 \\ 5\overline{)\,3\,5\,4\,9} \\ 3\,5\,0\,0 \\ \hline 4\,9 \\ 4\,5 \\ \hline 4 \end{array}$$

The answer is 709 R 4.

81.

83. Observe that $-32 \times 59 = -1888$ and $88 - 29 = 59$. Thus we have

$$-32 \boxed{\times} (88 \boxed{-} 29) = -1888.$$

85. $3(x+3) + 2(x-7) = 3x + 9 + 2x - 14 = 5x - 5$

87. $2(3-4a) + 5(a-7) = 6 - 8a + 5a - 35 = -3a - 29$

89. $-5(2+3x+4y) + 7(2x-y) =$
$-10 - 15x - 20y + 14x - 7y = -10 - x - 27y$

91. ***Familiarize***. First we will find the perimeter of each door and each window.

Translate.

The perimeter of each door is given by

$$P = 2(l+w) = 2(3\text{ ft} + 7\text{ ft}).$$

The perimeter of each window is given by

$$P = 2(l+w) = 2(3\text{ ft} + 4\text{ ft}).$$

Solve. First we calculate the perimeters.

For each door: $P = 2(3\text{ ft} + 7\text{ ft}) = 2 \cdot 10\text{ ft} = 20\text{ ft}$

For each window: $P = 2(3\text{ ft} + 4\text{ ft}) = 2 \cdot 7\text{ ft} = 14\text{ ft}$

We multiply to find the perimeter of 3 doors and of 12 windows.

Doors: $3 \cdot 20\text{ ft} = 60\text{ ft}$

Windows: $12 \cdot 14\text{ ft} = 168\text{ ft}$

We add to find the total of the perimeters:

$$60\text{ ft} + 168\text{ ft} = 228\text{ ft}$$

Finally we divide to determine how many tubes of sealant are needed:

$$\begin{array}{r} 4 \\ 51\,\overline{)\,228} \\ 204 \\ \hline 24 \end{array}$$

The answer is 4 R 24. Since 24 ft will be left unsealed after 4 tubes are used, the homeowner should buy 5 tubes of sealant.

Check. We multiply 51 by 4 and add 24:

$$4 \cdot 51 + 24 = 204 + 24 = 228$$

The result checks.

State. The homeowner should buy 5 tubes of sealant.

Chapter 3

Fractional Notation: Multiplication and Division

Exercise Set 3.1

1. $1 \cdot 4 = 4$ $6 \cdot 4 = 24$
$2 \cdot 4 = 8$ $7 \cdot 4 = 28$
$3 \cdot 4 = 12$ $8 \cdot 4 = 32$
$4 \cdot 4 = 16$ $9 \cdot 4 = 36$
$5 \cdot 4 = 20$ $10 \cdot 4 = 40$

3. $1 \cdot 20 = 20$ $6 \cdot 20 = 120$
$2 \cdot 20 = 40$ $7 \cdot 20 = 140$
$3 \cdot 20 = 60$ $8 \cdot 20 = 160$
$4 \cdot 20 = 80$ $9 \cdot 20 = 180$
$5 \cdot 20 = 100$ $10 \cdot 20 = 200$

5. $1 \cdot 3 = 3$ $6 \cdot 3 = 18$
$2 \cdot 3 = 6$ $7 \cdot 3 = 21$
$3 \cdot 3 = 9$ $8 \cdot 3 = 24$
$4 \cdot 3 = 12$ $9 \cdot 3 = 27$
$5 \cdot 3 = 15$ $10 \cdot 3 = 30$

7. $1 \cdot 12 = 12$ $6 \cdot 12 = 72$
$2 \cdot 12 = 24$ $7 \cdot 12 = 84$
$3 \cdot 12 = 36$ $8 \cdot 12 = 96$
$4 \cdot 12 = 48$ $9 \cdot 12 = 108$
$5 \cdot 12 = 60$ $10 \cdot 12 = 120$

9. $1 \cdot 10 = 10$ $6 \cdot 10 = 60$
$2 \cdot 10 = 20$ $7 \cdot 10 = 70$
$3 \cdot 10 = 30$ $8 \cdot 10 = 80$
$4 \cdot 10 = 40$ $9 \cdot 10 = 90$
$5 \cdot 10 = 50$ $10 \cdot 10 = 100$

11. $1 \cdot 9 = 9$ $6 \cdot 9 = 54$
$2 \cdot 9 = 18$ $7 \cdot 9 = 63$
$3 \cdot 9 = 27$ $8 \cdot 9 = 72$
$4 \cdot 9 = 36$ $9 \cdot 9 = 81$
$5 \cdot 9 = 45$ $10 \cdot 9 = 90$

13. We divide 26 by 7.

$$\begin{array}{r} 3 \\ 7\,\overline{)\,26} \\ 21 \\ \hline 5 \end{array}$$

Since the remainder is not 0, 26 is not divisible by 7.

15. We divide 1880 by 8.

$$\begin{array}{r} 235 \\ 8\,\overline{)\,1880} \\ 1600 \\ \hline 280 \\ 240 \\ \hline 40 \\ 40 \\ \hline 0 \end{array}$$

The remainder of 0 indicates that 1880 is divisible by 8.

17. We divide 256 by 16.

$$\begin{array}{r} 16 \\ 16\,\overline{)\,256} \\ 160 \\ \hline 96 \\ 96 \\ \hline 0 \end{array}$$

The remainder of 0 indicates that 256 is divisible by 16.

19. We divide 4227 by 9.

$$\begin{array}{r} 469 \\ 9\,\overline{)\,4227} \\ 3600 \\ \hline 627 \\ 540 \\ \hline 87 \\ 81 \\ \hline 6 \end{array}$$

Since the remainder is not 0, 4227 is not divisible by 9.

21. We divide 8650 by 16.

$$\begin{array}{r} 540 \\ 16\,\overline{)\,8650} \\ 8000 \\ \hline 650 \\ 640 \\ \hline 10 \end{array}$$

Since the remainder is not 0, 8650 is not divisible by 16.

23. A number is divisible by 2 if its ones digit is even.

4$\underline{6}$ is divisible by 2 because $\underline{6}$ is even.
22$\underline{4}$ is divisible by 2 because $\underline{4}$ is even.
1$\underline{9}$ is not divisible by 2 because $\underline{9}$ is not even.
55$\underline{5}$ is not divisible by 2 because $\underline{5}$ is not even.
30$\underline{0}$ is divisible by 2 because $\underline{0}$ is even.
3$\underline{6}$ is divisible by 2 because $\underline{6}$ is even.
45,27$\underline{0}$ is divisible by 2 because $\underline{0}$ is even.
444$\underline{4}$ is divisible by 2 because $\underline{4}$ is even.
8$\underline{5}$ is not divisible by 2 because $\underline{5}$ is not even.
71$\underline{1}$ is not divisible by 2 because $\underline{1}$ is not even.
13,25$\underline{1}$ is not divisible by 2 because $\underline{1}$ is not even.
254,76$\underline{5}$ is not divisible by 2 because $\underline{5}$ is not even.

25. A number is divisible by 10 if its ones digit is 0.

Of the numbers under consideration, only 300 and 45,270 have one digits of 0. Therefore, only 300 and 45,270 are divisible by 10.

27. For a number to be divisible by 6, the sum of the digits must be divisible by 3 and the ones digit must be 0, 2, 4, 6 or 8 (even). It is most efficient to determine if the ones digit is even first and then, if so, to determine if the sum of the digits is divisible by 3.

46 is not divisible by 6 because 46 is not divisible by 3.

$4 + 6 = 10$
↑
Not divisible by 3

224 is not divisible by 6 because 224 is not divisible by 3.

$2 + 2 + 4 = 8$
↑
Not divisible by 3

19 is not divisible by 6 because 19 is not even.

19
↑
Not even

555 is not divisible by 6 because 555 is not even.

555
↑
Not even

300 is divisible by 6.

300 ↑ Even $3 + 0 + 0 = 3$ ↑ Divisible by 3

36 is divisible by 6.

36 ↑ Even $3 + 6 = 9$ ↑ Divisible by 3

45,270 is divisible by 6.

45,270 ↑ Even $4 + 5 + 2 + 7 + 0 = 18$ ↑ Divisible by 3

4444 is not divisible by 6 because 4444 is not divisible by 3.

$4 + 4 + 4 + 4 = 16$
↑
Not divisible by 3

85 is not divisible by 6 because 85 is not even.

85
↑
Not even

711 is not divisible by 6 because 711 is not even.

711
↑
Not even

13,251 is not divisible by 6 because 13,251 is not even.

13,251
↑
Not even

254,765 is not divisible by 6 because 254,765 is not even.

254,765
↑
Not even

29. A number is divisible by 3 if the sum of the digits is divisible by 3.

56 is not divisible by 3 because $5 + 6 = 11$ and 11 is not divisible by 3.

324 is divisible by 3 because $3 + 2 + 4 = 9$ and 9 is divisible by 3.

784 is not divisible by 3 because $7 + 8 + 4 = 19$ and 19 is not divisible by 3.

55,555 is not divisible by 3 because $5 + 5 + 5 + 5 + 5 = 25$ and 25 is not divisible by 3.

200 is not divisible by 3 because $2 + 0 + 0 = 2$ and 2 is not divisible by 3.

42 is divisible by 3 because $4 + 2 = 6$ and 6 is divisible by 3.

501 is divisible by 3 because $5 + 0 + 1 = 6$ and 6 is divisible by 3.

3009 is divisible by 3 because $3 + 0 + 0 + 9 = 12$ and 12 is divisible by 3.

75 is divisible by 3 because $7 + 5 = 12$ and 12 is divisible by 3.

812 is not divisible by 3 because $8 + 1 + 2 = 11$ and 11 is not divisible by 3.

2345 is not divisible by 3 because $2 + 3 + 4 + 5 = 14$ and 14 is not divisible by 3.

2001 is divisible by 3 because $2 + 0 + 0 + 1 = 3$ and 3 is divisible by 3.

31. A number is divisible by 5 if the ones digit is 0 or 5.

$5\underline{6}$ is not divisible by 5 because the ones digit (6) is not 0 or 5.

$32\underline{4}$ is not divisible by 5 because the ones digit (4) is not 0 or 5.

$78\underline{4}$ is not divisible by 5 because the ones digit (4) is not 0 or 5.

$55{,}55\underline{5}$ is divisible by 5 because the ones digit (5) is 5.

$20\underline{0}$ is divisible by 5 because the ones digit (0) is 0.

$4\underline{2}$ is not divisible by 5 because the ones digit (2) is not 0 or 5.

$50\underline{1}$ is not divisible by 5 because the ones digit (1) is not 0 or 5.

$300\underline{9}$ is not divisible by 5 because the ones digit (9) is not 0 or 5.

$7\underline{5}$ is divisible by 5 because the ones digit (5) is 5.

$81\underline{2}$ is not divisible by 5 because the ones digit (2) is not 0 or 5.

$234\underline{5}$ is divisible by 5 because the ones digit (5) is 5.

$200\underline{1}$ is not divisible by 5 because the ones digit (1) is not 0 or 5.

33. A number is divisible by 9 if the sum of the digits is divisible by 9.

56 is not divisible by 9 because $5 + 6 = 11$ and 11 is not divisible by 9.

324 is divisible by 9 because $3+2+4=9$ and 9 is divisible by 9.

784 is not divisible by 9 because $7+8+4=19$ and 19 is not divisible by 9.

55,555 is not divisible by 9 because $5+5+5+5+5=25$ and 25 is not divisible by 9.

200 is not divisible by 9 because $2+0+0=2$ and 2 is not divisible by 9.

42 is not divisible by 9 because $4+2=6$ and 6 is not divisible by 9.

501 is not divisible by 9 because $5+0+1=6$ and 6 is not divisible by 9.

3009 is not divisible by 9 because $3+0+0+9=12$ and 12 is not divisible by 9.

75 is not divisible by 9 because $7+5=12$ and 12 is not divisible by 9.

812 is not divisible by 9 because $8+1+2=11$ and 11 is not divisible by 9.

2345 is not divisible by 9 because $2+3+4+5=14$ and 14 is not divisible by 9.

2001 is not divisible by 9 because $2+0+0+1=3$ and 3 is not divisible by 9.

35. $16 \cdot t = 848$

$$\frac{16 \cdot t}{16} = \frac{848}{16} \quad \text{Dividing by 16 on both sides}$$

$$t = 53$$

The solution is 53.

37.
$$\begin{aligned} 56 + x &= 194 \\ 56 + x - 56 &= 194 - 56 \quad \text{Subtracting 56 on both sides} \\ x &= 138 \end{aligned}$$

The solution is 138.

39. ***Familiarize***. This is a multistep problem. Find the total cost of the shirts and the total cost of the pants and then find the sum of the two.

We let s = the total cost of the shirts and t = the total cost of the pants.

Translate. We write two equations.

$$\underbrace{\text{Number of shirts}}_{12} \ \underbrace{\text{times}}_{\cdot} \ \underbrace{\text{Cost of one shirt}}_{37} \ \underbrace{\text{is}}_{=} \ \underbrace{\text{Total cost of shirts}}_{s}$$

$$\underbrace{\text{Number of pairs of pants}}_{4} \ \underbrace{\text{times}}_{\cdot} \ \underbrace{\text{Cost of one pair}}_{59} \ \underbrace{\text{is}}_{=} \ \underbrace{\text{Total cost of pants}}_{t}$$

Solve. We carry out the multiplication.

$$\begin{aligned} 12 \cdot 37 &= s \\ 444 &= s \quad \text{Doing the multiplication} \end{aligned}$$

The total cost of the 12 shirts is \$444.

$$\begin{aligned} 4 \cdot 59 &= t \\ 236 &= t \quad \text{Doing the multiplication} \end{aligned}$$

The total cost of the 4 pairs of pants is \$236.

Now we find the total amount spent. We let a = this amount.

$$\underbrace{\text{Total cost of shirts}}_{444} \ \underbrace{\text{plus}}_{+} \ \underbrace{\text{Total cost of pants}}_{236} \ \underbrace{\text{is}}_{=} \ \underbrace{\text{Total amount spent}}_{a}$$

To solve the equation, carry out the addition.

$$\begin{array}{r} 444 \\ +\,236 \\ \hline 680 \end{array}$$

Check. We can repeat the calculations. The answer checks.

State. The total cost is \$680.

41. 

43. When we use a calculator to divide the largest five-digit number, 99,999, by 47 we get 2127.638298. This tells us that 99,999 is not divisible by 47 but that 2127×47, or 99,969, is divisible by 47 and that it is the largest such five-digit number.

45. We list multiples of 2, 3, and 5 and find the smallest number that is on all 3 lists.

Multiples of 2: $2, 4, 6, 8, 10, 12, 14, 16, 18, 20, 22, 24, 26, 28, \underline{30}, 32, \cdots$

Multiples of 3: $3, 6, 9, 12, 15, 18, 21, 24, 27, \underline{30}, 33, \cdots$

Multiples of 5: $5, 10, 15, 20, 25, \underline{30}, 35, \cdots$

The smallest number that is simultaneously a multiple of 2, 3, and 5 is 30.

(We might also have observed that since 2, 3, and 5 are all prime numbers, the number we are looking for is their product, $2 \cdot 3 \cdot 5$, or 30.)

47. We list multiples of 4, 6, and 10 and find the smallest number that is on all 3 lists.

Multiples of 4: $4, 8, 12, 16, 20, 24, 28, 32, 36, 40, 44, 48, 52, 56, \underline{60}, 64, \cdots$

Multiples of 6: $6, 12, 18, 24, 30, 36, 42, 48, 54, \underline{60}, 66, \cdots$

Multiples of 10: $10, 20, 30, 40, 50, \underline{60}, 70, \cdots$

The smallest number that is simultaneously a multiple of 4, 6, and 10 is 60.

(We might also have written the prime factorization of each number: $4 = 2 \cdot 2$, $6 = 2 \cdot 3$, $10 = 2 \cdot 5$. Then the number we are looking for must contain two factors of 2, one factor of 3, and one factor 5: $2 \cdot 2 \cdot 3 \cdot 5 = 60$.)

49. Since $15 = 3 \cdot 5$, a number is divisible by 15 if it is divisible by both 3 and 5. This means that the sum of its digits must be divisible by 3 and its ones digit must be 0 or 5.

Exercise Set 3.2

1. Since 16 is even, we know that 2 is a factor. Using other tests for divisibility we determine that 3, 5, 6, 9, and 10 are not factors. We now write some factorizations.

$16 = 1\cdot 16$ $\quad$ $16 = 4\cdot 4$
$16 = 2\cdot 8$

Factors: 1, 2, 4, 8, 16

3. Since 54 is even, we know that 2 is a factor. Since the sum of the digits is 9 and 9 is divisible by both 3 and 9, we know that 3 and 9 are both factors. Also, since 2 and 3 are factors, 6 is a factor as well. We now write some factorizations.

$54 = 1\cdot 54$ $\quad$ $54 = 3\cdot 18$
$54 = 2\cdot 27$ $\quad$ $54 = 6\cdot 9$

Factors: 1, 2, 3, 6, 9, 18, 27, 54

5. Since 4 is even, we know that 2 is a factor. We write some factorizations:

$4 = 1\cdot 4$ $\quad$ $4 = 2\cdot 2$

Factors: 1, 2, 4

7. The only factorization is $7 = 1\cdot 7$.

Factors: 1, 7

9. The only factorization is $1 = 1\cdot 1$.

Factor: 1

11. Since 98 is even, we now that 2 is a factor. Using other tests for divisibility, we determine that 3, 5, 6, 9, and 10 are not factors. We now write some factorizations:

$98 = 1\cdot 98$ $\quad$ $98 = 7\cdot 14$
$98 = 2\cdot 49$

Factors: 1, 2, 7, 14, 49, 98

13. Using tests for divisibility we determine that 2, 3, and 6 are factors. We write some factorizations.

$42 = 1\cdot 42$ $\quad$ $42 = 3\cdot 14$
$42 = 2\cdot 21$ $\quad$ $42 = 6\cdot 7$

Factors: 1, 2, 3, 6, 7, 14, 21, 42

15. Using tests for divisibility we determine that 5 is a factor. We write some factorizations.

$385 = 1\cdot 385$ $\quad$ $385 = 7\cdot 55$
$385 = 5\cdot 77$ $\quad$ $385 = 11\cdot 35$

Factors: 1, 5, 7, 11, 35, 55, 77, 385

17. Using tests for divisibility we determine that 2, 3, 6, and 9 are factors. We write some factorizations.

$36 = 1\cdot 36$ $\quad$ $36 = 4\cdot 9$
$36 = 2\cdot 18$ $\quad$ $36 = 6\cdot 6$
$36 = 3\cdot 12$

Factors: 1, 2, 3, 4, 6, 9, 12, 18, 36

19. Using tests for divisibility we determine that 3, 5, and 9 are factors. We write some factorizations.

$225 = 1\cdot 225$ $\quad$ $225 = 9\cdot 25$
$225 = 3\cdot 75$ $\quad$ $225 = 15\cdot 15$
$225 = 5\cdot 45$

Factors: 1, 3, 5, 9, 15, 25, 45, 75, 225

21. The number 17 is prime. It has only the factors 1 and 17.

23. The number 22 has factors 1, 2, 11, and 22. Since it has at least one factor other than itself and 1, it is composite.

25. The number 32 has factors 1, 2, 4, 8, 16, and 32. Since it has at least one factor other than itself and 1, it is composite.

27. The number 31 is prime. It has only the factors 1 and 31.

29. 1 is neither prime nor composite.

31. The number 9 has factors 1, 3, and 9.

Since it has at least one factor other than itself and 1, it is composite.

33. The number 13 is prime. It has only the factors 1 and 13.

35. The number 29 is prime. It has only the factors 1 and 29.

37.
2 ← 2 is prime
$2\,\overline{)\,4}$
$2\,\overline{)\,8}$

$8 = 2\cdot 2\cdot 2$

39.
7 ← 7 is prime
$2\,\overline{)\,14}$

$14 = 2\cdot 7$

41.
11 ← 11 is prime
$2\,\overline{)\,22}$

$22 = 2\cdot 11$

43.
5 ← 5 is prime
$5\,\overline{)\,25}$ (25 is not divisible by 2 or 3. We move to 5.)

$25 = 5\cdot 5$

45.
5 ← 5 is prime
$5\,\overline{)\,25}$ (25 is not divisible by 2 or 3. We
$2\,\overline{)\,50}$ move to 5.)

$50 = 2\cdot 5\cdot 5$

47.
13 ← 13 is prime
$13\,\overline{)\,169}$ (169 is not divisible by 2, 3, 5, 7 or 11. We move to 13.)

$169 = 13\cdot 13$

49.
5 ← 5 is prime
$5\,\overline{)\,25}$ (25 is not divisible by 2 or 3. We
$2\,\overline{)\,50}$ move to 5.)
$2\,\overline{)\,100}$

$100 = 2\cdot 2\cdot 5\cdot 5$

We can also use a factor tree.

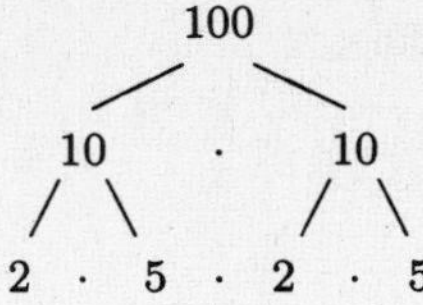

51.
$$\begin{array}{r|l} & 7 \quad \leftarrow \text{7 is prime} \\ 5 & 35 \quad \text{(35 is not divisible by 2 or 3. We move to 5.)} \end{array}$$

$35 = 5 \cdot 7$

53.
$$\begin{array}{r|l} & 13 \quad \leftarrow \text{13 is prime} \\ 3 & 39 \quad \text{(39 is not divisible by 2. We move} \\ 2 & 78 \quad \text{to 3.)} \end{array}$$

$78 = 2 \cdot 3 \cdot 13$

55.
$$\begin{array}{r|l} & 11 \quad \leftarrow \text{11 is prime} \\ 7 & 77 \quad \text{(77 is not divisible by 2, 3, or 5. We move to 7.)} \end{array}$$

$77 = 7 \cdot 11$

57.
$$\begin{array}{r|l} & 7 \quad \leftarrow \text{7 is prime} \\ 2 & 14 \\ 2 & 28 \\ 2 & 56 \\ 2 & 112 \end{array}$$

$112 = 2 \cdot 2 \cdot 2 \cdot 2 \cdot 7$

We can also use a factor tree.

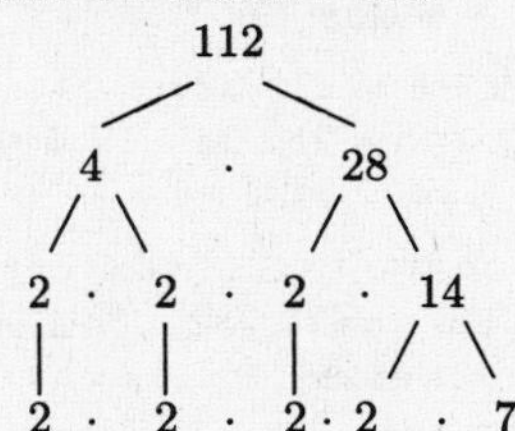

59.
$$\begin{array}{r|l} & 5 \quad \leftarrow \text{5 is prime} \\ 5 & 25 \\ 3 & 75 \\ 2 & 150 \\ 2 & 300 \end{array}$$

$300 = 2 \cdot 2 \cdot 3 \cdot 5 \cdot 5$

We can also use a factor tree.

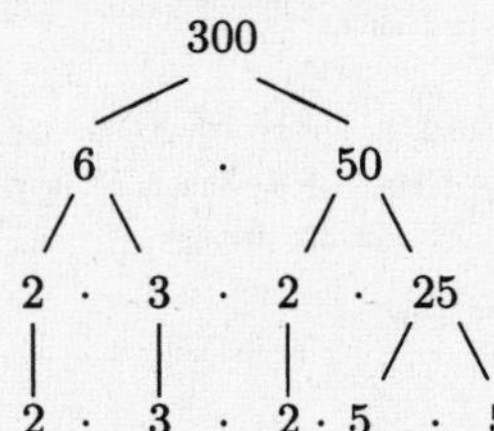

61. $-2 \cdot 13 = -26$ (The signs are different, so the answer is negative.)

63. $-17 + 25$ The absolute values are 17 and 25. The difference is 8. The positive number has the larger absolute value, so the answer is positive. $-17 + 25 = 8$

65. $0 \div 22 = 0$ (0 divided by a nonzero number is 0.)

67.

69. Use a calculator to perform a string of successive divisions by prime numbers.

$473,073,361 = 23 \cdot 31 \cdot 61 \cdot 73 \cdot 149$

71. We use a factor tree.

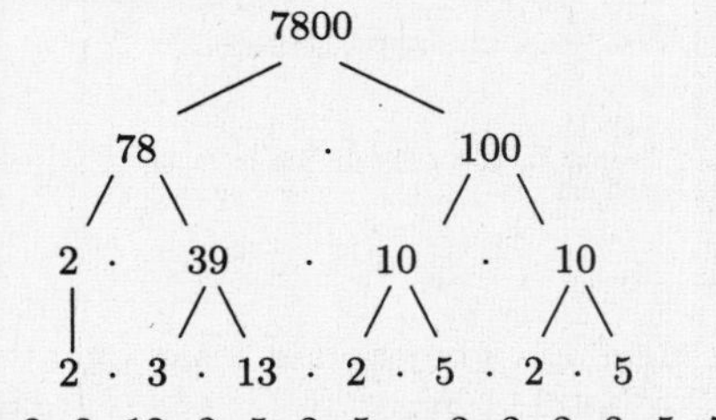

$7800 = 2 \cdot 3 \cdot 13 \cdot 2 \cdot 5 \cdot 2 \cdot 5$, or $2 \cdot 2 \cdot 2 \cdot 3 \cdot 5 \cdot 5 \cdot 13$

73.
$$\begin{array}{r|l} & 11 \quad \leftarrow \text{11 is prime} \\ 7 & 77 \\ 3 & 231 \\ 3 & 693 \\ 2 & 1386 \\ 2 & 2772 \end{array}$$

$2772 = 2 \cdot 2 \cdot 3 \cdot 3 \cdot 7 \cdot 11$

75. Since there are 2 factors a rectangular array is indicated. One factor gives the number of rows, the other the number of objects in each row. Thus, we have a rectangular array of 6 rows with 9 objects each or 9 rows with 6 objects each.

Exercise Set 3.3

1. The top number is the numerator, and the bottom number is the denominator.

$$\frac{3}{4} \begin{array}{l} \leftarrow \text{Numerator} \\ \leftarrow \text{Denominator} \end{array}$$

3. $\frac{7}{-9} \begin{array}{l} \leftarrow \text{Numerator} \\ \leftarrow \text{Denominator} \end{array}$

5. $\frac{2x}{3z} \begin{array}{l} \leftarrow \text{Numerator} \\ \leftarrow \text{Denominator} \end{array}$

7. The dollar is divided into 4 parts of the same size, and 2 of them are shaded. This is $2 \cdot \frac{1}{4}$ or $\frac{2}{4}$. Thus, $\$\frac{2}{4}$ (two-fourths of a dollar) is shaded.

9. We have 2 yards, each of which is divided into 8 parts of the same size. We take 2 of those parts. This is $2 \cdot \frac{1}{8}$, or $\frac{2}{8}$ yd.

11. We have 2 liters, each divided into three parts of the same size. We take 4 of those parts. This is $4 \cdot \frac{1}{3}$, or $\frac{4}{3}$ L.

13. The triangle is divided into 4 triangles of the same size, and 3 of them are shaded. This is $3 \cdot \frac{1}{4}$ or $\frac{3}{4}$. Thus, $\frac{3}{4}$ (three-fourths) of the triangle is shaded.

15. The pie is divided into 8 parts of the same size, and 4 of them are shaded. This is $4\cdot\frac{1}{8}$, or $\frac{4}{8}$. Thus, $\frac{4}{8}$ (four-eighths) of the pie is shaded.

17. The acre is divided into 12 parts of the same size, and 6 of them are shaded. This is $6\cdot\frac{1}{12}$, or $\frac{6}{12}$. Thus, $\frac{6}{12}$ acre (six-twelfths of the acre) is shaded.

19. There are 5 objects in the set, and 3 of the objects are shaded. Thus, $\frac{3}{5}$ of the set is shaded.

21. Remember: $\frac{0}{n}=0$, for any integer n that is not 0.

$\frac{0}{5}=0$

Think of dividing an object into 5 parts and taking none of them. We get 0.

23. Remember: $\frac{n}{1}=n$.

$\frac{15}{1}=15$

Think of taking 15 objects and dividing them into 1 part. (We do not divide them.) We have 15 objects.

25. Remember: $\frac{n}{n}=1$, for any integer n that is not 0.

$\frac{20}{20}=1$

If we divide an object into 20 parts and take 20 of them, we get all of the object (1 whole object).

27. Remember: $\frac{n}{n}=1$, for any integer n that is not 0.

$\frac{-14}{-14}=1$

29. Remember: $\frac{0}{n}=0$, for any integer n that is not 0.

$\frac{0}{-234}=0$

31. Remember: $\frac{n}{n}=1$, for any integer n that is not 0.

$\frac{3n}{3n}=1$

33. Remember: $\frac{n}{n}=1$, for any integer n that is not 0.

$\frac{9x}{9x}=1$

35. Remember: $\frac{n}{1}=n$

$\frac{-63}{1}=-63$

37. Remember: $\frac{0}{n}=0$, for any integer n that is not 0.

$\frac{0}{2a}=0$

39. Remember: $\frac{n}{0}$ is not defined.

$\frac{52}{0}$ is not defined.

41. Remember: $\frac{n}{1}=n$

$\frac{7n}{1}=7n$

43. $\frac{6}{7-7}=\frac{6}{0}$

Remember: $\frac{n}{0}$ is not defined. Thus, $\frac{6}{7-7}$ is not defined.

45. $-7(30)=-210$
(The signs are different, so the answer is negative.)

47. $(-71)(-12)0=-71\cdot 0=0$
(We might have observed at the outset that the answer is 0 since one of the factors is 0.)

49. ***Familiarize.*** We let $a=$ the amount by which the average income in Alaska exceeds the average income in Colorado.

Translate. This is "how-much-more" situation.

Average income in Colorado	plus	Excess income in Alaska	is	Average income in Alaska
↓	↓	↓	↓	↓
19,440	+	a	=	21,932

Solve. We solve the equation.

$$\begin{aligned}19,440+a&=21,932\\19,440+a-19,440&=21,932-19,440\\a&=2492\end{aligned}$$

Check. We add the answer to the average income in Colorado: $2492 + $19,440 = $21,932. This is the average income in Alaska, so our answer checks.

State. On average, people living in Alaska make $2492 more than those living in Colorado.

51.

53. The surface of the earth is divided into $3+1$, or 4 parts. Three of them are taken up by water, so $\frac{3}{4}$ is water. One of them is land, so $\frac{1}{4}$ is land.

55. The couple's 3 sons had a total of $3\cdot 3$, or 9 daughters. The 9 daughters had a total of $9\cdot 3$, or 27 sons. Altogether the couple had $3+9+27$, or 39 descendants. Nine of them were female, so $\frac{9}{39}$ are female.

Exercise Set 3.4

1. $3 \cdot \frac{1}{5} = \frac{3 \cdot 1}{5} = \frac{3}{5}$

3. $(-5) \times \frac{1}{6} = \frac{-5 \times 1}{6} = \frac{-5}{6}$, or $-\frac{5}{6}$

5. $\frac{2}{3} \cdot 7 = \frac{2 \cdot 7}{3} = \frac{14}{3}$

7. $(-1)\frac{7}{9} = \frac{(-1)7}{9} = \frac{-7}{9}$, or $-\frac{7}{9}$

9. $\frac{2}{5} \cdot x = \frac{2 \cdot x}{5} = \frac{2x}{5}$

11. $\frac{2}{5}(-3) = \frac{2(-3)}{5} = \frac{-6}{5}$, or $-\frac{6}{5}$

13. $a \cdot \frac{3}{4} = \frac{a \cdot 3}{4} = \frac{3a}{4}$

15. $17 \times \frac{m}{6} = \frac{17 \times m}{6} = \frac{17m}{6}$

17. $\frac{1}{2} \cdot \frac{1}{3} = \frac{1 \cdot 1}{2 \cdot 3} = \frac{1}{6}$

19. $\left(-\frac{1}{4}\right) \times \frac{1}{10} = -\frac{1 \times 1}{4 \times 10} = -\frac{1}{40}$

21. $\frac{2}{3} \times \frac{1}{5} = \frac{2 \times 1}{3 \times 5} = \frac{2}{15}$

23. $\frac{2}{y} \cdot \frac{x}{5} = \frac{2 \cdot x}{y \cdot 5} = \frac{2x}{5y}$

25. $\left(-\frac{3}{4}\right)\left(-\frac{3}{4}\right) = \frac{(-3)(-3)}{4 \cdot 4} = \frac{9}{16}$

27. $\frac{2}{3} \cdot \frac{7}{13} = \frac{2 \cdot 7}{3 \cdot 13} = \frac{14}{39}$

29. $\frac{1}{10}\left(\frac{-7}{10}\right) = \frac{1(-7)}{10 \cdot 10} = \frac{-7}{100}$, or $-\frac{7}{100}$

31. $\frac{7}{8} \cdot \frac{a}{8} = \frac{7 \cdot a}{8 \cdot 8} = \frac{7a}{64}$

33. $\frac{1}{y} \cdot \frac{1}{100} = \frac{1 \cdot 1}{y \cdot 100} = \frac{1}{100y}$

35. $\frac{-14}{15} \cdot \frac{13}{19} = \frac{-14 \cdot 13}{15 \cdot 19} = \frac{-182}{285}$, or $-\frac{182}{285}$

37. ***Familiarize***. Recall that area is length times width. We draw a picture. We will let A = the area of the cutting board.

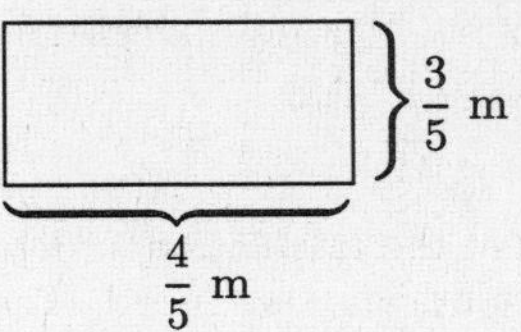

Translate. Then we translate.

$$\begin{array}{ccccc} \text{Area} & \text{is} & \text{length} & \text{times} & \text{width} \\ \downarrow & \downarrow & \downarrow & \downarrow & \downarrow \\ A & = & \frac{4}{5} & \times & \frac{3}{5} \end{array}$$

Solve. The sentence tells us what to do. We multiply.

$$\frac{4}{5} \times \frac{3}{5} = \frac{4 \times 3}{5 \times 5} = \frac{12}{25}$$

Check. We repeat the calculation. The answer checks.

State. The area is $\frac{12}{25}$ m^2.

39. ***Familiarize***. We draw a picture. We let n = the amount of kerosene the lamp holds when it is $\frac{1}{2}$ full.

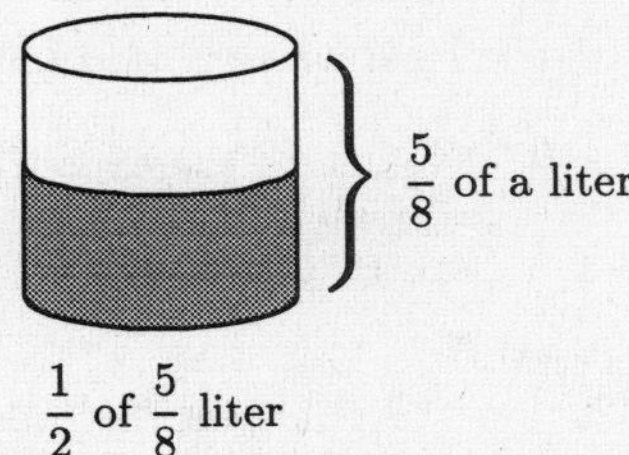

Translate. The multiplication sentence $\frac{1}{2} \cdot \frac{5}{8} = n$ corresponds to the situation.

Solve. We multiply:

$$\frac{1}{2} \cdot \frac{5}{8} = \frac{1 \cdot 5}{2 \cdot 8} = \frac{5}{16}$$

Check. We repeat the calculation. The answer checks.

State. The lamp will hold $\frac{5}{16}$ of a liter of kerosene when it is full.

41. ***Familiarize***. We draw a picture. We let m = the amount of molasses in $\frac{3}{4}$ of a batch.

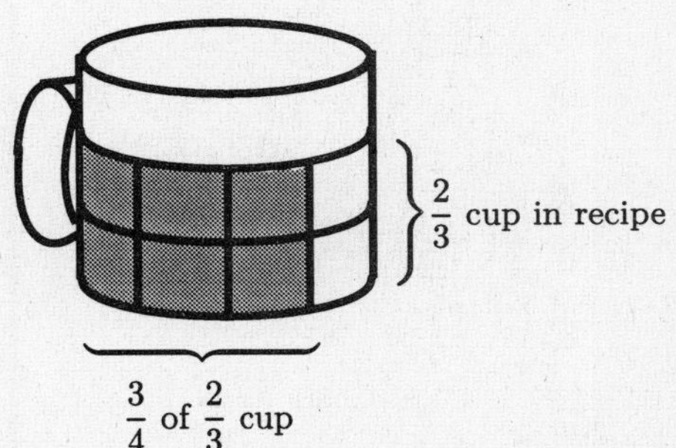

Translate. The multiplication sentence $\frac{3}{4} \cdot \frac{2}{3} = m$ corresponds to the situation.

Solve. We carry out the multiplication:

$$\frac{3}{4}\cdot\frac{2}{3}=\frac{3\cdot 2}{4\cdot 3}=\frac{6}{12}$$

Check. We repeat the calculation or determine what fractional part of the drawing is shaded. The answer checks.

State. $\frac{6}{12}$ cup of molasses is needed to make $\frac{3}{4}$ of a batch.

43. ***Familiarize***. Two of every 3 tons of municipal waste are dumped in landfills, so $\frac{2}{3}$ of all municipal waste is dumped in landfills. We also know that $\frac{1}{10}$ of the waste in landfills is yard trimmings. We let y = the fractional part of municipal waste that is yard trimmings.

Translate. The multiplication sentence $\frac{1}{10}\cdot\frac{2}{3}=y$ corresponds to this situation.

Solve. We carry out the multiplication.

$$\frac{1}{10}\cdot\frac{2}{3}=\frac{1\cdot 2}{10\cdot 3}=\frac{2}{30}$$

Check. We repeat the calculation. The answer checks.

State. $\frac{2}{30}$ of municipal waste is yard trimmings.

45. $5-3^2=5-9$ Evaluating the exponential expression

$=-4$ Subtracting

47. $8\cdot 12-(7+13)$

$=8\cdot 12-20$ Adding inside parentheses

$=96-20$ Multiplying

$=76$ Subtracting

49. 4, [6] 7 8, 9 5 2

The digit 6 means 6 hundred thousands.

51.

53. Use a calculator.

$$\frac{341}{517}\cdot\frac{209}{349}=\frac{341\cdot 209}{517\cdot 349}=\frac{71,269}{180,433}$$

55. $\left(\frac{2}{5}\right)^3\left(-\frac{7}{9}\right)=\frac{8}{125}\left(-\frac{7}{9}\right)$ Evaluating the exponential expression

$$=\frac{8(-7)}{125\cdot 9}$$

$$=\frac{-56}{1125},\text{ or }-\frac{56}{1125}$$

57. $$-\frac{2}{3}xy=-\frac{2}{3}\cdot\frac{2}{5}\left(-\frac{1}{7}\right)$$

$$=-\frac{4}{15}\left(-\frac{1}{7}\right)$$

$$=\frac{4}{105}$$

Exercise Set 3.5

1. Since $10\div 2=5$, we multiply by $\frac{5}{5}$.

$$\frac{1}{2}=\frac{1}{2}\cdot\frac{5}{5}=\frac{1\cdot 5}{2\cdot 5}=\frac{5}{10}$$

3. Since $-48\div 4=-12$, we multiply by $\frac{-12}{-12}$.

$$\frac{3}{4}=\frac{3}{4}\left(\frac{-12}{-12}\right)=\frac{3(-12)}{4(-12)}==\frac{-36}{-48}$$

5. Since $30\div 10=3$, we multiply by $\frac{3}{3}$.

$$\frac{9}{10}=\frac{9}{10}\cdot\frac{3}{3}=\frac{9\cdot 3}{10\cdot 3}=\frac{27}{30}$$

7. Since $30\div 5=6$, we multiply by $\frac{6}{6}$.

$$\frac{11}{5}=\frac{11}{5}\cdot\frac{6}{6}=\frac{11\cdot 6}{5\cdot 6}=\frac{66}{30}$$

9. Since $48\div 12=4$, we multiply by $\frac{4}{4}$.

$$\frac{5}{12}=\frac{5}{12}\cdot\frac{4}{4}=\frac{5\cdot 4}{12\cdot 4}=\frac{20}{48}$$

11. Since $54\div 18=3$, we multiply by $\frac{3}{3}$.

$$-\frac{17}{18}=-\frac{17}{18}\cdot\frac{3}{3}=-\frac{17\cdot 3}{18\cdot 3}=-\frac{51}{54}$$

13. Since $-25\div -5=5$, we multiply by $\frac{5}{5}$.

$$\frac{2}{-5}=\frac{2}{-5}\cdot\frac{5}{5}=\frac{2\cdot 5}{-5\cdot 5}=\frac{10}{-25}$$

15. Since $132\div 22=6$, we multiply by $\frac{6}{6}$.

$$\frac{-7}{22}=\frac{-7}{22}\cdot\frac{6}{6}=\frac{-7\cdot 6}{22\cdot 6}=\frac{-42}{132}$$

17. Since $8x\div 8=x$, we multiply by $\frac{x}{x}$.

$$\frac{5}{8}=\frac{5}{8}\cdot\frac{x}{x}=\frac{5x}{8x}$$

19. Since $11m\div 11=m$, we multiply by $\frac{m}{m}$.

$$\frac{7}{11}\cdot\frac{m}{m}=\frac{7m}{11m}$$

21. Since $9ab\div 9=ab$, we multiply by $\frac{ab}{ab}$.

$$\frac{4}{9}\cdot\frac{ab}{ab}=\frac{4ab}{9ab}$$

23. Since $27b\div 9=3b$, we multiply by $\frac{3b}{3b}$.

$$\frac{4}{9}=\frac{4}{9}\cdot\frac{3b}{3b}=\frac{12b}{27b}$$

25. $\dfrac{2}{4} = \dfrac{1\cdot 2}{2\cdot 2}$ ⟵ Factor the numerator / ⟵ Factor the denominator

$= \dfrac{1}{2}\cdot\dfrac{2}{2}$ ⟵ Factor the fraction

$= \dfrac{1}{2}\cdot 1$ ⟵ $\dfrac{2}{2}=1$

$= \dfrac{1}{2}$ ⟵ Removing a factor of 1

27. $\dfrac{-6}{8} = \dfrac{-3\cdot 2}{4\cdot 2}$ ⟵ Factor the numerator / ⟵ Factor the denominator

$= \dfrac{-3}{4}\cdot\dfrac{2}{2}$ ⟵ Factor the fraction

$= \dfrac{-3}{4}\cdot 1$ ⟵ $\dfrac{2}{2}=1$

$= \dfrac{-3}{4}$ ⟵ Removing a factor of 1

29. $\dfrac{2}{15} = \dfrac{1\cdot 3}{5\cdot 3}$ ⟵ Factor the numerator / ⟵ Factor the denominator

$= \dfrac{1}{5}\cdot\dfrac{3}{3}$ ⟵ Factor the fraction

$= \dfrac{1}{5}\cdot 1$ ⟵ $\dfrac{3}{3}=1$

$= \dfrac{1}{5}$ ⟵ Removing a factor of 1

31. $\dfrac{24}{-8} = \dfrac{3\cdot 8}{-1\cdot 8} = \dfrac{3}{-1}\cdot\dfrac{8}{8} = \dfrac{3}{-1}\cdot 1 = \dfrac{3}{-1} = -3$

33. $\dfrac{27}{36} = \dfrac{9\cdot 3}{9\cdot 4} = \dfrac{9}{9}\cdot\dfrac{3}{4} = 1\cdot\dfrac{3}{4} = \dfrac{3}{4}$

35. $-\dfrac{12}{10} = -\dfrac{6\cdot 2}{5\cdot 2} = -\dfrac{6}{5}\cdot\dfrac{2}{2} = -\dfrac{6}{5}\cdot 1 = -\dfrac{6}{5}$

37. $\dfrac{16}{48} = \dfrac{1\cdot 16}{3\cdot 16} = \dfrac{1}{3}\cdot\dfrac{16}{16} = \dfrac{1}{3}\cdot 1 = \dfrac{1}{3}$

39. $\dfrac{-17}{51} = \dfrac{-1\cdot 17}{3\cdot 17} = \dfrac{-1}{3}\cdot\dfrac{17}{17} = \dfrac{-1}{3}\cdot 1 = \dfrac{-1}{3}$

41. $\dfrac{420}{480} = \dfrac{2\cdot 2\cdot 3\cdot 5\cdot 7}{2\cdot 2\cdot 2\cdot 2\cdot 2\cdot 3\cdot 5}$

$= \dfrac{2}{2}\cdot\dfrac{2}{2}\cdot\dfrac{3}{3}\cdot\dfrac{5}{5}\cdot\dfrac{7}{2\cdot 2\cdot 2}$

$= 1\cdot 1\cdot 1\cdot 1\cdot\dfrac{7}{2\cdot 2\cdot 2}$

$= \dfrac{7}{2\cdot 2\cdot 2}$

$= \dfrac{7}{8}$

43. $\dfrac{5m}{7m} = \dfrac{5\cdot m}{7\cdot m} = \dfrac{5}{7}\cdot\dfrac{m}{m} = \dfrac{5}{7}\cdot 1 = \dfrac{5}{7}$

45. $\dfrac{3ab}{8ab} = \dfrac{3\cdot a\cdot b}{8\cdot a\cdot b} = \dfrac{3}{8}\cdot\dfrac{a}{a}\cdot\dfrac{b}{b} = \dfrac{3}{8}\cdot 1\cdot 1 = \dfrac{3}{8}$

47. $\dfrac{9xy}{6x} = \dfrac{3\cdot 3\cdot x\cdot y}{2\cdot 3\cdot x} = \dfrac{3\cdot y}{2}\cdot\dfrac{3}{3}\cdot\dfrac{x}{x} = \dfrac{3y}{2}\cdot 1\cdot 1 = \dfrac{3y}{2}$

49. ***Familiarize***. We make a drawing. We let A = the area.

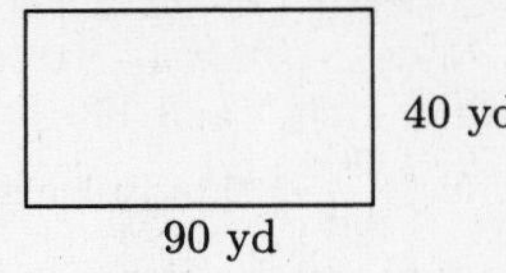

Translate. Using the formula for area, we have

$A = l\cdot w = 90\cdot 40.$

Solve. We carry out the multiplication.

$$\begin{array}{r} 4\,0 \\ \times\ \ 9\,0 \\ \hline 3\,6\,0\,0 \end{array}$$

Thus, $A = 3600$.

Check. We repeat the calculation. The answer checks.

State. The area is 3600 yd^2.

51. $30x = 150$

$\dfrac{30x}{30} = \dfrac{150}{30}$

$x = 5$

The solution is 5.

53. 

55. Use a calculator and the table of prime numbers on page 142 to find the prime factorizations of the numerator and denominator.

$\dfrac{2831}{2603} = \dfrac{19\cdot 149}{19\cdot 137} = \dfrac{19}{19}\cdot\dfrac{149}{137} = 1\cdot\dfrac{149}{137} = \dfrac{149}{137}$

57. Think of each person as $\dfrac{1}{10}$. First we write fractional notation for the part of the population that is shy. The multiplication sentence $4\cdot\dfrac{1}{10} = s$ corresponds to the situation. We multiply:

$$4\cdot\frac{1}{10} = \frac{4\cdot 1}{10} = \frac{4}{10}$$

Then we simplify:

$$\frac{4}{10} = \frac{2\cdot 2}{5\cdot 2} = \frac{2}{5}\cdot\frac{2}{2} = \frac{2}{5}\cdot 1 = \frac{2}{5}$$

Next we write fractional notation for the part of the population that is not shy. Since 4 of 10 people are shy, then 10 − 4, or 6, of 10 people are not shy. The multiplication sentence $6\cdot\dfrac{1}{10} = n$ corresponds to the situation. We multiply:

$$6\cdot\frac{1}{10} = \frac{6\cdot 1}{10} = \frac{6}{10}$$

Then we simplify:

$$\frac{6}{10} = \frac{2\cdot 3}{2\cdot 5} = \frac{2}{2}\cdot\frac{3}{5} = 1\cdot\frac{3}{5} = \frac{3}{5}$$

59. Of every 100 students, 18 gave a rating of "poor," 2 had no response, 4 gave a rating of "excellent," and 32 gave a rating of "good." Now $18 + 2 + 4 + 32 = 56$, so 56 gave a response other than "fair," and $100 - 56$, or 44 gave a rating of "fair." Thus $\frac{44}{100}$ of the students gave a rating of "fair." We can simplify this fraction:

$$\frac{44}{100} = \frac{4 \cdot 11}{4 \cdot 25} = \frac{4}{4} \cdot \frac{11}{25} = 1 \cdot \frac{11}{25} = \frac{11}{25}$$

Exercise Set 3.6

1. $\frac{3}{8} \cdot \frac{1}{3} = \frac{3 \cdot 1}{8 \cdot 3} = \frac{3}{3} \cdot \frac{1}{8} = 1 \cdot \frac{1}{8} = \frac{1}{8}$

3. $\frac{7}{8} \cdot \frac{-1}{7} = \frac{7(-1)}{8 \cdot 7} = \frac{7}{7} \cdot \frac{-1}{8} = \frac{-1}{8}$, or $-\frac{1}{8}$

5. $\frac{1}{8} \cdot \frac{4}{5} = \frac{1 \cdot 4}{8 \cdot 5} = \frac{1 \cdot 4}{2 \cdot 4 \cdot 5} = \frac{4}{4} \cdot \frac{1}{2 \cdot 5} = \frac{1}{2 \cdot 5} = \frac{1}{10}$

7. $\frac{1}{6} \cdot \frac{2}{3} = \frac{1 \cdot 2}{6 \cdot 3} = \frac{1 \cdot 2}{2 \cdot 3 \cdot 3} = \frac{2}{2} \cdot \frac{1}{3 \cdot 3} = \frac{1}{3 \cdot 3} = \frac{1}{9}$

9. $\frac{12}{-5} \cdot \frac{9}{8} = \frac{12 \cdot 9}{-5 \cdot 8} = \frac{4 \cdot 3 \cdot 9}{-5 \cdot 2 \cdot 4} = \frac{4}{4} \cdot \frac{3 \cdot 9}{-5 \cdot 2} = \frac{3 \cdot 9}{-5 \cdot 2} =$
$\frac{27}{-10}$, or $-\frac{27}{10}$

11. $\frac{5x}{9} \cdot \frac{7}{5} = \frac{5x \cdot 7}{9 \cdot 5} = \frac{5 \cdot x \cdot 7}{9 \cdot 5} = \frac{5}{5} \cdot \frac{x \cdot 7}{9} = \frac{7x}{9}$

13. $\frac{1}{4} \cdot 8 = \frac{1 \cdot 8}{4} = \frac{8}{4} = \frac{4 \cdot 2}{4 \cdot 1} = \frac{4}{4} \cdot \frac{2}{1} = \frac{2}{1} = 2$

15. $15 \cdot \frac{1}{3} = \frac{15 \cdot 1}{3} = \frac{15}{3} = \frac{3 \cdot 5}{3 \cdot 1} = \frac{3}{3} \cdot \frac{5}{1} = \frac{5}{1} = 5$

17. $-12 \cdot \frac{3}{4} = \frac{-12 \cdot 3}{4} = \frac{4(-3) \cdot 3}{4 \cdot 1} = \frac{4}{4} \cdot \frac{-3 \cdot 3}{1} =$
$\frac{-3 \cdot 3}{1} = \frac{-9}{1} = -9$

19. $\frac{3}{8} \cdot 8a = \frac{3 \cdot 8a}{8} = \frac{3 \cdot 8 \cdot a}{8 \cdot 1} = \frac{8}{8} \cdot \frac{3 \cdot a}{1} = \frac{3a}{1} = 3a$

21. $13\left(\frac{-2}{5}\right) = \frac{13(-2)}{5} = \frac{-26}{5}$, or $-\frac{26}{5}$

23. $\frac{m}{10} \cdot 28 = \frac{m \cdot 28}{10} = \frac{m \cdot 2 \cdot 14}{2 \cdot 5} = \frac{2}{2} \cdot \frac{m \cdot 14}{5} = \frac{14m}{5}$

25. $\frac{1}{6} \cdot 360x = \frac{1 \cdot 360x}{6} = \frac{360x}{6} = \frac{6 \cdot 60 \cdot x}{6 \cdot 1} = \frac{6}{6} \cdot \frac{60 \cdot x}{1} =$
$\frac{60x}{1} = 60x$

27. $240\left(\frac{1}{-8}\right) = \frac{240 \cdot 1}{-8} = \frac{240}{-8} = \frac{8 \cdot 30}{8(-1)} = \frac{8}{8} \cdot \frac{30}{-1} =$
$\frac{30}{-1} = -30$

29. $9 \cdot \frac{1}{9} = \frac{9 \cdot 1}{9} = \frac{9 \cdot 1}{9 \cdot 1} = 1$

31. $-\frac{1}{3} \cdot 3 = -\frac{1 \cdot 3}{3} = -\frac{1 \cdot 3}{1 \cdot 3} = -1$

33. $\frac{7}{10} \cdot \frac{10}{7} = \frac{7 \cdot 10}{10 \cdot 7} = \frac{7 \cdot 10}{7 \cdot 10} = 1$

35. $\frac{m}{n} \cdot \frac{n}{m} = \frac{m \cdot n}{n \cdot m} = \frac{m \cdot n}{m \cdot n} = 1$

37. $\frac{4}{10} \cdot \frac{5}{10} = \frac{4 \cdot 5}{10 \cdot 10} = \frac{2 \cdot 2 \cdot 5 \cdot 1}{2 \cdot 5 \cdot 2 \cdot 5} = \frac{2 \cdot 2 \cdot 5}{2 \cdot 2 \cdot 5} \cdot \frac{1}{5} = \frac{1}{5}$

39. $\frac{8}{10} \cdot \frac{45}{100} = \frac{8 \cdot 45}{10 \cdot 100} = \frac{2 \cdot 2 \cdot 2 \cdot 5 \cdot 9}{2 \cdot 5 \cdot 2 \cdot 2 \cdot 5 \cdot 5}$
$= \frac{2 \cdot 2 \cdot 2 \cdot 5}{2 \cdot 2 \cdot 2 \cdot 5} \cdot \frac{9}{5 \cdot 5} = \frac{9}{5 \cdot 5} = \frac{9}{25}$

41. $\left(-\frac{11}{24}\right)\frac{3}{5} = -\frac{11 \cdot 3}{24 \cdot 5} = -\frac{11 \cdot 3}{3 \cdot 8 \cdot 5} = \frac{3}{3}\left(-\frac{11}{8 \cdot 5}\right) =$
$-\frac{11}{8 \cdot 5} = -\frac{11}{40}$

43. $\frac{10a}{21} \cdot \frac{3}{4a} = \frac{10a \cdot 3}{21 \cdot 4a} = \frac{2 \cdot 5 \cdot a \cdot 3}{3 \cdot 7 \cdot 2 \cdot 2 \cdot a}$
$= \frac{2 \cdot 3 \cdot a}{2 \cdot 3 \cdot a} \cdot \frac{5}{2 \cdot 7} = \frac{5}{2 \cdot 7} = \frac{5}{14}$

45. ***Familiarize.*** We visualize the situation. We let a = the amount received for working $\frac{3}{4}$ of a day.

1 day \$56	
3/4 day a	

Translate. We write an equation.

Pay for 3/4 of a day is $\frac{3}{4}$ of \$56

$$a = \frac{3}{4} \cdot 56$$

Solve. We carry out the multiplication.

$$a = \frac{3}{4} \cdot 56 = \frac{3 \cdot 56}{4}$$
$$= \frac{3 \cdot 14 \cdot 4}{1 \cdot 4} = \frac{3 \cdot 14}{1} \cdot \frac{4}{4}$$
$$= 42$$

Check. We can repeat the calculation. We can also determine that the answer seems reasonable since we multiplied 56 by a number less than 1 and the result is less than 56. The answer checks.

State. \$42 is received for working $\frac{3}{4}$ of a day.

47. ***Familiarize.*** We visualize the situation. We let n = the number of addresses that will be incorrect after one year.

Mailing list 2500 addresses			
1/4 of the addresses n			

Translate.

Number incorrect is $\frac{1}{4}$ of Number of addresses

$$n = \frac{1}{4} \cdot 2500$$

Solve. We carry out the multiplication.

$$n = \frac{1}{4} \cdot 2500 = \frac{1 \cdot 2500}{4} = \frac{2500}{4}$$
$$= \frac{4 \cdot 625}{4 \cdot 1} = \frac{4}{4} \cdot \frac{625}{1}$$
$$= 625$$

Check. We can repeat the calculation. We can also determine that the answer seems reasonable since we multiplied 2500 by a number less than 1 and the result is less than 2500. The answer checks.

State. After one year 625 addresses will be incorrect.

49. ***Familiarize.*** We draw a picture.

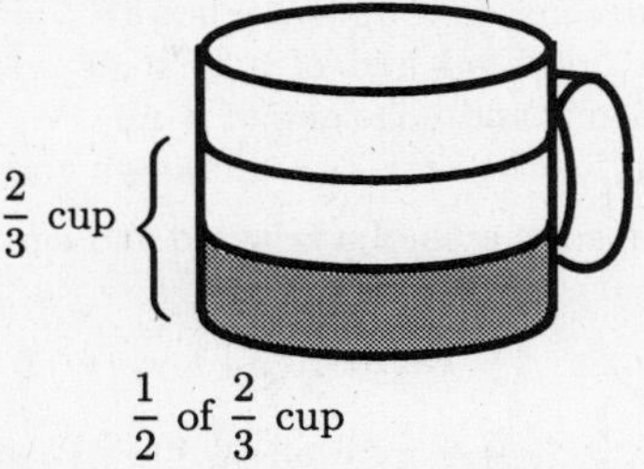

We let n = the amount of flour the chef should use.

Translate. The multiplication sentence

$$\frac{1}{2} \cdot \frac{2}{3} = n$$

corresponds to the situation.

Solve. We multiply and simplify:

$$n = \frac{1}{2} \cdot \frac{2}{3} = \frac{1 \cdot 2}{2 \cdot 3} = \frac{2}{2} \cdot \frac{1}{3} = \frac{1}{3}$$

Check. We can repeat the calculation. We can also determine that the answer seems reasonable since we multiplied $\frac{2}{3}$ by a number less than 1 and the result is less than $\frac{2}{3}$. The answer checks.

State. The chef should use $\frac{1}{3}$ cup of flour.

51. ***Familiarize.*** We visualize the situation. Let a = the amount of the loan.

Tuition $2400	
2/3 of the tuition $a	

Translate. We write an equation.

Amount of loan is $\frac{2}{3}$ of the tuition

$$a = \frac{2}{3} \cdot 2400$$

Solve. We carry out the multiplication.

$$a = \frac{2}{3} \cdot 2400 = \frac{2 \cdot 2400}{3}$$
$$= \frac{2 \cdot 3 \cdot 800}{3 \cdot 1} = \frac{3}{3} \cdot \frac{2 \cdot 800}{1}$$
$$= 1600$$

Check. We can repeat the calculation. We can also determine that the answer seems reasonable since we multiplied 2400 by a number less than 1 and the result is less than 2400. The answer checks.

State. The loan was $1600.

53. ***Familiarize.*** We draw a picture.

$\frac{2}{3}$ in.

1 in.
240 miles

We let n = the number of miles represented by $\frac{2}{3}$ in.

Translate. The multiplication sentence

$$n = \frac{2}{3} \cdot 240$$

corresponds to the situation.

Solve. We multiply and simplify:

$$n = \frac{2}{3} \cdot 240 = \frac{2 \cdot 240}{3} = \frac{2 \cdot 3 \cdot 80}{1 \cdot 3}$$
$$= \frac{3}{3} \cdot \frac{2 \cdot 80}{1} = \frac{2 \cdot 80}{1}$$
$$= 160$$

Check. We can repeat the calculation. We can also determine that the answer seems reasonable since we multiplied 240 by a number less than 1 and the result is less than 240.

State. $\frac{2}{3}$ in. on the map represents 160 miles.

55. ***Familiarize.*** This is a multistep problem. First we find the amount of each of the given expenses. Then we find the total of these expenses and take it away from the annual income to find how much is spent for other expenses.

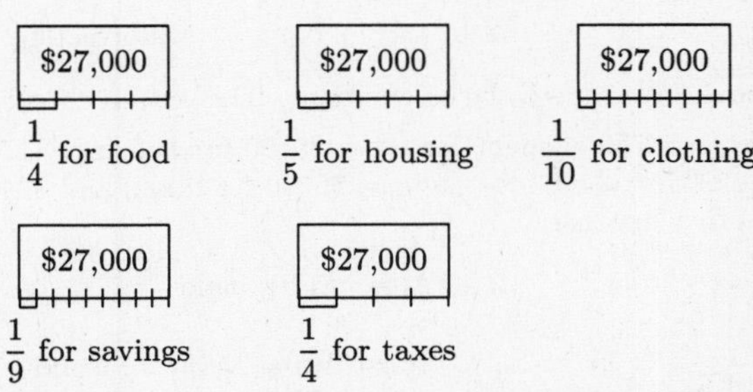

We let f, h, c, s, and t represent the amounts spent on food, housing, clothing, savings, and taxes, respectively.

Translate. The following multiplication sentences correspond to the situation.

$\frac{1}{4} \cdot 27,000 = f$ $\quad \frac{1}{9} \cdot 27,000 = s$

$\frac{1}{5} \cdot 27,000 = h$ $\quad \frac{1}{4} \cdot 27,000 = t$

$\frac{1}{10} \cdot 27,000 = c$

Solve. We multiply and simplify.

$$f = \frac{1}{4} \cdot 27,000 = \frac{27,000}{4} = \frac{4 \cdot 6750}{4 \cdot 1} = \frac{4}{4} \cdot \frac{6750}{1} = 6750$$

$$h = \frac{1}{5} \cdot 27,000 = \frac{27,000}{5} = \frac{5 \cdot 5400}{5 \cdot 1} = \frac{5}{5} \cdot \frac{5400}{1} = 5400$$

$$c = \frac{1}{10} \cdot 27,000 = \frac{27,000}{10} = \frac{10 \cdot 2700}{10 \cdot 1} = \frac{10}{10} \cdot \frac{2700}{1} = 2700$$

$$s = \frac{1}{9} \cdot 27,000 = \frac{27,000}{9} = \frac{9 \cdot 3000}{9 \cdot 1} = \frac{9}{9} \cdot \frac{3000}{1} = 3000$$

$$t = \frac{1}{4} \cdot 27,000 = \frac{27,000}{4} = \frac{4 \cdot 6750}{4 \cdot 1} = \frac{4}{4} \cdot \frac{6750}{1} = 6750$$

We add to find the total of these expenses.

```
   2 1
 $ 6 7 5 0
   5 4 0 0
   2 7 0 0
   3 0 0 0
   6 7 5 0
 $ 2 4, 6 0 0
```

We let m = the amount spent on other expenses and subtract to find this amount.

Annual income	minus	Total of itemized expenses	is	Total spent on other expenses	
↓	↓	↓	↓	↓	
\$27,000	−	\$24,600	=	m	
		\$2400	=	m	Subtracting

Check. We repeat the calculations. The results check.

State. \$6750 is spent for food, \$5400 for housing, \$2700 for clothing, \$3000 for savings, \$6750 for taxes, and \$2400 for other expenses.

57. $A = \frac{1}{2} \cdot b \cdot h$ Area of a triangle

$A = \frac{1}{2} \cdot 15 \text{ in.} \cdot 8 \text{ in.}$ Substituting 15 in. for b and 8 in. for h

$A = \frac{15 \cdot 8}{2} \text{ in}^2$

$A = 60 \text{ in}^2$

59. $A = \frac{1}{2} \cdot b \cdot h$ Area of a triangle

$A = \frac{1}{2} \cdot \frac{17}{5} \text{ km} \cdot 4 \text{ km}$ Substituting $\frac{17}{5}$ km for b and 4 km for h

$A = \frac{17 \cdot 4}{2 \cdot 5} \text{ km}^2$

$A = \frac{34}{5} \text{ km}^2$

61. $A = \frac{1}{2} \cdot b \cdot h$ Area of a triangle

$A = \frac{1}{2} \cdot 4 \text{ m} \cdot \frac{7}{2} \text{ m}$ Substituting 4 m for b and $\frac{7}{2}$ m for h

$A = \frac{4 \cdot 7}{2 \cdot 2} \text{ m}^2$

$A = 7 \text{ m}^2$

63. ***Familiarize***. We look for figures whose areas we can calculate using area formulas we already know.

Translate. The figure consists of a rectangle with a length of 10 mi and a width of 8 mi and of a triangle with a base of 13 − 10, or 3 mi, and a height of 8 mi. We use the formula $A = l \cdot w$ for the area of a rectangle and the formula $A = \frac{1}{2} \cdot b \cdot h$ for the area of a triangle and add the two areas.

Solve. For the rectangle: $A = l \cdot w = 10 \text{ mi} \cdot 8 \text{ mi} = 80 \text{ mi}^2$

For the triangle: $A = \frac{1}{2} \cdot b \cdot h = \frac{1}{2} \cdot 3 \text{ mi} \cdot 8 \text{ mi} = 12 \text{ mi}^2$

Then we add: $80 \text{ mi}^2 + 12 \text{ mi}^2 = 92 \text{ mi}^2$

Check. We repeat the calculations.

State. The area of the figure is 92 mi^2.

65. ***Familiarize***. We look for figures whose areas we can calculate using area formulas we already know.

Translate. Each of the 2 ends of the building consists of a rectangle with a length of 50 ft and a width of 25 ft and of a triangle with a base of 50 ft and a height of 11 ft. Each of the two sides is a rectangle with a length of 75 ft and a width of 25 ft. We use the formula $A = l \cdot w$ for the area of a rectangle twice and the formula $A = \frac{1}{2} \cdot b \cdot h$ for the area of a triangle, multiply each area by 2, and then add the two results.

Solve.

For each end:

Rectangle: $A = l \cdot w = 50 \text{ ft} \cdot 25 \text{ ft} = 1250 \text{ ft}^2$

Triangle: $A = \frac{1}{2} \cdot b \cdot h = \frac{1}{2} \cdot 50 \text{ ft} \cdot 11 \text{ ft} = 275 \text{ ft}^2$

For each side:

$A = l \cdot w = 75 \text{ ft} \cdot 25 \text{ ft} = 1875 \text{ ft}^2$

We multiply each area by 2:

$2 \cdot 1250 \text{ ft}^2 = 2500 \text{ ft}^2$

$2 \cdot 275 \text{ ft}^2 = 550 \text{ ft}^2$

$2 \cdot 1875 \text{ ft}^2 = 3750 \text{ ft}^2$

Now we add:

$$2500 \text{ ft}^2 + 550 \text{ ft}^2 + 3750 \text{ ft}^2 = 6800 \text{ ft}^2$$

Check. We repeat the calculations.

State. The total area of the sides and ends of the building is 6800 ft^2.

67. $48 \cdot t = 1680$

$$\frac{48 \cdot t}{48} = \frac{1680}{48}$$

$$t = 35$$

The solution is 35.

69. $747 = x + 270$

$$747 - 270 = x + 270 - 270$$

$$477 = x$$

The solution is 477.

71. We add absolute values and make the result negative.

$$(-39) + (-72) = -111$$

73. 

75. Use a calculator and the table of prime numbers on page 142 to find factors that are common to the numerator and denominator of the product.

$$\frac{201}{535} \cdot \frac{4601}{6499} = \frac{201 \cdot 4601}{535 \cdot 6499} = \frac{3 \cdot 67 \cdot 43 \cdot 107}{5 \cdot 107 \cdot 67 \cdot 97} = \frac{67 \cdot 107}{67 \cdot 107} \cdot \frac{3 \cdot 43}{5 \cdot 97} = \frac{3 \cdot 43}{5 \cdot 97} = \frac{129}{485}$$

77. ***Familiarize***. We look for figures whose areas we can calculate using area formulas we already know.

Translate. Each of the 2 ends of the box is a triangle with a base of 30 mm and a height of 26 mm. Each of the 3 sides is a rectangle with a length of 140 mm and a width of 30 mm. We use the formula $A = \frac{1}{2} \cdot b \cdot h$ for the area of a triangle and multiply the area by 2. We use the formula $A = l \cdot w$ for the area of a rectangle and multiply that area by 3. Then we add the results.

Solve.

For each end:

$$A = \frac{1}{2} \cdot b \cdot h = \frac{1}{2} \cdot 30 \text{ mm} \cdot 26 \text{ mm} = 390 \text{ mm}^2$$

We multiply this area by 2: $2 \cdot 390 \text{ mm}^2 = 780 \text{ mm}^2$

For each side:

$$A = l \cdot w = 140 \text{ mm} \cdot 30 \text{ mm} = 4200 \text{ mm}^2$$

We multiply this area by 3: $3 \cdot 4200 \text{ mm}^2 = 12,600 \text{ mm}^2$

Finally we add the areas of the ends and sides:

$$780 \text{ mm}^2 + 12,600 \text{ mm}^2 = 13,380 \text{ mm}^2$$

Check. We repeat the calculations.

State. The surface area of the candy box is 13,380 mm^2.

79. ***Familiarize***. We are told that $\frac{2}{3}$ of $\frac{7}{8}$ of the students are high school graduates who are older than 20, and $\frac{1}{7}$ of this fraction are left-handed. Thus, we want to find $\frac{1}{7}$ of $\frac{2}{3}$ of $\frac{7}{8}$. We let f represent this fraction.

Translate. The multiplication sentence

$$f = \frac{1}{7} \cdot \frac{2}{3} \cdot \frac{7}{8}$$

corresponds to this situation.

Solve. We multiply and simplify.

$$f = \frac{1}{7} \cdot \frac{2}{3} \cdot \frac{7}{8} = \frac{1 \cdot 2}{7 \cdot 3} \cdot \frac{7}{8} = \frac{1 \cdot 2 \cdot 7}{7 \cdot 3 \cdot 8} = \frac{1 \cdot 2 \cdot 7}{7 \cdot 3 \cdot 2 \cdot 4} = \frac{2 \cdot 7}{2 \cdot 7} \cdot \frac{1}{3 \cdot 4} = \frac{1}{3 \cdot 4} = \frac{1}{12}$$

Check. We repeat the calculation. The result checks.

State. $\frac{1}{12}$ of the students are left-handed high school graduates over the age of 20.

81. ***Familiarize***. If we divide the group of entering students into 8 equal parts and take 7 of them, we have the fractional part of the students that completed high school. Then the 1 part remaining, or $\frac{1}{8}$ of the students, did not graduate from high school. Similarly, if we divide the group of entering students into 3 equal parts and take 2 of them, we have the fractional part of the students that is older than 20. Then the 1 part remaining, or $\frac{1}{3}$ of the students, are 20 years old or younger. From Exercise 79 we know that $\frac{1}{7}$ of the students are left-handed. Thus, we want to find $\frac{1}{7}$ of $\frac{1}{3}$ of $\frac{1}{8}$. We let f = this fraction.

Translate. The multiplication sentence

$$f = \frac{1}{7} \cdot \frac{1}{3} \cdot \frac{1}{8}$$

corresponds to this situation.

Solve. We multiply.

$$f = \frac{1}{7} \cdot \frac{1}{3} \cdot \frac{1}{8} = \frac{1 \cdot 1 \cdot 1}{7 \cdot 3 \cdot 8} = \frac{1}{168}$$

Check. We repeat the calculation. The result checks.

State. $\frac{1}{168}$ of the students did not graduate from high school, are 20 years old or younger, and are left-handed.

Exercise Set 3.7

1. $\frac{6}{5}$ Interchange the numerator and denominator.

The reciprocal of $\frac{6}{5}$ is $\frac{5}{6}$. $\left(\frac{6}{5} \cdot \frac{5}{6} = \frac{30}{30} = 1\right)$

3. Think of 4 as $\frac{4}{1}$.

$\frac{4}{1}$ Interchange the numerator and denominator.

The reciprocal of 4 is $\frac{1}{4}$. $\left(\frac{4}{1}\cdot\frac{1}{4}=\frac{4}{4}=1\right)$

5. $\frac{1}{6}$ Interchange the numerator and denominator.

The reciprocal of $\frac{1}{6}$ is 6. $\left(\frac{6}{1}=6;\ \frac{1}{6}\cdot\frac{6}{1}=\frac{6}{6}=1\right)$

7. $-\frac{10}{3}$ Interchange the numerator and denominator.

The reciprocal of $-\frac{10}{3}$ is $-\frac{3}{10}$. $\left(-\frac{10}{3}\left(-\frac{3}{10}\right)=\frac{30}{30}=1\right)$

9. $\frac{2}{21}$ Interchange the numerator and denominator.

The reciprocal of $\frac{2}{21}$ is $\frac{21}{2}$. $\left(\frac{2}{21}\cdot\frac{21}{2}=\frac{42}{42}=1\right)$

11. $\frac{2x}{y}$ Interchange the numerator and denominator.

The reciprocal of $\frac{2x}{y}$ is $\frac{y}{2x}$. $\left(\frac{2x}{y}\cdot\frac{y}{2x}=\frac{2xy}{2xy}=1\right)$

13. $\frac{7}{-15}$ Interchange the numerator and denominator.

The reciprocal of $\frac{7}{-15}$ is $\frac{-15}{7}$. $\left(\frac{7}{-15}\left(\frac{-15}{7}\right)=\frac{-105}{-105}=1\right)$

15. Think of $7m$ as $\frac{7m}{1}$.

$\frac{7m}{1}$ Interchange the numerator and denominator.

The reciprocal of $7m$ is $\frac{1}{7m}$. $\left(\frac{7m}{1}\cdot\frac{1}{7m}=\frac{7m}{7m}=1\right)$

17. $\frac{3}{5}\div\frac{3}{4}=\frac{3}{5}\cdot\frac{4}{3}$ Multiplying the dividend $\left(\frac{3}{5}\right)$ by the reciprocal of the divisor $\left(\text{The reciprocal of } \frac{3}{4} \text{ is } \frac{4}{3}.\right)$

$=\frac{3\cdot 4}{5\cdot 3}$ Multiplying numerators and denominators

$=\frac{3}{3}\cdot\frac{4}{5}=\frac{4}{5}$ Simplifying

19. $\frac{3}{5}\div\frac{-9}{4}=\frac{3}{5}\cdot\frac{4}{-9}$ Multiplying the dividend $\left(\frac{3}{5}\right)$ by the reciprocal of the divisor $\left(\text{The reciprocal of } \frac{-9}{4} \text{ is } \frac{4}{-9}.\right)$

$=\frac{3\cdot 4}{5(-9)}$ Multiplying numerators and denominators

$=\frac{3\cdot 4}{5\cdot 3(-3)}$

$=\frac{3}{3}\cdot\frac{4}{5(-3)}$ Simplifying

$=\frac{4}{5(-3)}=\frac{4}{-15}$, or $-\frac{4}{15}$

21. $\frac{4}{3}\div\frac{1}{3}=\frac{4}{3}\cdot 3=\frac{4\cdot 3}{3}=\frac{3}{3}\cdot 4=4$

23. $\left(-\frac{1}{3}\right)\div\frac{1}{6}=-\frac{1}{3}\cdot 6=-\frac{1\cdot 6}{3}=-\frac{1\cdot 2\cdot 3}{1\cdot 3}=$

$-\frac{1\cdot 3}{1\cdot 3}\cdot 2=-2$

25. $\frac{3}{8}\div 3=\frac{3}{8}\cdot\frac{1}{3}=\frac{3\cdot 1}{8\cdot 3}=\frac{3}{3}\cdot\frac{1}{8}=\frac{1}{8}$

27. $\frac{12}{7}\div 4x=\frac{12}{7}\cdot\frac{1}{4x}=\frac{12\cdot 1}{7\cdot 4x}=\frac{4\cdot 3\cdot 1}{7\cdot 4\cdot x}=\frac{4}{4}\cdot\frac{3\cdot 1}{7\cdot x}=$

$\frac{3\cdot 1}{7\cdot x}=\frac{3}{7x}$

29. $(-12)\div\frac{3}{2}=-12\cdot\frac{2}{3}=-\frac{12\cdot 2}{3}=-\frac{3\cdot 4\cdot 2}{3\cdot 1}$

$=-\frac{3}{3}\cdot\frac{4\cdot 2}{1}=-\frac{4\cdot 2}{1}=-\frac{8}{1}=-8$

31. $28\div\frac{4}{5a}=28\cdot\frac{5a}{4}=\frac{28\cdot 5a}{4}=\frac{4\cdot 7\cdot 5\cdot a}{4\cdot 1}=\frac{4}{4}\cdot\frac{7\cdot 5\cdot a}{1}$

$=\frac{7\cdot 5\cdot a}{1}=35a$

33. $\left(-\frac{5}{8}\right)\div\left(-\frac{5}{8}\right)=-\frac{5}{8}\left(-\frac{8}{5}\right)=\frac{5\cdot 8}{8\cdot 5}=\frac{5\cdot 8}{5\cdot 8}=1$

35. $\frac{-8}{15}\div\frac{4}{5}=\frac{-8}{15}\cdot\frac{5}{4}=\frac{-8\cdot 5}{15\cdot 4}=\frac{-2\cdot 4\cdot 5}{3\cdot 5\cdot 4}=$

$\frac{4\cdot 5}{4\cdot 5}\cdot\frac{-2}{3}=\frac{-2}{3}$, or $-\frac{2}{3}$

37. $\frac{9}{5}\div\frac{4}{5}=\frac{9}{5}\cdot\frac{5}{4}=\frac{9\cdot 5}{5\cdot 4}=\frac{5}{5}\cdot\frac{9}{4}=\frac{9}{4}$

39. $120a\div\frac{5}{6}=120a\cdot\frac{6}{5}=\frac{120a\cdot 6}{5}=\frac{5\cdot 24\cdot a\cdot 6}{5\cdot 1}$

$=\frac{5}{5}\cdot\frac{24\cdot a\cdot 6}{1}=\frac{24\cdot a\cdot 6}{1}=144a$

41. ***Familiarize***. We make a drawing. Repeated subtraction, or division, will work here.

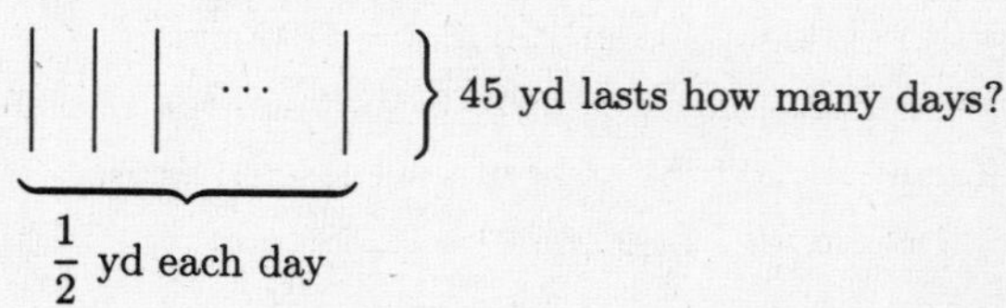

We let $n =$ the number of days the container will last.

Translate. The problem can be translated to the following equation:

$$n = 45 \div \frac{1}{2}$$

Solve. We carry out the division.

$$\begin{aligned} n &= 45 \div \frac{1}{2} \\ &= 45 \cdot 2 \qquad \text{Multiplying by the reciprocal} \\ &= 90 \end{aligned}$$

Check. If $\frac{1}{2}$ yd of dental floss is used on each of 90 days, a total of

$$\frac{1}{2} \cdot 90 = \frac{1 \cdot 90}{2} = \frac{1 \cdot 2 \cdot 45}{2} = 1 \cdot 45,$$

or 45 yd of dental floss is used. Since the problem states that Joy's container holds 45 yd, our answer checks.

State. The container will last 90 days.

43. ***Familiarize***. We make a drawing.

How long for $\frac{3}{4}$ mi?

$\frac{1}{12}$ mi each day

We let $d =$ the number of days it will take to repave a $\frac{3}{4}$-mi stretch of road.

Translate. The problem translates to the following equation:

$$d = \frac{3}{4} \div \frac{1}{12}$$

Solve. We carry out the division.

$$\begin{aligned} d &= \frac{3}{4} \div \frac{1}{12} \\ &= \frac{3}{4} \cdot \frac{12}{1} = \frac{3 \cdot 12}{4 \cdot 1} \\ &= \frac{3 \cdot 3 \cdot 4}{4 \cdot 1} = \frac{4}{4} \cdot \frac{3 \cdot 3}{1} \\ &= 9 \end{aligned}$$

Check. If the crew repaves $\frac{1}{12}$ mi of road each day for 9 days, a total of

$$\frac{1}{12} \cdot 9 = \frac{1 \cdot 9}{12} = \frac{1 \cdot 3 \cdot 3}{3 \cdot 4} = \frac{1 \cdot 3}{4},$$

or $\frac{3}{4}$ mi of road will be repaved. Our answer checks.

State. It will take 9 days to repave a $\frac{3}{4}$-mi stretch of road.

45. ***Familiarize***. We make a drawing.

6 lb feeds how many people?

$\frac{3}{8}$ lb per person

We let $p =$ the number of people who can attend the luncheon.

Translate. The problem translates to the following equation:

$$p = 6 \div \frac{3}{8}$$

Solve. We carry out the division.

$$\begin{aligned} p &= 6 \div \frac{3}{8} \\ &= 6 \cdot \frac{3}{8} = \frac{6 \cdot 8}{3} \\ &= \frac{2 \cdot 3 \cdot 8}{3 \cdot 1} = \frac{3}{3} \cdot \frac{2 \cdot 8}{1} \\ &= 16 \end{aligned}$$

Check. If each of 16 people is allotted $\frac{3}{8}$ lb of cold cuts, a total of

$$16 \cdot \frac{3}{8} = \frac{16 \cdot 3}{8} = \frac{2 \cdot 8 \cdot 3}{8} = 2 \cdot 3,$$

or 6 lb of cold cuts are used. Our answer checks.

State. 16 people can attend the luncheon.

47. ***Familiarize***. We make a drawing.

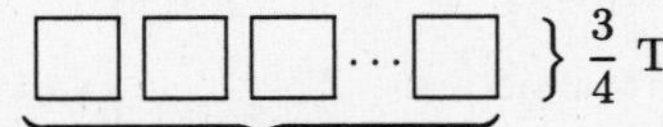

$\frac{3}{4}$ T

How much for each of 6 departments?

We let $c =$ the number of tons of clay each art department will receive.

Translate. The problem translates to the following equation:

$$c = \frac{3}{4} \div 6$$

Solve. We carry out the division.

$$\begin{aligned} c &= \frac{3}{4} \div 6 \\ &= \frac{3}{4} \cdot \frac{1}{6} = \frac{3 \cdot 1}{4 \cdot 6} \\ &= \frac{3 \cdot 1}{4 \cdot 2 \cdot 3} = \frac{3}{3} \cdot \frac{1}{4 \cdot 2} \\ &= \frac{1}{8} \end{aligned}$$

Check. If each of 6 art departments gets $\frac{1}{8}$ T of clay, a total of

$$6 \cdot \frac{1}{8} = \frac{6 \cdot 1}{8} = \frac{2 \cdot 3 \cdot 1}{2 \cdot 4} = \frac{3 \cdot 1}{4},$$

or $\frac{3}{4}$ T of clay is distributed. Our answer checks.

State. Each art department will receive $\frac{1}{8}$ T of clay.

49. ***Familiarize***. We draw a picture.

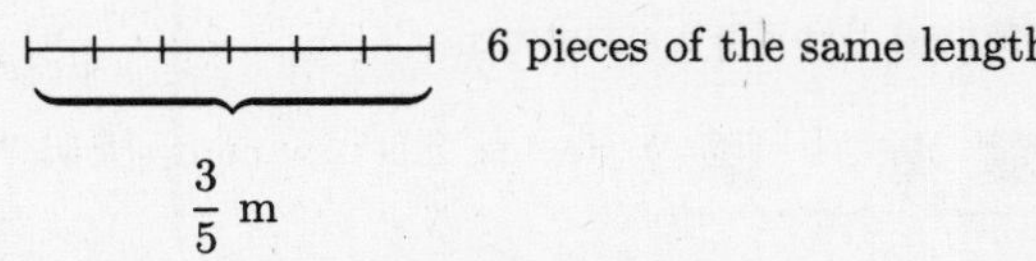

$\frac{3}{5}$ m

We let $n =$ the length of each piece.

Translate. The problem translates to the following equation:

$$n = \frac{3}{5} \div 6$$

Solve. We carry out the division.

$$\begin{aligned} n &= \frac{3}{5} \div 6 \\ &= \frac{3}{5} \cdot \frac{1}{6} = \frac{3 \cdot 1}{5 \cdot 6} \\ &= \frac{3 \cdot 1}{5 \cdot 2 \cdot 3} = \frac{3}{3} \cdot \frac{1}{5 \cdot 2} \\ &= \frac{1}{10} \end{aligned}$$

Check. If each of 6 pieces of wire is $\frac{1}{10}$ m long, there is a total of

$$6 \cdot \frac{1}{10} = \frac{6 \cdot 1}{10} = \frac{2 \cdot 3 \cdot 1}{2 \cdot 5} = \frac{2}{2} \cdot \frac{3 \cdot 1}{5},$$

or $\frac{3}{5}$ m of wire. Our answer checks.

State. Each piece is $\frac{1}{10}$ m.

51. *Familiarize*. We draw a picture.

} 24 yd makes how many pairs?

$\frac{3}{4}$ yd per pair

We let s = the number of pairs of basketball shorts that can be made.

Translate. The problem translates to the following equation:

$$s = 24 \div \frac{3}{4}$$

Solve. We carry out the division.

$$\begin{aligned} s &= 24 \div \frac{3}{4} \\ &= 24 \cdot \frac{4}{3} = \frac{24 \cdot 4}{3} \\ &= \frac{3 \cdot 8 \cdot 4}{1 \cdot 3} = \frac{3}{3} \cdot \frac{8 \cdot 4}{1} \\ &= 32 \end{aligned}$$

Check. If each of 32 pairs of shorts requires $\frac{3}{4}$ yd of nylon, a total of

$$32 \cdot \frac{3}{4} = \frac{32 \cdot 3}{4} = \frac{4 \cdot 8 \cdot 3}{4} = 8 \cdot 3,$$

or 24 yd of nylon is needed. Our answer checks.

State. 32 pairs of basketball shorts can be made from 24 yd of nylon.

53. *Familiarize*. We draw a picture.

} 16 cups fills how many bowls?

$\frac{2}{3}$ cup per bowl

We let b = the number of sugar bowls that can be filled.

Translate. The problem translates to the following equation:

$$b = 16 \div \frac{2}{3}$$

Solve. We carry out the division.

$$\begin{aligned} b &= 16 \div \frac{2}{3} \\ &= 16 \cdot \frac{3}{2} = \frac{16 \cdot 3}{2} \\ &= \frac{2 \cdot 8 \cdot 3}{1 \cdot 2} = \frac{2}{2} \cdot \frac{8 \cdot 3}{1} \\ &= 24 \end{aligned}$$

Check. If $\frac{2}{3}$ cup of sugar is put in each of 24 bowls, a total of

$$\frac{2}{3} \cdot 24 = \frac{2 \cdot 24}{3} = \frac{2 \cdot 3 \cdot 8}{3} = 2 \cdot 8,$$

or 16 cups of sugar are needed.

State. 24 sugar bowls can be filled.

55. $(-17)(-30) = 510$ (The signs are the same, so the answer is positive.)

57. $x^3 = 3^3 = 3 \cdot 3 \cdot 3 = 27$
$x^3 = (-3)^3 = (-3)(-3)(-3) = -27$

59. $3x^2 = 3 \cdot 7^2 = 3 \cdot 49 = 147$
$3x^2 = 3(-7)^2 = 3 \cdot 49 = 147$

61.

63. Use a calculator.

$$\begin{aligned} \frac{711}{1957} \div \frac{10,033}{13,081} &= \frac{711}{1957} \cdot \frac{13,081}{10,033} \\ &= \frac{711 \cdot 13,081}{1957 \cdot 10,033} \\ &= \frac{3 \cdot 3 \cdot 79 \cdot 103 \cdot 127}{19 \cdot 103 \cdot 79 \cdot 127} \\ &= \frac{79 \cdot 103 \cdot 127}{79 \cdot 103 \cdot 127} \cdot \frac{3 \cdot 3}{19} \\ &= \frac{9}{19} \end{aligned}$$

65. $\left(\frac{9}{10}\div\frac{2}{5}\div\frac{3}{8}\right)^2=\left(\frac{9}{10}\cdot\frac{5}{2}\div\frac{3}{8}\right)^2$

$=\left(\frac{9\cdot 5}{10\cdot 2}\div\frac{3}{8}\right)^2$

$=\left(\frac{9\cdot 5}{2\cdot 5\cdot 2}\div\frac{3}{8}\right)^2$

$=\left(\frac{9}{2\cdot 2}\div\frac{3}{8}\right)^2$

$=\left(\frac{9}{2\cdot 2}\cdot\frac{8}{3}\right)^2$

$=\left(\frac{9\cdot 8}{2\cdot 2\cdot 3}\right)^2$

$=\left(\frac{3\cdot 3\cdot 2\cdot 2\cdot 2}{2\cdot 2\cdot 3\cdot 1}\right)^2$

$=\left(\frac{3\cdot 2}{1}\right)^2$

$=6^2$

$=36$

Exercise Set 3.8

1. We can multiply $7x$ and 21 by any nonzero number to find an equivalent equation.

If both sides are multiplied by 2, we have

$2\cdot 7x=2\cdot 21$, or $14x=42$.

If both sides are multiplied by $\frac{1}{7}$, we have

$\frac{1}{7}\cdot 7x=\frac{1}{7}\cdot 21$, or $1x=3$, or $x=3$.

3. We can multiply $5a$ and 9 by any nonzero number to find an equivalent equation.

If both sides are multiplied by 3, we have

$3\cdot 5a=3\cdot 9$, or $15a=27$.

If both sides are multiplied by $\frac{1}{5}$, we have

$\frac{1}{5}\cdot 5a=\frac{1}{5}\cdot 9$, or $1a=\frac{9}{5}$, or $a=\frac{9}{5}$.

5. We can multiply $2x$ and 46 by any nonzero number to find an equivalent equation.

If both sides are multiplied by 2, we have

$2\cdot 2x=2\cdot 46$, or $4x=92$.

If both sides are multiplied by $\frac{1}{2}$, we have

$\frac{1}{2}\cdot 2x=\frac{1}{2}\cdot 46$, or $1x=23$, or $x=23$.

7. We can multiply $-4a$ and 22 by any nonzero number to find an equivalent equation.

If both sides are multiplied by -1, we have

$-1(-4a)=-1\cdot 22$, or $4a=-22$.

If both sides are multiplied by 3, we have

$3(-4a)=3\cdot 22$, or $-12a=66$.

9. We can multiply -42 and $3x$ by any nonzero number to find an equivalent equation.

If both sides are multiplied by 3, we have

$3(-42)=3\cdot 3x$, or $-126=9x$.

If both sides are multiplied by $\frac{1}{3}$, we have

$\frac{1}{3}(-42)=\frac{1}{3}\cdot 3x$, or $-14=1x$, or $-14=x$.

11. We can multiply $\frac{2}{3}a$ and 8 by any nonzero number to find an equivalent equation.

If both sides are multiplied by 3, we have

$3\cdot\frac{2}{3}a=3\cdot 8$, or $2a=24$.

If both sides are multiplied by $\frac{3}{2}$, we have

$\frac{3}{2}\cdot\frac{2}{3}a=\frac{3}{2}\cdot 8$, or $1a=12$, or $a=12$.

13. $\frac{4}{5}x=16$

$\frac{5}{4}\cdot\frac{4}{5}x=\frac{5}{4}\cdot 16$ The reciprocal of $\frac{4}{5}$ is $\frac{5}{4}$.

$1x=\frac{5\cdot 4\cdot 4}{4}$

$x=20$ Removing the factor $\frac{4}{4}$

Check: $\frac{4}{5}x=16$

$\frac{4}{5}\cdot 20 \;?\; 16$

$\frac{4\cdot 5\cdot 4}{5}$

$\frac{5}{5}\cdot\frac{4\cdot 4}{1}$ | 16 TRUE

The solution is 20.

15. $\frac{7}{3}a=21$

$\frac{3}{7}\cdot\frac{7}{3}a=\frac{3}{7}\cdot 21$ The reciprocal of $\frac{7}{3}$ is $\frac{3}{7}$.

$1a=\frac{3\cdot 3\cdot 7}{7}$

$a=9$ Removing the factor $\frac{7}{7}$

Check: $\frac{7}{3}a=21$

$\frac{7}{3}\cdot 9 \;?\; 21$

$\frac{7\cdot 3\cdot 3}{3}$

$\frac{3}{3}\cdot\frac{7\cdot 3}{1}$ | 21 TRUE

The solution is 9.

17. $\frac{2}{7}x = -16$

$\frac{7}{2}\cdot\frac{2}{7}x = \frac{7}{2}(-16)$ The reciprocal of $\frac{2}{7}$ is $\frac{7}{2}$.

$x = -\frac{7\cdot 2\cdot 8}{2}$

$x = -56$ Removing the factor $\frac{2}{2}$

Check:

$$\begin{array}{c|c} \frac{2}{7}x & -16 \\ \hline \frac{2}{7}(-56) & ?\ -16 \\ -\frac{2\cdot 7\cdot 8}{7} & \\ -\frac{2\cdot 8}{1}\cdot\frac{7}{7} & -16 \quad \text{TRUE} \end{array}$$

The solution is -56.

19. $-10 = \frac{2}{9}a$

$\frac{9}{2}(-10) = \frac{9}{2}\cdot\frac{2}{9}a$ The reciprocal of $\frac{2}{9}$ is $\frac{9}{2}$.

$-\frac{9\cdot 2\cdot 5}{2} = a$

$-45 = a$ Removing the factor $\frac{2}{2}$

Check:

$$\begin{array}{c|c} -10 & \frac{2}{9}a \\ \hline -10\ ? & \frac{2}{9}(-45) \\ & -\frac{2\cdot 9\cdot 5}{9} \\ -10 & -\frac{2\cdot 5}{1}\cdot\frac{9}{9} \quad \text{TRUE} \end{array}$$

The solution is -45.

21. $\frac{3}{5}x = \frac{2}{7}$

$\frac{5}{3}\cdot\frac{3}{5}x = \frac{5}{3}\cdot\frac{2}{7}$

$x = \frac{5\cdot 2}{3\cdot 7}$

$x = \frac{10}{21}$

$\frac{10}{21}$ checks and is the solution.

23. $\frac{3}{2}t = \frac{5}{7}$

$\frac{2}{3}\cdot\frac{3}{2}t = \frac{2}{3}\cdot\frac{5}{7}$

$t = \frac{2\cdot 5}{3\cdot 7}$

$t = \frac{10}{21}$

$\frac{10}{21}$ checks and is the solution.

25. $10 = \frac{4}{5}a$

$\frac{5}{4}\cdot 10 = \frac{5}{4}\cdot\frac{4}{5}a$

$\frac{5\cdot 2\cdot 5}{2\cdot 2} = a$

$\frac{25}{2} = a$

$\frac{25}{2}$ checks and is the solution.

27. $\frac{9}{5}x = \frac{3}{10}$

$\frac{5}{9}\cdot\frac{9}{5}x = \frac{5}{9}\cdot\frac{3}{10}$

$x = \frac{5\cdot 3\cdot 1}{3\cdot 3\cdot 2\cdot 5}$

$x = \frac{1}{6}$

$\frac{1}{6}$ checks and is the solution.

29. $-\frac{3}{10}x = 8$

$-\frac{10}{3}\left(-\frac{3}{10}x\right) = -\frac{10}{3}\cdot 8$

$x = -\frac{10\cdot 8}{3}$

$x = -\frac{80}{3}$

$-\frac{80}{3}$ checks and is the solution.

31. $a\cdot\frac{9}{7} = -\frac{3}{14}$

$a\cdot\frac{9}{7}\cdot\frac{7}{9} = -\frac{3}{14}\cdot\frac{7}{9}$

$a\cdot 1 = -\frac{3\cdot 7\cdot 1}{2\cdot 7\cdot 3\cdot 3}$

$a = -\frac{1}{6}$

$-\frac{1}{6}$ checks and is the solution.

33. $-\frac{15}{7} = \frac{3}{2}t$

$\frac{2}{3}\left(-\frac{15}{7}\right) = \frac{2}{3}\cdot\frac{3}{2}t$

$-\frac{2\cdot 3\cdot 5}{3\cdot 7} = t$

$-\frac{10}{7} = t$

$-\frac{10}{7}$ checks and is the solution.

35.
$$x \cdot \frac{5}{16} = \frac{15}{14}$$
$$x \cdot \frac{5}{16} \cdot \frac{16}{5} = \frac{15}{14} \cdot \frac{16}{5}$$
$$x \cdot 1 = \frac{3 \cdot 5 \cdot 2 \cdot 8}{2 \cdot 7 \cdot 5}$$
$$x = \frac{24}{7}$$

$\frac{24}{7}$ checks and is the solution.

37.
$$-\frac{3}{20}x = -\frac{21}{10}$$
$$-\frac{20}{3}\left(-\frac{3}{20}x\right) = -\frac{20}{3}\left(-\frac{21}{10}\right)$$
$$x = \frac{2 \cdot 10 \cdot 3 \cdot 7}{3 \cdot 10}$$
$$x = 14$$

14 checks and is the solution.

39.
$$-\frac{25}{17} = -\frac{35}{34}a$$
$$-\frac{34}{35}\left(-\frac{25}{17}\right) = -\frac{34}{35}\left(-\frac{35}{34}a\right)$$
$$\frac{2 \cdot 17 \cdot 5 \cdot 5}{5 \cdot 7 \cdot 17} = a$$
$$\frac{10}{7} = a$$

$\frac{10}{7}$ checks and is the solution.

41.
$$\begin{aligned} 36 \div (-3)^2 \times (7-2) &= 36 \div (-3)^2 \times 5 \\ &= 36 \div 9 \times 5 \\ &= 4 \times 5 \\ &= 20 \end{aligned}$$

43.
$$\begin{aligned} -13 + 3^2 &= -13 + 9 \\ &= -4 \end{aligned}$$

45. 

47. ***Familiarize***. This is a multistep problem. First we find the length of the total trip. Then we find how many kilometers were left to drive. We draw a picture. We let $n =$ the length of the total trip.

$\frac{5}{8}$ of the trip

180 km

n km

Translate. We translate to an equation.

Fraction of trip completed	times	Total length of trip	is	Amount already traveled
↓	↓	↓	↓	↓
$\frac{5}{8}$	$\cdot$	n	$=$	180

Solve. We solve the equation.
$$\frac{5}{8} \cdot n = 180$$
$$\frac{8}{5} \cdot \frac{5}{8}n = \frac{8}{5} \cdot 180$$
$$n = \frac{8 \cdot 5 \cdot 36}{5}$$
$$n = 288$$

The total trip was 288 km.

Now we find how many kilometers were left to travel. Let $t =$ this number.

Length of total trip	minus	Distance traveled	is	Distance left to travel
↓	↓	↓	↓	↓
288	$-$	180	$=$	t

We carry out the subtraction:
$$288 - 180 = t$$
$$108 = t$$

Check. We repeat the calculations. The results check.

State. The total trip was 288 km. There were 108 km left to travel.

49. ***Familiarize***. We draw a picture. We let $p =$ the number of pounds the package could hold when completely filled.

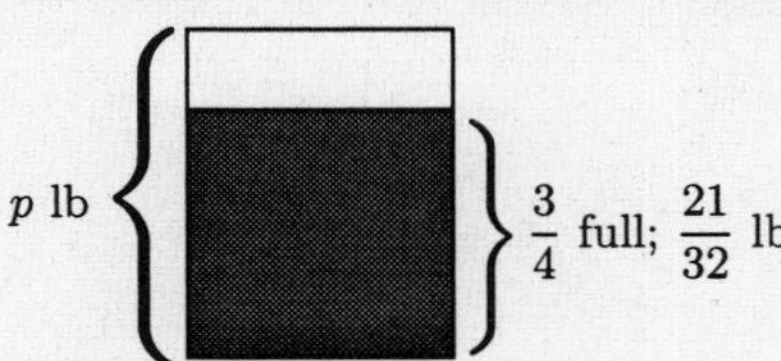

Translate. We translate to an equation.

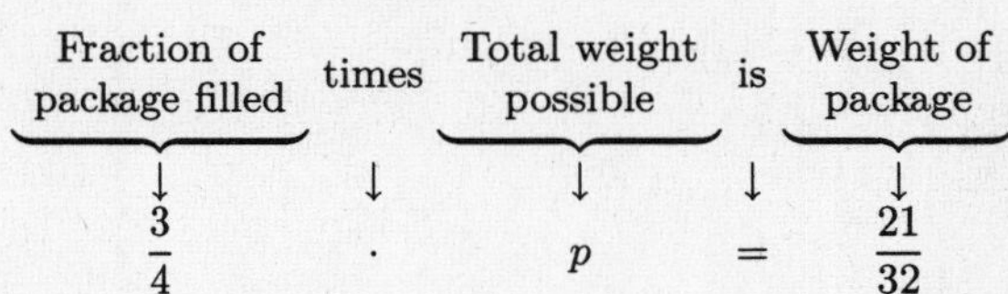

Solve. We solve the equation.
$$\frac{3}{4} \cdot p = \frac{21}{32}$$
$$\frac{4}{3} \cdot \frac{3}{4} \cdot p = \frac{4}{3} \cdot \frac{21}{32}$$
$$p = \frac{4 \cdot 3 \cdot 7}{3 \cdot 4 \cdot 8}$$
$$p = \frac{7}{8}$$

Check. $\frac{3}{4}$ of $\frac{7}{8}$ lb is $\frac{3}{4} \cdot \frac{7}{8}$, or $\frac{21}{32}$ lb. Our answer checks.

State. The package could hold $\frac{7}{8}$ lb when filled.

Chapter 4

Fractional Notation: Addition and Subtraction

Exercise Set 4.1

In this section we will find the LCM using the multiples method in Exercises 1 - 21 and the factorization method in Exercises 23 - 43.

1. 1. 4 is a multiple of 2, so it is the LCM.

The LCM is 4.

3. 1. 25 is not a multiple of 10.

2. Check multiples:

$2 \cdot 25 = 50$ A multiple of 10

3. The LCM is 50.

5. 1. 40 is a multiple of 20, so it is the LCM.

The LCM is 40.

7. 1. 27 is not a multiple of 18.

2. Check multiples:

$2 \cdot 27 = 54$ A multiple of 18

3. The LCM is 54.

9. 1. 50 is not a multiple of 30.

2. Check multiples:

$2 \cdot 50 = 100$ Not a multiple of 30
$3 \cdot 50 = 150$ A multiple of 30

3. The LCM is 150.

11. 1. 40 is not a multiple of 30.

2. Check multiples:

$2 \cdot 40 = 80$ Not a multiple of 30
$3 \cdot 40 = 120$ A multiple of 30

3. The LCM is 120.

13. 1. 48 is a multiple of 8, so it is the LCM.

The LCM is 48.

15. 1. 50 is a multiple of 5, so it is the LCM.

The LCM is 50.

17. 1. 24 is not a multiple of 18.

2. Check multiples:

$2 \cdot 24 = 48$ Not a multiple of 18
$3 \cdot 24 = 72$ A multiple of 18

3. The LCM is 72.

19. 1. 70 is not a multiple of 60.

2. Check multiples:

$2 \cdot 70 = 140$ Not a multiple of 60
$3 \cdot 70 = 210$ Not a multiple of 60
$4 \cdot 70 = 280$ Not a multiple of 60
$5 \cdot 70 = 350$ Not a multiple of 60
$6 \cdot 70 = 420$ A multiple of 60

3. The LCM is 420.

21. 1. 36 is not a multiple of 16.

2. Check multiples:

$2 \cdot 36 = 72$ Not a multiple of 16
$3 \cdot 36 = 108$ Not a multiple of 16
$4 \cdot 36 = 144$ A multiple of 16

3. The LCM is 144.

23. 1. Write the prime factorization of each number.

$32 = 2 \cdot 2 \cdot 2 \cdot 2 \cdot 2$
$36 = 2 \cdot 2 \cdot 3 \cdot 3$

2. a) Neither factorization contains the other.

b) We multiply the factorization of 32, $2 \cdot 2 \cdot 2 \cdot 2 \cdot 2$, by any prime factors of 36 that are missing. In this case, two factors of 3 are needed. The LCM is $2 \cdot 2 \cdot 2 \cdot 2 \cdot 2 \cdot 3 \cdot 3$.

3. To check, note that 2 and 3 appear in the LCM the greatest number of times that each appears as a factor of 32 or 36. The LCM is $2 \cdot 2 \cdot 2 \cdot 2 \cdot 2 \cdot 3 \cdot 3$, or 288.

25. 1. Write the prime factorization of each number. Because 11 and 13 are both prime we write $11 = 11$ and $13 = 13$.

2. a) Neither factorization contains the other.

b) We multiply the factorization of 11 by any prime factors of 13 that are missing. In this case, a factor of 13 is needed. The LCM is $11 \cdot 13$.

3. To check, note that 11 and 13 appear in the LCM the greatest number of times that each appears as a factor of 11 or 13. The LCM is $11 \cdot 13$, or 143.

27. 1. Write the prime factorization of each number.

$12 = 2 \cdot 2 \cdot 3$
$35 = 5 \cdot 7$

2. a) Neither factorization contains the other.

b) We multiply the factorization of 12 by any prime factors of 35 that are missing. In this case, one factor of 5 and one factor of 7 are needed. The LCM is $2 \cdot 2 \cdot 3 \cdot 5 \cdot 7$.

3. To check, note that 2, 3, 5, and 7 appear in the LCM the greatest number of times that each appears as a factor of 12 or 35. The LCM is $2 \cdot 2 \cdot 3 \cdot 5 \cdot 7$, or 420.

29. 1. Write the prime factorization of each number.

$54 = 2 \cdot 3 \cdot 3 \cdot 3$
$63 = 3 \cdot 3 \cdot 7$

2. a) Neither factorization contains the other.

b) We multiply the factorization of 54 by any prime factors of 63 that are missing. In this case, one factor of 7 is needed. The LCM is $2 \cdot 3 \cdot 3 \cdot 3 \cdot 7$, or 378.

3. The result checks.

31. 1. Write the prime factorization of each number.

$81 = 3 \cdot 3 \cdot 3 \cdot 3$
$90 = 2 \cdot 3 \cdot 3 \cdot 5$

2. a) Neither factorization contains the other.

b) We multiply the factorization of 90 by any prime factors of 81 that are missing. In this case, two factors of 3 are needed. The LCM is $2 \cdot 3 \cdot 3 \cdot 5 \cdot 3 \cdot 3$, or 810.

3. The result checks.

33. 1. Write the prime factorization of each number. Because 2, 3, and 5 are all prime we write $2 = 2$, $3 = 3$, and $5 = 5$.

2. a) None of the factorizations contains the other two.

b) We begin with 2. Since 3 contains a factor of 3, we multiply by 3:

$2 \cdot 3$

Next we multiply $2 \cdot 3$ by 5, the factor of 5 that is missing:

$2 \cdot 3 \cdot 5$

The LCM is $2 \cdot 3 \cdot 5$, or 30.

3. The result checks.

35. 1. Write the prime factorization of each number. Because 3, 5, and 7 are all prime we write $3 = 3$, $5 = 5$, and $7 = 7$.

2. a) None of the factorizations contains the other two.

b) We begin with 3. Since 5 contains a factor of 5, we multiply by 5:

$3 \cdot 5$

Next we multiply $3 \cdot 5$ by 7, the factor of 7 that is missing:

$3 \cdot 5 \cdot 7$

The LCM is $3 \cdot 5 \cdot 7$, or 105.

3. The result checks.

37. 1. Write the prime factorization of each number.

$24 = 2 \cdot 2 \cdot 2 \cdot 3$
$36 = 2 \cdot 2 \cdot 3 \cdot 3$
$12 = 2 \cdot 2 \cdot 3$

2. a) None of the factorizations contains the other two.

b) We begin with the factorization of 24, $2 \cdot 2 \cdot 2 \cdot 3$. Since 36 contains a second factor of 3, we multiply by another factor of 3:

$2 \cdot 2 \cdot 2 \cdot 3 \cdot 3$

Next we look for factors of 12 that are still missing. There are none. The LCM is

$2 \cdot 2 \cdot 2 \cdot 3 \cdot 3$, or 72.

3. The result checks.

39. 1. Write the prime factorization of each number.

$5 = 5$
$12 = 2 \cdot 2 \cdot 3$
$15 = 3 \cdot 5$

2. a) None of the factorizations contains the other two.

b) We begin with the factorization of 12, $2 \cdot 2 \cdot 3$. Since 5 contains a factor of 5, we multiply by 5:

$2 \cdot 2 \cdot 3 \cdot 5$

Next we look for factors of 15 that are still missing. There are none. The LCM is $2 \cdot 2 \cdot 3 \cdot 5$, or 60.

3. The result checks.

41. 1. Write the prime factorization of each number.

$9 = 3 \cdot 3$
$12 = 2 \cdot 2 \cdot 3$
$6 = 2 \cdot 3$

2. a) None of the factorizations contains the other two.

b) We begin with the factorization of 12, $2 \cdot 2 \cdot 3$. Since 9 contains a second factor of 3, we multiply by another factor of 3:

$2 \cdot 2 \cdot 3 \cdot 3$

Next we look for factors of 6 that are still missing. There are none. The LCM is $2 \cdot 2 \cdot 3 \cdot 3$, or 36.

3. The result checks.

43. 1. Write the prime factorization of each number.

$3 = 3$
$6 = 2 \cdot 3$
$8 = 2 \cdot 2 \cdot 2$

2. a) None of the factorizations contains the other two.

b) We begin with the factorization of 8, $2 \cdot 2 \cdot 2$. Since 3 contains a factor of 3, we multiply by 3:

$$2 \cdot 2 \cdot 2 \cdot 3$$

Next we look for factors of 6 that are still missing. There are none. The LCM is $2 \cdot 2 \cdot 2 \cdot 3$, or 24.

3. The result checks.

45. ***Familiarize***. We draw a picture. Repeated addition applies here.

$3250			
$13	$13	···	$13

Translate. We must determine how many 13's there are in 3250. This number is the number of seats in the facility. We let x = the number of seats in the facility.

$$\underbrace{\$13}_{13}\ \underbrace{\text{times}}_{\cdot}\ \underbrace{\text{number of seats}}_{x}\ \underbrace{\text{is}}_{=}\ \underbrace{\$3250}_{3250}$$

Solve. To solve the equation, we divide on both sides by 13.

$$13 \cdot x = 3250$$
$$\frac{13 \cdot x}{13} = \frac{3250}{13}$$
$$x = 250$$

Check. If 250 seats are sold at $13 each, the total receipts are $250 \cdot \$13$, or $3250. The result checks.

State. The facility contains 250 seats.

47. $\dfrac{-4}{5} \cdot \dfrac{10}{12} = \dfrac{-4 \cdot 10}{5 \cdot 12} = \dfrac{-40}{60} = \dfrac{-2 \cdot 20}{3 \cdot 20} = \dfrac{-2}{3} \cdot \dfrac{20}{20} = \dfrac{-2}{3} \cdot 1 = \dfrac{-2}{3}$, or $-\dfrac{2}{3}$

49. 

51. 1. 324 is not a multiple of 288.

2. Check multiples, using a calculator.

$2 \cdot 324 = 648$	Not a multiple of 288
$3 \cdot 324 = 972$	Not a multiple of 288
$4 \cdot 324 = 1296$	Not a multiple of 288
$5 \cdot 324 = 1620$	Not a multiple of 288
$6 \cdot 324 = 1944$	Not a multiple of 288
$7 \cdot 324 = 2268$	Not a multiple of 288
$8 \cdot 324 = 2592$	A multiple of 288

3. The LCM is 2592.

53. 1. From Example 9 we know that the LCM of 27, 90, and 84 is $2 \cdot 3 \cdot 3 \cdot 5 \cdot 3 \cdot 2 \cdot 7$, so the LCM of 27, 90, 84, 210, 108, and 50 must contain at least these factors. We write the prime factorizations of 210, 108, and 50:

$210 = 2 \cdot 3 \cdot 5 \cdot 7$
$108 = 2 \cdot 2 \cdot 3 \cdot 3 \cdot 3$
$50 = 2 \cdot 5 \cdot 5$

2. a) Neither of the four factorizations above contains the other three.

b) Begin with the LCM of 27, 90, and 84, $2 \cdot 3 \cdot 3 \cdot 5 \cdot 3 \cdot 2 \cdot 7$. Neither 210 nor 108 contains any factors that are missing in this factorization. Next we look for factors of 50 that are missing. Since 50 contains a second factor of 5, we multiply by 5:

$$2 \cdot 3 \cdot 3 \cdot 5 \cdot 3 \cdot 2 \cdot 7 \cdot 5$$

The LCM is $2 \cdot 3 \cdot 3 \cdot 5 \cdot 3 \cdot 2 \cdot 7 \cdot 5$, or 18,900.

3. The result checks.

55. The width of the smallest carton will be the width of one box, 5 in. The length of the carton must be a multiple of both 6 and 8. The length of the smallest carton will be the least common multiple of 6 and 8.

$6 = 2 \cdot 3$
$8 = 2 \cdot 2 \cdot 2$

LCM is $2 \cdot 2 \cdot 2 \cdot 3$, or 24.

The length of the smallest carton is 24 in.

Exercise Set 4.2

1. $\dfrac{2}{9} + \dfrac{4}{9} = \dfrac{2+4}{9} = \dfrac{6}{9} = \dfrac{2 \cdot 3}{3 \cdot 3} = \dfrac{2}{3} \cdot \dfrac{3}{3} = \dfrac{2}{3} \cdot 1 = \dfrac{2}{3}$

3. $\dfrac{1}{8} + \dfrac{5}{8} = \dfrac{1+5}{8} = \dfrac{6}{8} = \dfrac{3 \cdot 2}{4 \cdot 2} = \dfrac{3}{4} \cdot \dfrac{2}{2} = \dfrac{3}{4} \cdot 1 = \dfrac{3}{4}$

5. $\dfrac{-7}{11} + \dfrac{5}{11} = \dfrac{-7+5}{11} = \dfrac{-2}{11}$, or $-\dfrac{2}{11}$

7. $\dfrac{9}{a} + \dfrac{2}{a} = \dfrac{9+2}{a} = \dfrac{11}{a}$

9. $-\dfrac{3}{x} + \left(-\dfrac{7}{x}\right) = \dfrac{-3}{x} + \dfrac{-7}{x} = \dfrac{-3+(-7)}{x} = \dfrac{-10}{x}$, or $-\dfrac{10}{x}$

11. $\frac{1}{8}+\frac{1}{6}$ $8 = 2\cdot 2\cdot 2$ and $6 = 2\cdot 3$, so the LCD is $2\cdot 2\cdot 2\cdot 3$, or 24

$= \underbrace{\frac{1}{8}\cdot\frac{3}{3}} + \underbrace{\frac{1}{6}\cdot\frac{4}{4}}$

Think: $6\times\square = 24$. The answer is 4, so we multiply by 1, using $\frac{4}{4}$.

Think: $8\times\square = 24$. The answer is 3, so we multiply by 1, using $\frac{3}{3}$.

$=\frac{3}{24}+\frac{4}{24}$

$=\frac{7}{24}$

13. $\frac{-4}{5}+\frac{7}{10}$ 5 is a factor of 10, so the LCD is 10.

$= \underbrace{\frac{-4}{5}\cdot\frac{2}{2}} + \frac{7}{10}$ ← This fraction already has the LCD as denominator.

Think: $5\times\square = 10$. The answer is 2, so we multiply by 1, using $\frac{2}{2}$.

$=\frac{-8}{10}+\frac{7}{10}$

$=\frac{-1}{10}$, or $-\frac{1}{10}$

15. $\frac{5}{12}+\frac{3}{8}$ $12 = 2\cdot 2\cdot 3$ and $8 = 2\cdot 2\cdot 2$, so the LCD is $2\cdot 2\cdot 2\cdot 3$, or 24.

$= \underbrace{\frac{5}{12}\cdot\frac{2}{2}} + \underbrace{\frac{3}{8}\cdot\frac{3}{3}}$

Think: $8\times\square = 24$. The answer is 3, so we multiply by 1, using $\frac{3}{3}$.

Think: $12\times\square = 24$. The answer is 2, so we multiply by 1, using $\frac{2}{2}$.

$=\frac{10}{24}+\frac{9}{24}=\frac{19}{24}$

17. $\frac{3}{20}+4$

$=\frac{3}{20}+\frac{4}{1}$ Rewriting 4 in fractional notation

$=\frac{3}{20}+\frac{4}{1}\cdot\frac{20}{20}$ The LCD is 20.

$=\frac{3}{20}+\frac{80}{20}$

$=\frac{83}{20}$

19. $\frac{5}{-8}+\frac{5}{6}$

$=\frac{-5}{8}+\frac{5}{6}$ Recall that $\frac{m}{-n}=\frac{-m}{n}$. The LCD is 24. (See Exercise 11.)

$=\frac{-5}{8}\cdot\frac{3}{3}+\frac{5}{6}\cdot\frac{4}{4}$

$=\frac{-15}{24}+\frac{20}{24}$

$=\frac{5}{24}$

21. $\frac{3}{10}+\frac{1}{100}$ 10 is a factor of 100, so the LCD is 100.

$=\frac{3}{10}\cdot\frac{10}{10}+\frac{1}{100}$

$=\frac{30}{100}+\frac{1}{100}=\frac{31}{100}$

23. $\frac{5}{12}+\frac{4}{15}$ $12 = 2\cdot 2\cdot 3$ and $15 = 3\cdot 5$, so the LCM is $2\cdot 2\cdot 3\cdot 5$, or 60.

$=\frac{5}{12}\cdot\frac{5}{5}+\frac{4}{15}\cdot\frac{4}{4}$

$=\frac{25}{60}+\frac{16}{60}=\frac{41}{60}$

25. $\frac{9}{10}+\frac{-99}{100}$ 10 is a factor of 100, so the LCD is 100.

$=\frac{9}{10}\cdot\frac{10}{10}+\frac{-99}{100}$

$=\frac{90}{100}+\frac{-99}{100}=\frac{-9}{100}$, or $-\frac{9}{100}$

27. $5+\frac{7}{12}$

$=\frac{5}{1}+\frac{7}{12}$ The LCD is 12.

$=\frac{5}{1}\cdot\frac{12}{12}+\frac{7}{12}$

$=\frac{60}{12}+\frac{7}{12}$

$=\frac{67}{12}$

29. $-4+\frac{2}{9}$

$=\frac{-4}{1}+\frac{2}{9}$ The LCD is 9.

$=\frac{-4}{1}\cdot\frac{9}{9}+\frac{2}{9}$

$=\frac{-36}{9}+\frac{2}{9}$

$=\frac{-34}{9}$, or $-\frac{34}{9}$

31. $-\frac{5}{12}+\frac{7}{24}$

$\frac{-5}{12}+\frac{7}{24}$ 12 is a factor of 24, so the LCD is 24.

$=\frac{-5}{12}\cdot\frac{2}{2}+\frac{7}{24}$

$=\frac{-10}{24}+\frac{7}{24}=\frac{-3}{24}$

$=\frac{-1}{8}$, or $-\frac{1}{8}$

33. $\frac{4}{10}+\frac{3}{100}+\frac{7}{1000}$ 10 and 100 are factors of 1000, so the LCD is 1000.

$=\frac{4}{10}\cdot\frac{100}{100}+\frac{3}{100}\cdot\frac{10}{10}+\frac{7}{1000}$

$=\frac{400}{1000}+\frac{30}{1000}+\frac{7}{1000}$

$=\frac{437}{1000}$

35. $\frac{3}{10}+\frac{5}{12}+\frac{8}{15}$

$=\frac{3}{2\cdot 5}+\frac{5}{2\cdot 2\cdot 3}+\frac{8}{3\cdot 5}$ Factoring the denominators

The LCD is $2\cdot 5\cdot 2\cdot 3$.

$=\frac{3}{2\cdot 5}\cdot\frac{2\cdot 3}{2\cdot 3}+\frac{5}{2\cdot 2\cdot 3}\cdot\frac{5}{5}+\frac{8}{3\cdot 5}\cdot\frac{2\cdot 2}{2\cdot 2}$

In each case we multiply by 1 to obtain the LCD.

$=\frac{3\cdot 2\cdot 3}{2\cdot 5\cdot 2\cdot 3}+\frac{5\cdot 5}{2\cdot 2\cdot 3\cdot 5}+\frac{8\cdot 2\cdot 2}{3\cdot 5\cdot 2\cdot 2}$

$=\frac{18}{2\cdot 5\cdot 2\cdot 3}+\frac{25}{2\cdot 5\cdot 2\cdot 3}+\frac{32}{2\cdot 5\cdot 2\cdot 3}$

$=\frac{75}{2\cdot 5\cdot 2\cdot 3}$

$=\frac{3\cdot 5\cdot 5}{2\cdot 5\cdot 2\cdot 3}=\frac{3\cdot 5}{3\cdot 5}\cdot\frac{5}{2\cdot 2}$

$=\frac{5}{4}$

37. $\frac{5}{6}+\frac{25}{52}+\frac{7}{4}$

$=\frac{5}{2\cdot 3}+\frac{25}{2\cdot 2\cdot 13}+\frac{7}{2\cdot 2}$ LCD is $2\cdot 3\cdot 2\cdot 13$.

$=\frac{5}{2\cdot 3}\cdot\frac{2\cdot 13}{2\cdot 13}+\frac{25}{2\cdot 2\cdot 13}\cdot\frac{3}{3}+\frac{7}{2\cdot 2}\cdot\frac{3\cdot 13}{3\cdot 13}$

$=\frac{5\cdot 2\cdot 13}{2\cdot 3\cdot 2\cdot 13}+\frac{25\cdot 3}{2\cdot 2\cdot 13\cdot 3}+\frac{7\cdot 3\cdot 13}{2\cdot 2\cdot 3\cdot 13}$

$=\frac{130}{2\cdot 3\cdot 2\cdot 13}+\frac{75}{2\cdot 3\cdot 2\cdot 13}+\frac{273}{2\cdot 3\cdot 2\cdot 13}$

$=\frac{478}{2\cdot 3\cdot 2\cdot 13}$

$=\frac{2\cdot 239}{2\cdot 3\cdot 2\cdot 13}=\frac{2}{2}\cdot\frac{239}{3\cdot 2\cdot 13}$

$=\frac{239}{78}$

39. $\frac{2}{9}+\frac{7}{10}+\frac{-4}{15}$

$=\frac{2}{3\cdot 3}+\frac{7}{2\cdot 5}+\frac{-4}{3\cdot 5}$ LCD is $3\cdot 3\cdot 2\cdot 5$.

$=\frac{2}{3\cdot 3}\cdot\frac{2\cdot 5}{2\cdot 5}+\frac{7}{2\cdot 5}\cdot\frac{3\cdot 3}{3\cdot 3}+\frac{-4}{3\cdot 5}\cdot\frac{3\cdot 2}{3\cdot 2}$

$=\frac{2\cdot 2\cdot 5}{3\cdot 3\cdot 2\cdot 5}+\frac{7\cdot 3\cdot 3}{2\cdot 5\cdot 3\cdot 3}+\frac{-4\cdot 3\cdot 2}{3\cdot 5\cdot 3\cdot 2}$

$=\frac{20}{3\cdot 3\cdot 2\cdot 5}+\frac{63}{3\cdot 3\cdot 2\cdot 5}+\frac{-24}{3\cdot 3\cdot 2\cdot 5}$

$=\frac{59}{3\cdot 3\cdot 2\cdot 5}$

$=\frac{59}{90}$

41. Since there is a common denominator, compare the numerators.

$5<6$, so $\frac{5}{8}<\frac{6}{8}$.

43. The LCD is 12. We multiply by 1 to make the denominators the same.

$\frac{1}{3}\cdot\frac{4}{4}=\frac{4}{12}$

$\frac{1}{4}\cdot\frac{3}{3}=\frac{3}{12}$

Since $4>3$, it follows that $\frac{4}{12}>\frac{3}{12}$, so $\frac{1}{3}>\frac{1}{4}$.

45. The LCD is 21. We multiply by 1 to make the denominators the same.

$\frac{-2}{3}\cdot\frac{7}{7}=\frac{-14}{21}$

$\frac{-5}{7}\cdot\frac{3}{3}=\frac{-15}{21}$

Since $-14>-15$, it follows that $\frac{-14}{21}>\frac{-15}{21}$, so $\frac{-2}{3}>\frac{-5}{7}$.

47. The LCD is 30. We multiply by 1 to make the denominators the same.

$\frac{4}{5}\cdot\frac{6}{6}=\frac{24}{30}$

$\frac{5}{6}\cdot\frac{5}{5}=\frac{25}{30}$

Since $24<25$, it follows that $\frac{24}{30}<\frac{25}{30}$, so $\frac{4}{5}<\frac{5}{6}$.

49. The LCD is 20.

The denominator of $\frac{19}{20}$ is the LCD.

$\frac{4}{5}\cdot\frac{4}{4}=\frac{16}{20}$

Since $19>16$, it follows that $\frac{19}{20}>\frac{16}{20}$, so $\frac{19}{20}>\frac{4}{5}$.

51. The LCD is 20.

The denominator of $\frac{-19}{20}$ is the LCD.

$\frac{-9}{10} \cdot \frac{2}{2} = \frac{-18}{20}$

Since $-19 < -18$, it follows that $\frac{-19}{20} < \frac{-18}{20}$, so $\frac{-19}{20} < \frac{-9}{10}$.

53. The LCD is $21 \cdot 13$, or 273. We multiply by 1 to make the denominators the same.

$$\frac{31}{21} \cdot \frac{13}{13} = \frac{403}{273}$$
$$\frac{41}{13} \cdot \frac{21}{21} = \frac{861}{273}$$

Since $403 < 861$, it follows that $\frac{403}{273} < \frac{861}{273}$, so $\frac{31}{21} < \frac{41}{13}$.

55. The LCD is 49.

$$\frac{12}{7} \cdot \frac{7}{7} = \frac{84}{49}$$

The denominator of $\frac{132}{49}$ is the LCD.

Since $84 < 132$, it follows that $\frac{84}{49} < \frac{132}{49}$, so $\frac{12}{7} < \frac{132}{49}$.

57. ***Familiarize.*** We draw a picture. We let p = the number of pounds of tea Rose bought.

$\frac{1}{3}$ lb	$\frac{1}{2}$ lb
p	

Translate. The problem can be translated to an equation as follows:

Pounds of orange pekoe	plus	Pounds of English cinnamon	is	Total pounds of tea
↓	↓	↓	↓	↓
$\frac{1}{3}$	$+$	$\frac{1}{2}$	$=$	p

Solve. We carry out the addition. The LCM of the denominators is $3 \cdot 2$, or 6.

$$\frac{1}{3} + \frac{1}{2} = p$$
$$\frac{1}{3} \cdot \frac{2}{2} + \frac{1}{2} \cdot \frac{3}{3} = p$$
$$\frac{2}{6} + \frac{3}{6} = p$$
$$\frac{5}{6} = p$$

Check. We check by repeating the calculation. We also note that the sum is larger than either of the individual weights, so the answer seems reasonable.

State. Rose bought $\frac{5}{6}$ lb of tea.

59. ***Familiarize.*** We draw a picture. We let D = the total distance walked.

$\frac{7}{6}$ mi	$\frac{3}{4}$ mi
D	

Translate. The problem can be translated to an equation as follows:

Distance to friend's dorm	plus	Distance to class	is	Total distance
↓	↓	↓	↓	↓
$\frac{7}{6}$	$+$	$\frac{3}{4}$	$=$	D

Solve. To solve the equation, carry out the addition. Since $6 = 2 \cdot 3$ and $4 = 2 \cdot 2$, the LCM of the denominators is $2 \cdot 2 \cdot 3$, or 12.

$$\frac{7}{6} + \frac{3}{4} = D$$
$$\frac{7}{6} \cdot \frac{2}{2} + \frac{3}{4} \cdot \frac{3}{3} = D$$
$$\frac{14}{12} + \frac{9}{12} = D$$
$$\frac{23}{12} = D$$

Check. We repeat the calculation. We also note that the sum is larger than either of the original distances, so the answer seems reasonable.

State. The student walked $\frac{23}{12}$ mi.

61. ***Familiarize.*** We draw a picture and let c = the total number of cups of liquid ingredients.

$\frac{2}{3}$ cup	$\frac{1}{4}$ cup	$\frac{1}{8}$ cup
c		

Translate. The problem can be translated to an equation as follows:

Amount of water	plus	Amount of milk	plus	Amount of oil	is	Amount of liquid
↓	↓	↓	↓	↓	↓	↓
$\frac{2}{3}$	$+$	$\frac{1}{4}$	$+$	$\frac{1}{8}$	$=$	c

Solve. We carry out the addition. Since $3 = 3$, $4 = 2 \cdot 2$, and $8 = 2 \cdot 2 \cdot 2$, the LCM of the denominators is $3 \cdot 2 \cdot 2 \cdot 2$, or 24.

$$\frac{2}{3} + \frac{1}{4} + \frac{1}{8} = c$$
$$\frac{2}{3} \cdot \frac{8}{8} + \frac{1}{4} \cdot \frac{6}{6} + \frac{1}{8} \cdot \frac{3}{3} = c$$
$$\frac{16}{24} + \frac{6}{24} + \frac{3}{24} = c$$
$$\frac{25}{24} = c$$

Check. We repeat the calculation. We also note that the sum is larger than any of the individual amounts, as expected.

State. The recipe calls for $\frac{25}{24}$ cups of liquid ingredients.

63. **Familiarize**. This is a multistep problem. First we find the total weight of the cubic meter of concrete mix. We visualize the situation, letting w = the total weight.

420 kg	150 kg	120 kg
w		

Translate. We translate to an equation.

$$\underbrace{\text{Weight of cement}}_{\downarrow\; 420} \text{ plus } \underbrace{\text{Weight of stone}}_{\downarrow\; 150} \text{ plus } \underbrace{\text{Weight of sand}}_{\downarrow\; 120} \text{ is } \underbrace{\text{Total weight}}_{\downarrow\; w}$$

$$420 + 150 + 120 = w$$

Solve. We carry out the addition.

$$420 + 150 + 120 = w$$
$$690 = w$$

Since the mix contains 420 kg of cement, the part that is cement is $\frac{420}{690} = \frac{14\cdot 30}{23\cdot 30} = \frac{14}{23}\cdot\frac{30}{30} = \frac{14}{23}$.

Since the mix contains 150 kg of stone, the part that is stone is $\frac{150}{690} = \frac{5\cdot 30}{23\cdot 30} = \frac{5}{23}\cdot\frac{30}{30} = \frac{5}{23}$.

Since the mix contains 120 kg of sand, the part that is sand is $\frac{120}{690} = \frac{4\cdot 30}{23\cdot 30} = \frac{4}{23}\cdot\frac{30}{30} = \frac{4}{23}$.

We add these amounts: $\frac{14}{23} + \frac{5}{23} + \frac{4}{23} =$

$\frac{14+5+4}{23} = \frac{23}{23} = 1.$

Check. We repeat the calculations. We also note that since the total of the fractional parts is 1, the answer is probably correct.

State. The total weight of the cubic meter of concrete mix is 690 kg. Of this, $\frac{14}{23}$ is cement, $\frac{5}{23}$ is stone, and $\frac{4}{23}$ is sand. The result when we add these amounts is 1.

65. **Familiarize**. We draw a picture and let d = the number of miles the triathlete covers.

Running	Canoeing	Swimming
$\frac{7}{8}$ mi	$\frac{1}{3}$ mi	$\frac{1}{6}$ mi
d		

Translate. We translate to an equation as follows:

$$\underbrace{\text{Miles run}}_{\downarrow\; \frac{7}{8}} \text{ plus } \underbrace{\text{Miles canoed}}_{\downarrow\; \frac{1}{3}} \text{ plus } \underbrace{\text{Miles swum}}_{\downarrow\; \frac{1}{6}} \text{ is } \underbrace{\text{Total miles}}_{\downarrow\; d}$$

$$\frac{7}{8} + \frac{1}{3} + \frac{1}{6} = d$$

Solve. We carry out the addition. Since $8 = 2\cdot 2\cdot 2$, $3 = 3$, and $6 = 2\cdot 3$, the LCM of the denominators is $2\cdot 2\cdot 2\cdot 3$, or 24.

$$\frac{7}{8} + \frac{1}{3} + \frac{1}{6} = d$$
$$\frac{7}{8}\cdot\frac{3}{3} + \frac{1}{3}\cdot\frac{8}{8} + \frac{1}{6}\cdot\frac{4}{4} = d$$
$$\frac{21}{24} + \frac{8}{24} + \frac{4}{24} = d$$
$$\frac{33}{24} = d$$
$$\frac{11}{8} = d \quad \text{Simplifying}$$

Check. We repeat the calculation. We also note that the sum is larger than any of the individual distances, as expected.

State. The triathlete covers a total of $\frac{11}{8}$ mi.

67. **Familiarize**. We draw a picture. We let t = the total thickness.

Layer	Thickness
Board	$\frac{5}{8}$ in.
Glue	$\frac{3}{32}$ in.
Board	$\frac{7}{8}$ in.
Total	t

Translate. We translate to an equation.

$$\underbrace{\text{Thickness of one board}}_{\downarrow\; \frac{5}{8}} \text{ plus } \underbrace{\text{Thickness of glue}}_{\downarrow\; \frac{3}{32}} \text{ plus } \underbrace{\text{Thickness of second board}}_{\downarrow\; \frac{7}{8}} \text{ is } \underbrace{\text{Total thickness}}_{\downarrow\; t}$$

$$\frac{5}{8} + \frac{3}{32} + \frac{7}{8} = t$$

Solve. We carry out the addition. The LCD is 32 since 8 is a factor of 32.

$$\frac{5}{8} + \frac{3}{32} + \frac{7}{8} = t$$
$$\frac{5}{8}\cdot\frac{4}{4} + \frac{3}{32} + \frac{7}{8}\cdot\frac{4}{4} = t$$
$$\frac{20}{32} + \frac{3}{32} + \frac{28}{32} = t$$
$$\frac{51}{32} = t$$

Check. We repeat the calculation. We also note that the sum is larger than any of the individual thicknesses, as expected.

State. The result is $\frac{51}{32}$ in. thick.

69. $-7-6=-7+(-6)=-13$

71. $9-17=9+(-17)=-8$

73. $\dfrac{x-y}{3}=\dfrac{7-(-3)}{3}=\dfrac{7+3}{3}=\dfrac{10}{3}$

75. 

77. Use a calculator to do this exercise. First, add on the left.

$$\frac{12}{97}+\frac{67}{139}=\frac{8167}{13,483}$$

Now compare $\dfrac{8167}{13,483}$ and $\dfrac{8167}{13,289}$.

Since the numerators are the same, the fraction with the larger denominator is smaller. (Think of two objects of the same size. Divide one into 13,483 equal parts and the other into 13,289 equal parts. Then 8167 of the 13,483 smaller parts are smaller than 8167 of the 13,289 larger parts.)

Thus, we have $\dfrac{8167}{13,483}<\dfrac{8167}{13,289}$, so $\dfrac{12}{97}+\dfrac{67}{139}<\dfrac{8167}{13,289}$.

79. ***Familiarize.*** First we find the fractional part of the band's pay that the guitarist received. We let f = this fraction.

Translate. We translate to an equation.

One-third	of	one-half	plus	one-fifth	of	one-half	is	fractional part
↓	↓	↓	↓	↓	↓	↓	↓	↓
$\frac{1}{3}$	$\cdot$	$\frac{1}{2}$	$+$	$\frac{1}{5}$	$\cdot$	$\frac{1}{2}$	$=$	f

Solve. We carry out the calculation.

$$\frac{1}{3}\cdot\frac{1}{2}+\frac{1}{5}\cdot\frac{1}{2}=f$$

$$\frac{1}{6}+\frac{1}{10}=f \quad \text{LCD is 30.}$$

$$\frac{1}{6}\cdot\frac{5}{5}+\frac{1}{10}\cdot\frac{3}{3}=f$$

$$\frac{5}{30}+\frac{3}{30}=f$$

$$\frac{8}{30}=f$$

$$\frac{4}{15}=f$$

Now we find how much of the \$1200 received by the band was paid to the guitarist. We let p = the amount.

Four-fifteenths	of	\$1200	=	guitarist's pay
↓	↓	↓	↓	↓
$\frac{4}{15}$	$\cdot$	1200	$=$	p

We solve the equation.

$$\frac{4}{15}\cdot 1200=p$$

$$\frac{4\cdot 1200}{15}=p$$

$$\frac{4\cdot 3\cdot 5\cdot 80}{3\cdot 5}=p$$

$$320=p$$

Check. We repeat the calculations.

State. The guitarist received $\dfrac{4}{15}$ of the band's pay. This was \$320.

81. $\dfrac{8}{15a}+\dfrac{2}{15a}=\dfrac{10}{15a}=\dfrac{2\cdot 5}{3\cdot 5\cdot a}=\dfrac{5}{5}\cdot\dfrac{2}{3\cdot a}=\dfrac{2}{3a}$

Exercise Set 4.3

1. When denominators are the same, subtract the numerators and keep the denominator.

$$\frac{5}{6}-\frac{1}{6}=\frac{5-1}{6}=\frac{4}{6}=\frac{2\cdot 2}{2\cdot 3}=\frac{2}{2}\cdot\frac{2}{3}=\frac{2}{3}$$

3. When denominators are the same, subtract the numerators and keep the denominator.

$$\frac{11}{16}-\frac{15}{16}=\frac{11-15}{16}=\frac{-4}{16}=\frac{-1\cdot 4}{4\cdot 4}=\frac{-1}{4}\cdot\frac{4}{4}=\frac{-1}{4},$$

or $-\dfrac{1}{4}$

5. The LCM of 3 and 9 is 9.

$$\frac{2}{3}-\frac{1}{9}=\frac{2}{3}\cdot\frac{3}{3}-\frac{1}{9}$$

←This fraction already has the LCM as the denominator.

Think: $3\times\square=9$. The answer is 3, so we multiply by 1, using $\dfrac{3}{3}$.

$$=\frac{6}{9}-\frac{1}{9}=\frac{5}{9}$$

7. The LCM of 8 and 12 is 24.

$$\frac{1}{8}-\frac{1}{12}=\frac{1}{8}\cdot\frac{3}{3}-\frac{1}{12}\cdot\frac{2}{2}$$

Think: $12\times\square=24$. The answer is 2, so we multiply by 1, using $\dfrac{2}{2}$.

Think: $8\times\square=24$. The answer is 3, so we multiply by 1, using $\dfrac{3}{3}$.

$$=\frac{3}{24}-\frac{2}{24}=\frac{1}{24}$$

9. The LCM of 3 and 6 is 6.

$$\frac{4}{3}-\frac{5}{6}=\frac{4}{3}\cdot\frac{2}{2}-\frac{5}{6}$$
$$=\frac{8}{6}-\frac{5}{6}=\frac{3}{6}$$
$$=\frac{1\cdot 3}{2\cdot 3}=\frac{1}{2}\cdot\frac{3}{3}$$
$$=\frac{1}{2}$$

11. The LCM of 6 and 2 is 6.

$$\frac{5}{6}-\frac{1}{2}=\frac{5}{6}-\frac{1}{2}\cdot\frac{3}{3}$$
$$=\frac{5}{6}-\frac{3}{6}=\frac{2}{6}$$
$$=\frac{2\cdot 1}{2\cdot 3}=\frac{2}{3}\cdot\frac{1}{3}$$
$$=\frac{1}{3}$$

13. The LCM of 15 and 5 is 15.

$$\frac{2}{15}-\frac{2}{5}=\frac{2}{15}-\frac{2}{5}\cdot\frac{3}{3}$$
$$=\frac{2}{15}-\frac{6}{15}=\frac{2-6}{15}$$
$$=\frac{-4}{15},\text{ or }-\frac{4}{15}$$

15. The LCM of 4 and 20 is 20.

$$\frac{3}{4}-\frac{1}{20}=\frac{3}{4}\cdot\frac{5}{5}-\frac{1}{20}$$
$$=\frac{15}{20}-\frac{1}{20}=\frac{14}{20}$$
$$=\frac{2\cdot 7}{2\cdot 10}=\frac{2}{2}\cdot\frac{7}{10}$$
$$=\frac{7}{10}$$

17. The LCM of 15 and 12 is 60.

$$\frac{2}{15}-\frac{5}{12}=\frac{2}{15}\cdot\frac{4}{4}-\frac{5}{12}\cdot\frac{5}{5}$$
$$=\frac{8}{60}-\frac{25}{60}=\frac{8-25}{60}$$
$$=\frac{-17}{60},\text{ or }-\frac{17}{60}$$

19. The LCM of 10 and 100 is 100.

$$\frac{6}{10}-\frac{7}{100}=\frac{6}{10}\cdot\frac{10}{10}-\frac{7}{100}$$
$$=\frac{60}{100}-\frac{7}{100}=\frac{53}{100}$$

21. The LCM of 15 and 25 is 75.

$$\frac{7}{15}-\frac{3}{25}=\frac{7}{15}\cdot\frac{5}{5}-\frac{3}{25}\cdot\frac{3}{3}$$
$$=\frac{35}{75}-\frac{9}{75}=\frac{26}{75}$$

23. The LCM of 10 and 100 is 100.

$$\frac{69}{100}-\frac{9}{10}=\frac{69}{100}-\frac{9}{10}\cdot\frac{10}{10}$$
$$=\frac{69}{100}-\frac{90}{100}=\frac{69-90}{100}$$
$$=\frac{-21}{100},\text{ or }-\frac{21}{100}$$

25. The LCM of 3 and 8 is 24.

$$\frac{2}{3}-\frac{1}{8}=\frac{2}{3}\cdot\frac{8}{8}-\frac{1}{8}\cdot\frac{3}{3}$$
$$=\frac{16}{24}-\frac{3}{24}$$
$$=\frac{13}{24}$$

27. The LCM of 5 and 2 is 10.

$$\frac{3}{5}-\frac{1}{2}=\frac{3}{5}\cdot\frac{2}{2}-\frac{1}{2}\cdot\frac{5}{5}$$
$$=\frac{6}{10}-\frac{5}{10}$$
$$=\frac{1}{10}$$

29. The LCM of 12 and 8 is 24.

$$\frac{-5}{12}-\frac{3}{8}=\frac{-5}{12}\cdot\frac{2}{2}-\frac{3}{8}\cdot\frac{3}{3}$$
$$=\frac{-10}{24}-\frac{9}{24}=\frac{-10-9}{24}$$
$$=\frac{-19}{24},\text{ or }-\frac{19}{24}$$

31. The LCM of 8 and 16 is 16.

$$\frac{7}{8}-\frac{1}{16}=\frac{7}{8}\cdot\frac{2}{2}-\frac{1}{16}$$
$$=\frac{14}{16}-\frac{1}{16}$$
$$=\frac{13}{16}$$

33. The LCM of 25 and 15 is 75.

$$\frac{17}{25}-\frac{4}{15}=\frac{17}{25}\cdot\frac{3}{3}-\frac{4}{15}\cdot\frac{5}{5}$$
$$=\frac{51}{75}-\frac{20}{75}$$
$$=\frac{31}{75}$$

35. The LCM of 25 and 150 is 150.

$$-\frac{23}{25}-\frac{112}{150} = -\frac{23}{25}\cdot\frac{6}{6}-\frac{112}{150}$$
$$= -\frac{138}{150}-\frac{112}{150} = \frac{-138-112}{150}$$
$$= \frac{-250}{150}$$
$$= \frac{-5\cdot 50}{3\cdot 50} = \frac{-5}{3}\cdot\frac{50}{50}$$
$$= \frac{-5}{3},\text{ or } -\frac{5}{3}$$

37. $x-\frac{2}{9}=\frac{4}{9}$

$x-\frac{2}{9}+\frac{2}{9}=\frac{4}{9}+\frac{2}{9}$ Adding $\frac{2}{9}$ on both sides

$x+0=\frac{6}{9}$

$x=\frac{6}{9}$

$x=\frac{2\cdot 3}{3\cdot 3}=\frac{2}{3}\cdot\frac{3}{3}$

$x=\frac{2}{3}$ Simplifying

The solution is $\frac{2}{3}$.

39. $a+\frac{2}{11}=\frac{8}{11}$

$a+\frac{2}{11}-\frac{2}{11}=\frac{8}{11}-\frac{2}{11}$ Subtracting $\frac{2}{11}$ on both sides

$a+0=\frac{6}{11}$

$a=\frac{6}{11}$

The solution is $\frac{6}{11}$.

41. $n+\frac{3}{10}=\frac{7}{10}$

$n+\frac{3}{10}-\frac{3}{10}=\frac{7}{10}-\frac{3}{10}$ Subtracting $\frac{3}{10}$ on both sides

$n+0=\frac{4}{10}$

$n=\frac{2\cdot 2}{2\cdot 5}=\frac{2}{2}\cdot\frac{2}{5}$

$n=\frac{2}{5}$

The solution is $\frac{2}{5}$.

43. $x-\frac{2}{5}=\frac{4}{5}$

$x-\frac{2}{5}+\frac{2}{5}=\frac{4}{5}+\frac{2}{5}$ Adding $\frac{2}{5}$ on both sides

$x+0=\frac{6}{5}$

$x=\frac{6}{5}$

The solution is $\frac{6}{5}$.

45. $a+\frac{1}{2}=\frac{7}{8}$

$a+\frac{1}{2}-\frac{1}{2}=\frac{7}{8}-\frac{1}{2}$ Subtracting $\frac{1}{2}$ on both sides

$a+0=\frac{7}{8}-\frac{1}{2}\cdot\frac{4}{4}$ The LCD is 8. We multiply by 1 to get the LCD.

$a=\frac{7}{8}-\frac{4}{8}=\frac{3}{8}$

The solution is $\frac{3}{8}$.

47. $x-\frac{3}{10}=\frac{2}{5}$

$x-\frac{3}{10}+\frac{3}{10}=\frac{2}{5}+\frac{3}{10}$ Adding $\frac{3}{10}$ on both sides

$x+0=\frac{2}{5}\cdot\frac{2}{2}+\frac{3}{10}$ The LCD is 10. We multiply by 1 to get the LCD.

$x=\frac{4}{10}+\frac{3}{10}=\frac{7}{10}$

The solution is $\frac{7}{10}$.

49. $\frac{2}{3}+x=\frac{4}{5}$

$\frac{2}{3}+x-\frac{2}{3}=\frac{4}{5}-\frac{2}{3}$ Subtracting $\frac{2}{3}$ on both sides

$x+0=\frac{4}{5}\cdot\frac{3}{3}-\frac{2}{3}\cdot\frac{5}{5}$ The LCD is 15. We multiply by 1 to get the LCD.

$x=\frac{12}{15}-\frac{10}{15}=\frac{2}{15}$

The solution is $\frac{2}{15}$.

51. $\frac{3}{8}+a=\frac{1}{12}$

$\frac{3}{8}+a-\frac{3}{8}=\frac{1}{12}-\frac{3}{8}$ Subtracting $\frac{3}{8}$ on both sides

$a+0=\frac{1}{12}\cdot\frac{2}{2}-\frac{3}{8}\cdot\frac{3}{3}$ The LCD is 24. We multiply by 1 to get the LCD.

$a=\frac{2}{24}-\frac{9}{24}=\frac{2-9}{24}$

$a=\frac{-7}{24}$, or $-\frac{7}{24}$

The solution is $-\frac{7}{24}$.

53. $n-\frac{1}{10}=-\frac{1}{30}$

$n-\frac{1}{10}+\frac{1}{10}=-\frac{1}{30}+\frac{1}{10}$ Adding $\frac{1}{10}$ on both sides

$n+0=-\frac{1}{30}+\frac{1}{10}\cdot\frac{3}{3}$ The LCD is 30. We multiply by 1 to get the LCD.

$n=-\frac{1}{30}+\frac{3}{30}$

$n=\frac{-1}{30}+\frac{3}{30}=\frac{2}{30}$

$n=\frac{2\cdot 1}{2\cdot 15}=\frac{2}{2}\cdot\frac{1}{15}$

$n=\frac{1}{15}$

The solution is $\frac{1}{15}$.

55. $x+\frac{3}{4}=-\frac{1}{2}$

$x+\frac{3}{4}-\frac{3}{4}=-\frac{1}{2}-\frac{3}{4}$ Subtracting $\frac{3}{4}$ on both sides

$x+0=-\frac{1}{2}\cdot\frac{2}{2}-\frac{3}{4}$ The LCD is 4. We multiply by 1 to get the LCD.

$x=-\frac{2}{4}-\frac{3}{4}=\frac{-2}{4}-\frac{3}{4}$

$x=\frac{-2-3}{4}$

$x=\frac{-5}{4}$, or $-\frac{5}{4}$

The solution is $-\frac{5}{4}$.

57. ***Familiarize***. We visualize the situation. Let t = the number of hours Monica listened to Brahms.

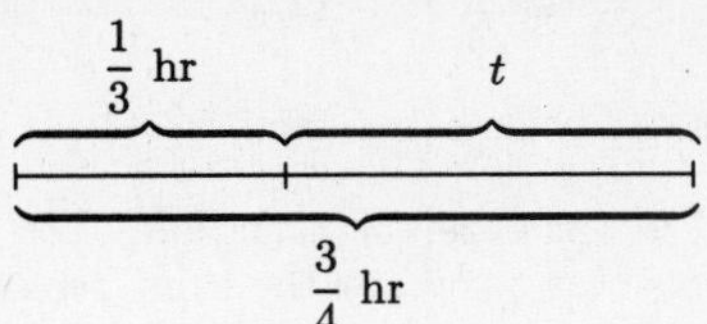

Translate. This is a "how much more" situation that can be translated as follows:

Time spent listening to Beethoven	plus	Time spent listening to Brahms	is	Total listening time
↓	↓	↓	↓	↓
$\frac{1}{3}$	$+$	t	$=$	$\frac{3}{4}$

Solve. We subtract $\frac{1}{3}$ on both sides of the equation.

$\frac{1}{3}+t-\frac{1}{3}=\frac{3}{4}-\frac{1}{3}$

$t+0=\frac{3}{4}\cdot\frac{3}{3}-\frac{1}{3}\cdot\frac{4}{4}$ The LCD is 12. We multiply by 1 to get the LCD.

$t=\frac{9}{12}-\frac{4}{12}=\frac{5}{12}$

Check. We return to the original problem and add.

$\frac{1}{3}+\frac{5}{12}=\frac{1}{3}\cdot\frac{4}{4}+\frac{5}{12}=\frac{4}{12}+\frac{5}{12}=\frac{9}{12}=\frac{3}{3}\cdot\frac{3}{4}=\frac{3}{4}$

State. Monica spent $\frac{5}{12}$ hr listening to Brahms.

59. ***Familiarize***. We visualize the situation. Let d = the distance that remains to be walked.

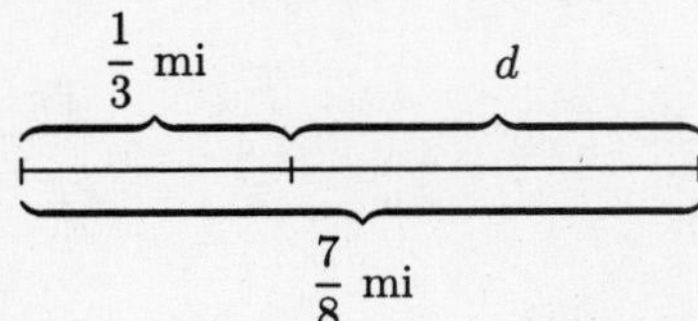

Translate. This is a "how much more" situation that can be translate as follows:

Distance already walked	plus	Distance remaining to be walked	is	Total distance
↓	↓	↓	↓	↓
$\frac{1}{3}$	$+$	d	$=$	$\frac{7}{8}$

Solve. We subtract $\frac{1}{3}$ on both sides of the equation.

$\frac{1}{3}+d-\frac{1}{3}=\frac{7}{8}-\frac{1}{3}$

$d+0=\frac{7}{8}\cdot\frac{3}{3}-\frac{1}{3}\cdot\frac{8}{8}$ The LCD is 24. We multiply by 1 to get the LCD.

$d=\frac{21}{24}-\frac{8}{24}=\frac{13}{24}$

Check. We return to the original problem and add.

$$\frac{1}{3}+\frac{13}{24}=\frac{1}{3}\cdot\frac{8}{8}+\frac{13}{14}=\frac{8}{24}+\frac{13}{24}=\frac{21}{24}=\frac{3}{3}\cdot\frac{7}{8}=\frac{7}{8}$$

State. Hugo should walk $\frac{13}{24}$ mi farther.

61. ***Familiarize***. We visualize the situation. Let x = the portion of the business owned by the third person.

$\frac{7}{12}$ | $\frac{1}{6}$ | x

1 entire business

Translate. This is a "how much more" situation.

First owner's portion	plus	Second owner's portion	plus	Third owner's portion	is	Entire business
$\frac{7}{12}$	$+$	$\frac{1}{6}$	$+$	x	$=$	1

Solve. We begin by adding the fractions on the left side of the equation.

$$\frac{7}{12}+\frac{1}{6}\cdot\frac{2}{2}+x=1 \qquad \text{The LCD is 12.}$$
$$\frac{7}{12}+\frac{2}{12}+x=1$$
$$\frac{9}{12}+x=1$$
$$\frac{3}{4}+x=1 \qquad \text{Simplifying } \frac{9}{12}$$
$$\frac{3}{4}+x-\frac{3}{4}=1-\frac{3}{4} \qquad \text{Subtracting } \frac{3}{4} \text{ on both sides}$$
$$x+0=1\cdot\frac{4}{4}-\frac{3}{4} \qquad \text{The LCD is 4.}$$
$$x=\frac{4}{4}-\frac{3}{4}$$
$$x=\frac{1}{4}$$

Check. We return to the original problem and add.

$$\frac{7}{12}+\frac{1}{6}+\frac{1}{4}=\frac{7}{12}+\frac{1}{6}\cdot\frac{2}{2}+\frac{1}{4}\cdot\frac{3}{3}=$$
$$\frac{7}{12}+\frac{2}{12}+\frac{3}{12}=\frac{12}{12}=1$$

State. The third person owned $\frac{1}{4}$ of the business.

63. $\frac{9}{10}\div\frac{3}{5}=\frac{9}{10}\cdot\frac{5}{3}=\frac{9\cdot 5}{10\cdot 3}=\frac{3\cdot 3\cdot 5}{2\cdot 5\cdot 3}=\frac{3\cdot 5}{3\cdot 5}\cdot\frac{3}{2}=\frac{3}{2}$

65. $(-7)\div\frac{1}{3}=-7\cdot\frac{3}{1}=\frac{-7\cdot 3}{1}=\frac{-21}{1}=-21$

67. ***Familiarize***. We visualize the situation. Repeated addition will work here.

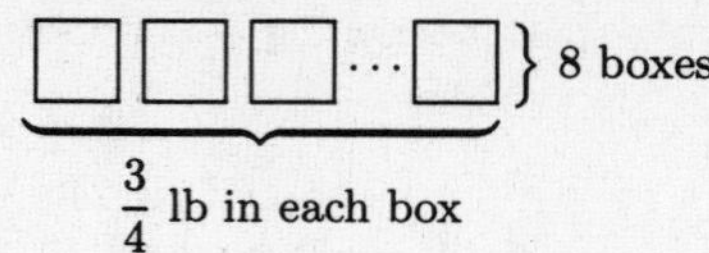

We let w = the weight of 8 boxes.

Translate. The problem translates to the following equation:

$$w=8\cdot\frac{3}{4}$$

Solve. We carry out the multiplication.

$$w=8\cdot\frac{3}{4}$$
$$w=\frac{8\cdot 3}{4}=\frac{2\cdot 4\cdot 3}{4\cdot 1}$$
$$w=\frac{4}{4}\cdot\frac{2\cdot 3}{1}$$
$$w=6$$

Check. We repeat the calculation. We an also observe that we are multiplying 8 by a number less than 1, so the product will be less than 8. Since 6 is less than 8, our answer seems reasonable.

State. The weight of 8 small boxes of cornflakes is 6 lb.

69. 

71. Use a calculator.

$$x-\frac{16}{323}=\frac{10}{187}$$
$$x-\frac{16}{323}+\frac{16}{323}=\frac{10}{187}+\frac{16}{323}$$
$$x+0=\frac{10}{11\cdot 17}+\frac{16}{17\cdot 19}$$
$$x=\frac{10}{11\cdot 17}\cdot\frac{19}{19}+\frac{16}{17\cdot 19}\cdot\frac{11}{11} \qquad \text{The LCD is } 11\cdot 17\cdot 19.$$
$$x=\frac{190}{11\cdot 17\cdot 19}+\frac{176}{17\cdot 19\cdot 11}$$
$$x=\frac{366}{11\cdot 17\cdot 19}$$
$$x=\frac{366}{3553}$$

The solution is $\frac{366}{3553}$.

73. ***Familiarize***. We visualize the situation. We let h = the elevation at which the climber finished.

$\frac{3}{5}$ km ↑ $\frac{1}{4}$ km ↓ $\frac{1}{3}$ km ↑ $\frac{1}{7}$ km ↓ } h

Sea level

Translate.

First climb	minus	First descent	plus	Second climb	minus
↓	↓	↓	↓	↓	↓
$\frac{3}{5}$	$-$	$\frac{1}{4}$	$+$	$\frac{1}{3}$	$-$

Second descent	is	Final elevation
↓	↓	↓
$\frac{1}{7}$	$=$	h

Solve. We carry out the calculation. The LCD is $5\cdot 4\cdot 3\cdot 7$, or 420.

$$\frac{3}{5}-\frac{1}{4}+\frac{1}{3}-\frac{1}{7}=h$$

$$\frac{3}{5}\cdot\frac{4\cdot 3\cdot 7}{4\cdot 3\cdot 7}-\frac{1}{4}\cdot\frac{5\cdot 3\cdot 7}{5\cdot 3\cdot 7}+\frac{1}{3}\cdot\frac{5\cdot 4\cdot 7}{5\cdot 4\cdot 7}-$$
$$\frac{1}{7}\cdot\frac{5\cdot 4\cdot 3}{5\cdot 4\cdot 3}=h$$

$$\frac{252}{5\cdot 4\cdot 3\cdot 7}-\frac{105}{5\cdot 4\cdot 3\cdot 7}+\frac{140}{5\cdot 4\cdot 3\cdot 7}-\frac{60}{5\cdot 4\cdot 3\cdot 7}=h$$

$$\frac{252-105+140-60}{5\cdot 4\cdot 3\cdot 7}=h$$

$$\frac{227}{5\cdot 4\cdot 3\cdot 7}=h$$

$$\frac{227}{420}=h$$

Check. We repeat the calculation.

State. The climber's final elevation is $\frac{227}{420}$ km.

75. $\frac{12}{5x}-\frac{7}{5x}=\frac{5}{5x}=\frac{5\cdot 1}{5\cdot x}=\frac{5}{5}\cdot\frac{1}{x}=\frac{1}{x}$

77. $\left(\frac{2}{3}\right)^2+\left(\frac{3}{4}\right)^2=\frac{4}{9}+\frac{9}{16}=\frac{4}{9}\cdot\frac{16}{16}+\frac{9}{16}\cdot\frac{9}{9}=$

$\frac{64}{144}+\frac{81}{144}=\frac{145}{144}$

79. ***Familiarize.*** First we find how far the athlete swam. We let s = this distance. We visualize the situation.

$\longrightarrow \longrightarrow \cdots \longrightarrow$ } 10 laps

$\frac{3}{80}$ km in each lap

Translate. We translate to the following equation:

$$s=10\cdot\frac{3}{80}$$

Solve. We carry out the multiplication.

$$s=10\cdot\frac{3}{80}=\frac{10\cdot 3}{80}$$

$$s=\frac{10\cdot 3}{10\cdot 8}=\frac{10}{10}\cdot\frac{3}{8}$$

$$s=\frac{3}{8}$$

Now we find the distance the athlete must walk. We let w = the distance.

Distance swum	plus	Distance walked	is	$\frac{9}{10}$ km
↓	↓	↓	↓	↓
$\frac{3}{8}$	$+$	w	$=$	$\frac{9}{10}$

We solve the equation.

$$\frac{3}{8}+w=\frac{9}{10}$$

$$\frac{3}{8}+w-\frac{3}{8}=\frac{9}{10}-\frac{3}{8}$$

$$w+0=\frac{9}{10}\cdot\frac{4}{4}-\frac{3}{8}\cdot\frac{5}{5}\quad\text{The LCD is 40.}$$

$$w=\frac{36}{40}-\frac{15}{40}$$

$$w=\frac{21}{40}$$

Check. We add the distance swum and the distance walked:

$\frac{3}{8}+\frac{21}{40}=\frac{3}{8}\cdot\frac{5}{5}+\frac{21}{40}=\frac{15}{40}+\frac{21}{40}=\frac{36}{40}=\frac{9\cdot 4}{10\cdot 4}=$

$\frac{9}{10}\cdot\frac{4}{4}=\frac{9}{10}$

State. The athlete must walk $\frac{21}{40}$ km after swimming 10 laps.

Exercise Set 4.4

1.

$$37+x=89$$
$$37+x-37=89-37\quad\text{Subtracting 37 on both sides}$$
$$x+0=52$$
$$x=52$$

Check:

$37+x=89$	
$37+52\ ?\ 89$	
$89 \mid 89$	TRUE

The solution is 52.

3.

$$\frac{4}{7}a=60$$

$$\frac{7}{4}\cdot\frac{4}{7}a=\frac{7}{4}\cdot 60\quad\text{Multiplying by } \frac{7}{4} \text{ on both sides}$$

$$1a=\frac{7\cdot 60}{4}$$

$$a=105$$

Check:

$\frac{4}{7}a=60$		
$\frac{4}{7}\cdot 105\ ?\ 60$		
$\frac{4\cdot 7\cdot 15}{7\cdot 1}$		
$\frac{4\cdot 15}{1}$	60	TRUE

The solution is 105.

5.
$$\begin{aligned} 84 &= x - 17 \\ 84 + 17 &= x - 17 + 17 \quad \text{Adding 17 on both sides} \\ 101 &= x + 0 \\ 101 &= x \end{aligned}$$

Check: $84 = x - 17$

$84 \ ? \ 101 - 17$

$84 \mid 84$ TRUE

The solution is 101.

7.
$$\begin{aligned} \frac{9}{7} &= 6a \\ \frac{1}{6} \cdot \frac{9}{7} &= \frac{1}{6} \cdot 6a \quad \text{Multiplying by } \frac{1}{6} \\ \frac{1 \cdot 9}{6 \cdot 7} &= 1a \\ \frac{3 \cdot 3}{2 \cdot 3 \cdot 7} &= a \\ \frac{3}{14} &= a \end{aligned}$$

Check: $\frac{9}{7} = 6a$

$\frac{9}{7} \ ? \ 6 \cdot \frac{3}{14}$

$\quad \mid \frac{2 \cdot 3 \cdot 3}{2 \cdot 7}$

$\frac{9}{7} \mid \frac{9}{7}$ TRUE

The solution is $\frac{3}{14}$.

9.
$$\begin{aligned} x + \frac{9}{7} &= \frac{10}{3} \\ x + \frac{9}{7} - \frac{9}{7} &= \frac{10}{3} - \frac{9}{7} \quad \text{Using the addition principle} \\ x + 0 &= \frac{10}{3} \cdot \frac{7}{7} - \frac{9}{7} \cdot \frac{3}{3} \quad \text{The LCD is 21.} \\ x &= \frac{70}{21} - \frac{27}{21} \\ x &= \frac{43}{21} \end{aligned}$$

The number $\frac{43}{21}$ checks and is the solution.

11.
$$\begin{aligned} -\frac{19}{5} &= \frac{3}{10}a \\ \frac{10}{3}\left(-\frac{19}{5}\right) &= \frac{10}{3} \cdot \frac{3}{10}a \quad \text{Using the multiplication principle} \\ -\frac{10 \cdot 19}{3 \cdot 5} &= 1a \\ -\frac{2 \cdot 5 \cdot 19}{3 \cdot 5} &= a \\ -\frac{2 \cdot 19}{3} \cdot \frac{5}{5} &= a \\ -\frac{38}{3} &= a \end{aligned}$$

The number $-\frac{38}{3}$ checks and is the solution.

13.
$$\begin{aligned} -\frac{4}{5} &= x - \frac{7}{10} \\ -\frac{4}{5} + \frac{7}{10} &= x - \frac{7}{10} + \frac{7}{10} \quad \text{Using the addition principle} \\ -\frac{4}{5} \cdot \frac{2}{2} + \frac{7}{10} &= x + 0 \\ -\frac{8}{10} + \frac{7}{10} &= x \\ -\frac{1}{10} &= x \end{aligned}$$

The number $-\frac{1}{10}$ checks and is the solution.

15.
$$\begin{aligned} -19 &= 23 + a \\ -19 - 23 &= 23 + a - 23 \quad \text{Using the addition principle} \\ -42 &= a + 0 \\ -42 &= a \end{aligned}$$

The number -42 checks and is the solution.

17.
$$\begin{aligned} -\frac{9}{4}x &= \frac{2}{3} \\ -\frac{4}{9}\left(-\frac{9}{4}x\right) &= -\frac{4}{9} \cdot \frac{2}{3} \quad \text{Using the multiplication principle} \\ 1x &= -\frac{4 \cdot 2}{9 \cdot 3} \\ x &= -\frac{8}{27} \end{aligned}$$

The number $-\frac{8}{27}$ checks and is the solution.

19.
$$\begin{aligned} -\frac{41}{10} + x &= \frac{7}{2} \\ -\frac{41}{10} + x + \frac{41}{10} &= \frac{7}{2} + \frac{41}{10} \quad \text{Using the addition principle} \\ x + 0 &= \frac{7}{2} \cdot \frac{5}{5} + \frac{41}{10} \\ x &= \frac{35}{10} + \frac{41}{10} \\ x &= \frac{76}{10} = \frac{2 \cdot 38}{2 \cdot 5} \\ x &= \frac{38}{5} \end{aligned}$$

The number $\frac{38}{5}$ checks and is the solution.

21.
$$\begin{aligned} 6x - 4 &= 14 \\ 6x - 4 + 4 &= 14 + 4 \quad \text{Using the addition principle} \\ 6x + 0 &= 18 \\ 6x &= 18 \\ \frac{1}{6} \cdot 6x &= \frac{1}{6} \cdot 18 \quad \text{Using the multiplication principle} \\ 1x &= \frac{18}{6} \\ x &= 3 \end{aligned}$$

Check: $6x - 4 = 14$

$6 \cdot 3 - 4 \ ? \ 14$

$18 - 4 \mid$

$14 \mid 14$ TRUE

The solution is 3.

23.
$$\begin{aligned} 3a+8&=23 \\ 3a+8-8&=23-8 \quad \text{Using the addition principle} \\ 3a+0&=15 \\ 3a&=15 \\ \frac{1}{3}\cdot 3a&=\frac{1}{3}\cdot 15 \quad \text{Using the multiplication principle} \\ 1a&=\frac{15}{3} \\ a&=5 \end{aligned}$$

Check:
$$\begin{array}{c|c} \multicolumn{2}{c}{3a+8=23} \\ \hline 3\cdot 5+8 \ ? & 23 \\ 15+8 & \\ 23 & 23 \end{array} \quad \text{TRUE}$$

The solution is 5.

25.
$$\begin{aligned} 19&=2x-7 \\ 19+7&=2x-7+7 \quad \text{Using the addition principle} \\ 26&=2x+0 \\ 26&=2x \\ \frac{1}{2}\cdot 26&=\frac{1}{2}\cdot 2x \quad \text{Using the multiplication principle} \\ \frac{26}{2}&=1x \\ 13&=x \end{aligned}$$

Check:
$$\begin{array}{c|c} \multicolumn{2}{c}{19=2x-7} \\ \hline 19 \ ? & 2\cdot 13-7 \\ & 26-7 \\ 19 & 19 \end{array} \quad \text{TRUE}$$

The solution is 13.

27.
$$\begin{aligned} -5t+4&=39 \\ -5t+4-4&=39-4 \quad \text{Using the addition principle} \\ -5t+0&=35 \\ -5t&=35 \\ -\frac{1}{5}(-5t)&=-\frac{1}{5}\cdot 35 \quad \text{Using the multiplication principle} \\ 1t&=-\frac{35}{5} \\ t&=-7 \end{aligned}$$

Check:
$$\begin{array}{c|c} \multicolumn{2}{c}{-5t+4=39} \\ \hline -5(-7)+4 \ ? & 39 \\ 35+4 & \\ 39 & 39 \end{array} \quad \text{TRUE}$$

The solution is -7.

29.
$$\begin{aligned} 3x+4&=-11 \\ 3x+4-4&=-11-4 \quad \text{Using the addition principle} \\ 3x+0&=-15 \\ 3x&=-15 \\ \frac{1}{3}\cdot 3x&=\frac{1}{3}(-15) \quad \text{Using the multiplication principle} \\ 1x&=-\frac{15}{3} \\ x&=-5 \end{aligned}$$

The number -5 checks and is the solution.

31.
$$\begin{aligned} 2a-9&=-7 \\ 2a-9+9&=-7+9 \quad \text{Using the addition principle} \\ 2a+0&=2 \\ 2a&=2 \\ \frac{1}{2}\cdot 2a&=\frac{1}{2}\cdot 2 \quad \text{Using the multiplication principle} \\ 1a&=\frac{2}{2} \\ a&=1 \end{aligned}$$

The number 1 checks and is the solution.

33.
$$\begin{aligned} \frac{3}{2}x-3&=12 \\ \frac{3}{2}x-3+3&=12+3 \quad \text{Using the addition principle} \\ \frac{3}{2}x+0&=15 \\ \frac{3}{2}x&=15 \\ \frac{2}{3}\cdot\frac{3}{2}x&=\frac{2}{3}\cdot 15 \quad \text{Using the multiplication principle} \\ 1x&=\frac{30}{3} \\ x&=10 \end{aligned}$$

The number 10 checks and is the solution.

35.
$$\begin{aligned} \frac{3}{5}t-4&=8 \\ \frac{3}{5}t-4+4&=8+4 \quad \text{Using the addition principle} \\ \frac{3}{5}t+0&=12 \\ \frac{3}{5}t&=12 \\ \frac{5}{3}\cdot\frac{3}{5}t&=\frac{5}{3}\cdot 12 \quad \text{Using the multiplication principle} \\ 1t&=\frac{60}{3} \\ t&=20 \end{aligned}$$

The number 20 checks and is the solution.

37.
$$\begin{aligned} \frac{3}{4}x&=18 \\ \frac{4}{3}\cdot\frac{3}{4}x&=\frac{4}{3}\cdot 18 \quad \text{Using the multiplication principle} \\ 1x&=\frac{72}{3} \\ x&=24 \end{aligned}$$

The number 24 checks and is the solution.

39.

$$7 = a + \frac{14}{5}$$

$$7 - \frac{14}{5} = a + \frac{14}{5} - \frac{14}{5} \quad \text{Using the addition principle}$$

$$7 \cdot \frac{5}{5} - \frac{14}{5} = a + 0$$

$$\frac{35}{5} - \frac{14}{5} = a$$

$$\frac{21}{5} = a$$

The number $\frac{21}{5}$ checks and is the solution.

41.

$$\frac{13}{3}x + \frac{11}{2} = \frac{35}{4}$$

$$\frac{13}{3}x + \frac{11}{2} - \frac{11}{2} = \frac{35}{4} - \frac{11}{2} \quad \text{Using the addition principle}$$

$$\frac{13}{3}x + 0 = \frac{35}{4} - \frac{11}{2} \cdot \frac{2}{2}$$

$$\frac{13}{3}x = \frac{35}{4} - \frac{22}{4}$$

$$\frac{13}{3}x = \frac{13}{4}$$

$$\frac{3}{13} \cdot \frac{13}{3}x = \frac{3}{13} \cdot \frac{13}{4} \quad \text{Using the multiplication principle}$$

$$1x = \frac{3 \cdot 13}{13 \cdot 4}$$

$$x = \frac{3}{4}$$

The number $\frac{3}{4}$ checks and is the solution.

43.

$$-\frac{11}{5}t + \frac{7}{2} = \frac{36}{5}$$

$$-\frac{11}{5}t + \frac{7}{2} - \frac{7}{2} = \frac{36}{5} - \frac{7}{2} \quad \text{Using the addition principle}$$

$$-\frac{11}{5}t + 0 = \frac{36}{5} \cdot \frac{2}{2} - \frac{7}{2} \cdot \frac{5}{5}$$

$$-\frac{11}{5}t = \frac{72}{10} - \frac{35}{10}$$

$$-\frac{11}{5}t = \frac{37}{10}$$

$$-\frac{5}{11}\left(-\frac{11}{5}t\right) = -\frac{5}{11} \cdot \frac{37}{10} \quad \text{Using the multiplication principle}$$

$$1t = -\frac{5 \cdot 37}{11 \cdot 2 \cdot 5}$$

$$t = -\frac{37}{22}$$

The number $-\frac{37}{22}$ checks and is the solution.

45.

$$8 = \frac{3}{4}x + \frac{5}{6}$$

$$8 - \frac{5}{6} = \frac{3}{4}x - \frac{5}{6} + \frac{5}{6} \quad \text{Using the addition principle}$$

$$8 \cdot \frac{6}{6} - \frac{5}{6} = \frac{3}{4}x + 0$$

$$\frac{48}{6} - \frac{5}{6} = \frac{3}{4}x$$

$$\frac{43}{6} = \frac{3}{4}x$$

$$\frac{4}{3} \cdot \frac{43}{6} = \frac{4}{3} \cdot \frac{3}{4}x \quad \text{Using the multiplication principle}$$

$$\frac{2 \cdot 2 \cdot 43}{3 \cdot 2 \cdot 3} = 1x$$

$$\frac{86}{9} = x$$

The number $\frac{86}{9}$ checks and is the solution.

47.

$$-\frac{51}{4} = \frac{5}{2} + \frac{3}{2}a$$

$$-\frac{51}{4} - \frac{5}{2} = \frac{5}{2} + \frac{3}{2}a - \frac{5}{2} \quad \text{Using the addition principle}$$

$$-\frac{51}{4} - \frac{5}{2} \cdot \frac{2}{2} = \frac{3}{2}a + 0$$

$$-\frac{51}{4} - \frac{10}{4} = \frac{3}{2}a$$

$$-\frac{61}{4} = \frac{3}{2}a$$

$$\frac{2}{3}\left(-\frac{61}{4}\right) = \frac{2}{3} \cdot \frac{3}{2}a \quad \text{Using the multiplication principle}$$

$$-\frac{2 \cdot 61}{3 \cdot 2 \cdot 2} = 1a$$

$$-\frac{61}{6} = a$$

The number $-\frac{61}{6}$ checks and is the solution.

49.

$$\frac{22}{7}x - \frac{14}{5} = -\frac{3}{2}$$

$$\frac{22}{7}x - \frac{14}{5} + \frac{14}{5} = -\frac{3}{2} + \frac{14}{5} \quad \text{Using the addition principle}$$

$$\frac{22}{7}x + 0 = -\frac{3}{2} \cdot \frac{5}{5} + \frac{14}{5} \cdot \frac{2}{2}$$

$$\frac{22}{7}x = -\frac{15}{10} + \frac{28}{10}$$

$$\frac{22}{7}x = \frac{13}{10}$$

$$\frac{7}{22} \cdot \frac{22}{7}x = \frac{7}{22} \cdot \frac{13}{10} \quad \text{Using the multiplication principle}$$

$$1x = \frac{7 \cdot 13}{22 \cdot 10}$$

$$x = \frac{91}{220}$$

The number $\frac{91}{220}$ checks and is the solution.

51. -200 represents the \$200 withdrawal, 90 represents the \$90 deposit, and -40 represents the \$40 withdrawal. We add these numbers to find the change in the balance.

$$-200 + 90 + (-40) = -110 + (-40) = -150$$

The account balance decreased by \$150.

53. $\frac{10}{7} \div 2m = \frac{10}{7} \cdot \frac{1}{2m} = \frac{10 \cdot 1}{7 \cdot 2m} = \frac{2 \cdot 5 \cdot 1}{7 \cdot 2 \cdot m} = \frac{2}{2} \cdot \frac{5 \cdot 1}{7 \cdot m} = \frac{5}{7m}$

55. $3(a+b) = 3 \cdot a + 3 \cdot b = 3a + 3b$

57. ◈

59. We use a calculator.

$$\begin{aligned}
\frac{1081}{3599}x - \frac{17}{61} &= \frac{19}{59} \\
\frac{1081}{3599}x - \frac{17}{61} + \frac{17}{61} &= \frac{19}{59} + \frac{17}{61} \\
\frac{1081}{3599}x + 0 &= \frac{19}{59} \cdot \frac{61}{61} + \frac{17}{61} \cdot \frac{59}{59} \\
\frac{1081}{3599}x &= \frac{1159}{3599} + \frac{1003}{3599} \\
\frac{1081}{3599}x &= \frac{2162}{3599} \\
\frac{3599}{1081} \cdot \frac{1081}{3599}x &= \frac{3599}{1081} \cdot \frac{2162}{3599} \\
1x &= \frac{2162}{1081} \\
x &= 2
\end{aligned}$$

The solution is 2.

61.

$$\begin{aligned}
-\frac{a}{5} + \frac{31}{4} &= \frac{16}{3} \\
-\frac{1}{5}a + \frac{31}{4} &= \frac{16}{3} \qquad \left(-\frac{1}{5} \cdot a = -\frac{a}{5}\right) \\
-\frac{1}{5}a + \frac{31}{4} - \frac{31}{4} &= \frac{16}{3} - \frac{31}{4} \\
-\frac{1}{5}a + 0 &= \frac{16}{3} \cdot \frac{4}{4} - \frac{31}{4} \cdot \frac{3}{3} \\
-\frac{1}{5}a &= \frac{64}{12} - \frac{93}{12} \\
-\frac{1}{5}a &= -\frac{29}{12} \\
-5\left(-\frac{1}{5}a\right) &= -5\left(-\frac{29}{12}\right) \\
1a &= \frac{5 \cdot 29}{12} \\
a &= \frac{145}{12}
\end{aligned}$$

63.

$$\begin{aligned}
4 + \frac{3x}{7} &= -\frac{11}{5} \\
4 + \frac{3}{7}x &= -\frac{11}{5} \qquad \left(\frac{3}{7} \cdot x = \frac{3x}{7}\right) \\
4 + \frac{3}{7}x - 4 &= -\frac{11}{5} - 4 \\
\frac{3}{7}x + 0 &= -\frac{11}{5} - 4 \cdot \frac{5}{5} \\
\frac{3}{7}x &= -\frac{11}{5} - \frac{20}{5} \\
\frac{3}{7}x &= -\frac{31}{5} \\
\frac{7}{3} \cdot \frac{3}{7}x &= \frac{7}{3}\left(-\frac{31}{5}\right) \\
1x &= -\frac{7 \cdot 31}{3 \cdot 5} \\
x &= -\frac{217}{15}
\end{aligned}$$

65. ***Familiarize***. The perimeter P is the sum of the lengths of the sides, so we have $P = \frac{5}{4}x + x + \frac{5}{2} + 6 + 2$.

Translate. We substitute 15 for P.

$$\frac{5}{4}x + x + \frac{5}{2} + 6 + 2 = 15$$

Solve. We solve the equation. We begin by collecting like terms on the left side.

$$\begin{aligned}
\frac{5}{4}x + x + \frac{5}{2} + 6 + 2 &= 15 \\
\left(\frac{5}{4} + 1\right)x + \frac{5}{2} + 6 \cdot \frac{2}{2} + 2 \cdot \frac{2}{2} &= 15 \\
\left(\frac{5}{4} + \frac{4}{4}\right)x + \frac{5}{2} + \frac{12}{2} + \frac{4}{2} &= 15 \\
\frac{9}{4}x + \frac{21}{2} &= 15 \\
\frac{9}{4}x + \frac{21}{2} - \frac{21}{2} &= 15 - \frac{21}{2} \\
\frac{9}{4}x + 0 &= 15 \cdot \frac{2}{2} - \frac{21}{2} \\
\frac{9}{4}x &= \frac{30}{2} - \frac{21}{2} \\
\frac{9}{4}x &= \frac{9}{2} \\
\frac{4}{9} \cdot \frac{9}{4}x &= \frac{4}{9} \cdot \frac{9}{2} \\
1x &= \frac{2 \cdot 2 \cdot 9}{9 \cdot 2} \\
x &= 2
\end{aligned}$$

Check. $\frac{5}{4} \cdot 2 + 2 + \frac{5}{2} + 6 + 2 = \frac{5}{2} + 2 + \frac{5}{2} + 6 + 2 = 15$, so the result checks.

State. x is 2 cm.

Exercise Set 4.5

1. [b] [a] Multiply: $3 \cdot 5 = 15$.

$3\frac{2}{5} = \frac{17}{5}$ [b] Add: $15 + 2 = 17$.

[a] [c] Keep the denominator.

3. [b] [a] Multiply: $6 \cdot 4 = 24$.

$6\frac{1}{4} = \frac{25}{4}$ [b] Add: $24 + 1 = 25$.

[a] [c] Keep the denominator.

5. $-20\frac{1}{8} = -\frac{161}{8}$ $(20 \cdot 8 = 160; 160 + 1 = 161;$ include the negative sign)

7. $5\frac{1}{10} = \frac{51}{10}$ $(5 \cdot 10 = 50;\ 50 + 1 = 51)$

9. $20\frac{3}{5} = \frac{103}{5}$ $(20 \cdot 5 = 100;\ 100 + 3 = 103)$

11. $-9\frac{5}{6} = -\frac{59}{6}$ $(9 \cdot 6 = 54;\ 54 + 5 = 59;$ include the negative sign)

13. $8\frac{3}{10} = \frac{83}{10}$ $(8 \cdot 10 = 80;\ 80 + 3 = 83)$

15. $1\frac{3}{5} = \frac{8}{5}$ $(1 \cdot 5 = 5;\ 5 + 3 = 8)$

17. $-12\frac{3}{4} = -\frac{51}{4}$ $(12 \cdot 4 = 48;\ 48 + 3 = 51;$ include the negative sign)

19. $4\frac{3}{10} = \frac{43}{10}$ $(4 \cdot 10 = 40;\ 40 + 3 = 43)$

21. $7\frac{3}{100} = \frac{703}{100}$ $(7 \cdot 100 = 700;\ 700 + 3 = 703)$

23. $-6\frac{4}{15} = -\frac{94}{15}$ $(6 \cdot 15 = 90;\ 90 + 4 = 94;$ include the negative sign)

25. To convert $\frac{17}{4}$ to a mixed numeral, we divide.

$$\begin{array}{r} 4 \\ 4\overline{)17} \\ \underline{16} \\ 1 \end{array} \qquad \frac{17}{4} = 4\frac{1}{4}$$

27. To convert $\frac{14}{3}$ to a mixed numeral, we divide.

$$\begin{array}{r} 4 \\ 3\overline{)14} \\ \underline{12} \\ 2 \end{array} \qquad \frac{14}{3} = 4\frac{2}{3}$$

29.

$$\begin{array}{r} 4 \\ 6\overline{)27} \\ \underline{24} \\ 3 \end{array} \qquad \frac{27}{6} = 4\frac{3}{6} = 4\frac{1}{2}$$

Since $\frac{27}{6} = 4\frac{1}{2}$, we have $-\frac{27}{6} = -4\frac{1}{2}$.

31.

$$\begin{array}{r} 5 \\ 10\overline{)57} \\ \underline{50} \\ 7 \end{array} \qquad \frac{57}{10} = 5\frac{7}{10}$$

33.

$$\begin{array}{r} 7 \\ 7\overline{)53} \\ \underline{49} \\ 4 \end{array} \qquad \frac{53}{7} = 7\frac{4}{7}$$

35.

$$\begin{array}{r} 7 \\ 6\overline{)45} \\ \underline{42} \\ 3 \end{array} \qquad \frac{45}{6} = 7\frac{3}{6} = 7\frac{1}{2}$$

37.

$$\begin{array}{r} 11 \\ 4\overline{)46} \\ \underline{40} \\ 6 \\ \underline{4} \\ 2 \end{array} \qquad \frac{46}{4} = 11\frac{2}{4} = 11\frac{1}{2}$$

39.

$$\begin{array}{r} 1 \\ 8\overline{)12} \\ \underline{8} \\ 4 \end{array} \qquad \frac{12}{8} = 1\frac{4}{8} = 1\frac{1}{2}$$

Since $\frac{12}{8} = 1\frac{1}{2}$, we have $-\frac{12}{8} = -1\frac{1}{2}$.

41.

$$\begin{array}{r} 7 \\ 100\overline{)757} \\ \underline{700} \\ 57 \end{array} \qquad \frac{757}{100} = 7\frac{57}{100}$$

43.

$$\begin{array}{r} 43 \\ 8\overline{)345} \\ \underline{320} \\ 25 \\ \underline{24} \\ 1 \end{array} \qquad \frac{345}{8} = 43\frac{1}{8}$$

Since $\frac{345}{8} = 43\frac{1}{8}$, we have $-\frac{345}{8} = -43\frac{1}{8}$.

45. We first divide as usual.

$$\begin{array}{r} 108 \\ 8\overline{)869} \\ \underline{800} \\ 69 \\ \underline{64} \\ 5 \end{array}$$

The answer is 108 R 5. We write a mixed numeral for the quotient as follows: $108\frac{5}{8}$.

47. We first divide as usual.

$$\begin{array}{r} 906 \\ 7\overline{)\,6345} \\ \underline{6300} \\ 45 \\ \underline{42} \\ 3 \end{array}$$

The answer is 906 R 3. We write a mixed numeral for the quotient as follows: $906\frac{3}{7}$.

49.
$$\begin{array}{r} 40 \\ 21\overline{)\,852} \\ \underline{840} \\ 12 \end{array}$$

We get $40\frac{12}{21}$. This simplifies as $40\frac{4}{7}$.

51.
$$\begin{array}{r} 55 \\ 102\overline{)\,5612} \\ \underline{5100} \\ 512 \\ \underline{510} \\ 2 \end{array}$$

We get $55\frac{2}{102}$. This simplifies as $55\frac{1}{51}$.

53. First we divide, as usual.

$$\begin{array}{r} 20 \\ 15\overline{)\,302} \\ \underline{300} \\ 2 \\ \underline{0} \\ 2 \end{array} \qquad \frac{302}{15} = 20\frac{2}{15}$$

Because we divided 15 into <u>negative</u> 302, the answer is $-20\frac{2}{15}$.

55. First we divide, as usual.

$$\begin{array}{r} 22 \\ 21\overline{)\,471} \\ \underline{420} \\ 51 \\ \underline{42} \\ 9 \end{array} \qquad \frac{471}{21} = 22\frac{9}{21} = 22\frac{3}{7}$$

Because we divided 21 into <u>negative</u> 471, the answer is $-22\frac{3}{7}$.

57. $\frac{6}{5}\cdot 15 = \frac{6\cdot 15}{5} = \frac{6\cdot 3\cdot 5}{5\cdot 1} = \frac{5}{5}\cdot\frac{6\cdot 3}{1} = \frac{6\cdot 3}{1} = \frac{18}{1} = 18$

59. $\frac{7}{10}\cdot\frac{5}{14} = \frac{7\cdot 5}{10\cdot 14} = \frac{7\cdot 5\cdot 1}{2\cdot 5\cdot 2\cdot 7} = \frac{7\cdot 5}{7\cdot 5}\cdot\frac{1}{2\cdot 2} = \frac{1}{2\cdot 2} = \frac{1}{4}$

61.

63. Use a calculator.

$$\frac{128,236}{541} = 237\frac{19}{541}$$

65. $\frac{56}{7} + \frac{2}{3} = 8 + \frac{2}{3}$ $\qquad (56 \div 7 = 8)$

$= 8\frac{2}{3}$

67.
$$\begin{array}{r} 52 \\ 7\overline{)\,366} \\ \underline{350} \\ 16 \\ \underline{14} \\ 2 \end{array} \qquad \frac{366}{7} = 52\frac{2}{7}$$

Exercise Set 4.6

1.
$$\begin{array}{r} 2\,\frac{7}{8} \\ +3\,\frac{5}{8} \\ \hline 5\,\frac{12}{8} \end{array}$$

$5\frac{12}{8} = 5 + \frac{12}{8}$

$= 5 + 1\frac{1}{2}$

$= 6\frac{1}{2}$

To find a mixed numeral for $\frac{12}{8}$ we divide:

$$\begin{array}{r} 1 \\ 8\overline{)\,12} \\ \underline{8} \\ 4 \end{array} \qquad \frac{12}{8} = 1\frac{4}{8} = 1\frac{1}{2}$$

3. The LCD is 12.

$$\begin{array}{rcr} 1\boxed{\frac{1}{4}\cdot\frac{3}{3}} & = & 1\,\frac{3}{12} \\ +1\boxed{\frac{2}{3}\cdot\frac{4}{4}} & = & +1\,\frac{8}{12} \\ \hline & & 2\,\frac{11}{12} \end{array}$$

5. The LCD is 12.

$$\begin{array}{rcr} 8\boxed{\frac{3}{4}\cdot\frac{3}{3}} & = & 8\,\frac{9}{12} \\ +5\boxed{\frac{5}{6}\cdot\frac{2}{2}} & = & +5\,\frac{10}{12} \\ \hline & & 13\,\frac{19}{12} \end{array}$$

$13\frac{19}{12} = 13 + \frac{19}{12}$

$= 13 + 1\frac{7}{12}$

$= 14\frac{7}{12}$

7. The LCD is 10.

$$\begin{array}{rl} 3\boxed{\frac{2}{5}\cdot\frac{2}{2}} = & 3\,\frac{4}{10} \\ +8\,\frac{7}{10} = & +8\,\frac{7}{10} \\ \hline & 11\,\frac{11}{10} = 11 + \frac{11}{10} \\ & = 11 + 1\frac{1}{10} \\ & = 12\frac{1}{10} \end{array}$$

9. The LCD is 24.

$$\begin{array}{rl} 5\boxed{\frac{3}{8}\cdot\frac{3}{3}} = & 5\,\frac{9}{24} \\ +10\boxed{\frac{5}{6}\cdot\frac{4}{4}} = & +10\,\frac{20}{24} \\ \hline & 15\,\frac{29}{24} = 15 + \frac{29}{24} \\ & = 15 + 1\frac{5}{24} \\ & = 16\frac{5}{24} \end{array}$$

11. The LCD is 10.

$$\begin{array}{rl} 12\boxed{\frac{4}{5}\cdot\frac{2}{2}} = & 12\,\frac{8}{10} \\ +8\,\frac{7}{10} = & +8\,\frac{7}{10} \\ \hline & 20\,\frac{15}{10} = 20 + \frac{15}{10} \\ & = 20 + 1\frac{5}{10} \\ & = 21\frac{5}{10} \\ & = 21\frac{1}{2} \end{array}$$

13. The LCD is 8.

$$\begin{array}{rl} 14\,\frac{5}{8} = & 14\,\frac{5}{8} \\ +13\boxed{\frac{1}{4}\cdot\frac{2}{2}} = & +13\,\frac{2}{8} \\ \hline & 27\,\frac{7}{8} \end{array}$$

15.

$$\begin{array}{rl} 4\frac{1}{5} = & 3\frac{6}{5} \\ -2\frac{3}{5} = & -2\frac{3}{5} \\ \hline & 1\frac{3}{5} \end{array}$$

Since $\frac{1}{5}$ is smaller than $\frac{3}{5}$, we cannot subtract until we borrow:

$4\frac{1}{5} = 3+\frac{5}{5}+\frac{1}{5} = 3+\frac{6}{5} = 3\frac{6}{5}$

17. The LCD is 10.

$$\begin{array}{rl} 6\boxed{\frac{3}{5}\cdot\frac{2}{2}} = & 6\,\frac{6}{10} \\ -2\boxed{\frac{1}{2}\cdot\frac{5}{5}} = & -2\,\frac{5}{10} \\ \hline & 4\,\frac{1}{10} \end{array}$$

19. The LCD is 24.

$$\begin{array}{rcl} 34\boxed{\frac{1}{3}\cdot\frac{8}{8}} = 34\,\frac{8}{24} & = & 33\,\frac{32}{24} \\ -12\boxed{\frac{5}{8}\cdot\frac{3}{3}} = -12\,\frac{15}{24} & = & -12\,\frac{15}{24} \\ \hline & & 21\,\frac{17}{24} \end{array}$$

(Since $\frac{8}{24}$ is smaller than $\frac{15}{24}$, we cannot subtract until we borrow: $34\frac{8}{24} = 33 + \frac{24}{24} + \frac{8}{24} = 33 + \frac{32}{24} = 33\frac{32}{24}$.)

21.

$$\begin{array}{rl} 21 = & 20\frac{4}{4} \\ -\ 8\frac{3}{4} = & -\ 8\frac{3}{4} \\ \hline & 12\frac{1}{4} \end{array}$$

$\left(21 = 20 + 1 = 20 + \frac{4}{4} = 20\frac{4}{4}\right)$

23.

$$\begin{array}{rl} 34 = & 33\frac{8}{8} \\ -\ 18\frac{5}{8} = & -\ 18\frac{5}{8} \\ \hline & 15\frac{3}{8} \end{array}$$

$\left(34 = 33 + 1 = 33 + \frac{8}{8} = 33\frac{8}{8}\right)$

25. The LCD is 12.

$$\begin{array}{rcl} 21\boxed{\frac{1}{6}\cdot\frac{2}{2}} = 21\,\frac{2}{12} & = & 20\,\frac{14}{12} \\ -13\boxed{\frac{3}{4}\cdot\frac{3}{3}} = -13\,\frac{9}{12} & = & -13\,\frac{9}{12} \\ \hline & & 7\,\frac{5}{12} \end{array}$$

(Since $\frac{2}{12}$ is smaller than $\frac{9}{12}$, we cannot subtract until we borrow: $21\frac{2}{12} = 20 + \frac{12}{12} + \frac{2}{12} = 20 + \frac{14}{12} = 20\frac{14}{12}$.)

27. The LCD is 18.

$$25\boxed{\frac{1}{9}\cdot\frac{2}{2}} = 25\frac{2}{18} = 24\frac{20}{18}$$
$$-13\boxed{\frac{5}{6}\cdot\frac{3}{3}} = -13\frac{15}{18} = -13\frac{15}{18}$$
$$11\frac{5}{18}$$

$\left(\text{Since } \frac{2}{18} \text{ is smaller than } \frac{15}{18}, \text{ we cannot subtract until we borrow: } 25\frac{2}{18} = 24 + \frac{18}{18} + \frac{2}{18} = 24 + \frac{20}{18} = 24\frac{20}{18}.\right)$

29. $5\frac{3}{14}t + 3\frac{2}{21}t$

$= \left(5\frac{3}{14} + 3\frac{2}{21}\right)t$ Using the distributive law

$= \left(5\frac{9}{42} + 3\frac{4}{42}\right)t$ The LCD is 42.

$= 8\frac{13}{42}t$ Adding

31. $9\frac{1}{2}x - 7\frac{3}{8}x$

$= \left(9\frac{1}{2} - 7\frac{3}{8}\right)x$ Using the distributive law

$= \left(9\frac{4}{8} - 7\frac{3}{8}\right)x$ The LCD is 8.

$= 2\frac{1}{8}x$ Subtracting

33. $3\frac{7}{8}t + 4\frac{9}{10}t$

$= \left(3\frac{7}{8} + 4\frac{9}{10}\right)t$ Using the distributive law

$= \left(3\frac{35}{40} + 4\frac{36}{40}\right)t$ The LCD is 40.

$= 7\frac{71}{40}t = 8\frac{31}{40}t$

35. $37\frac{5}{9}t - 25\frac{4}{5}t$

$= \left(37\frac{5}{9} - 25\frac{4}{5}\right)t$ Using the distributive law

$= \left(37\frac{25}{45} - 25\frac{36}{45}\right)t$ The LCD is 45.

$= \left(36\frac{70}{45} - 25\frac{36}{45}\right)t$

$= 11\frac{34}{45}t$

37. $2\frac{5}{6}x + 3\frac{1}{3}x$

$= \left(2\frac{5}{6} + 3\frac{1}{3}\right)x$ Using the distributive law

$= \left(2\frac{5}{6} + 3\frac{2}{6}\right)x$ The LCD is 6.

$= 5\frac{7}{6}x = 6\frac{1}{6}x$

39. $4\frac{3}{11}x + 5\frac{2}{3}x$

$= \left(4\frac{3}{11} + 5\frac{2}{3}\right)x$ Using the distributive law

$= \left(4\frac{9}{33} + 5\frac{22}{33}\right)x$ The LCD is 33.

$= 9\frac{31}{33}x$

41. ***Familiarize.*** We let w = the total weight of the meat.

Translate. We write an equation.

Weight of one package	plus	Weight of second package	is	Total weight
↓	↓	↓	↓	↓
$1\frac{2}{3}$	$+$	$5\frac{3}{4}$	$=$	w

Solve. We carry out the addition. The LCD is 12.

$$1\boxed{\frac{2}{3}\cdot\frac{4}{4}} = 1\frac{8}{12}$$
$$+5\boxed{\frac{3}{4}\cdot\frac{3}{3}} = +5\frac{9}{12}$$
$$6\frac{17}{12} = 6 + \frac{17}{12}$$
$$= 6 + 1\frac{5}{12}$$
$$= 7\frac{5}{12}$$

Thus, $w = 7\frac{5}{12}$.

Check. We repeat the calculation. We also note that the answer is larger than either of the individual weights, so the answer seems reasonable.

State. The total weight of the meat was $7\frac{5}{12}$ lb.

43. ***Familiarize.*** We let h = Rocky's excess height.

Translate. We have a "how much more" situation.

Height of daughter	plus	How much more height	is	Rocky's height
↓	↓	↓	↓	↓
$180\frac{3}{4}$	$+$	h	$=$	$187\frac{1}{10}$

Solve. We solve the equation as follows:

$$h = 187\frac{1}{10} - 180\frac{3}{4}$$

$$187\boxed{\frac{1}{10}\cdot\frac{2}{2}} = 187\frac{2}{20}$$
$$180\boxed{\frac{3}{4}\cdot\frac{5}{5}} = 180\frac{15}{20}$$

$$\begin{array}{rcrcr} 187\frac{1}{10} & = & 187\frac{2}{20} & = & 186\frac{22}{20} \\ -180\frac{3}{4} & = & -180\frac{15}{20} & = & -180\frac{15}{20} \\ \hline & & & & 6\frac{7}{20} \end{array}$$

Thus, $h = 6\frac{7}{20}$.

Check. We add Rocky's excess height to his daughter's height:

$$180\frac{3}{4} + 6\frac{7}{20} = 180\frac{15}{20} + 6\frac{7}{20} = 186\frac{22}{20} = 187\frac{2}{20} = 187\frac{1}{10}$$

The answer checks.

State. Rocky is $6\frac{7}{20}$ cm taller.

45. ***Familiarize.*** We draw a picture, letting x = the amount of pipe that was used.

$$\vdash\!\!-\!\!-\, 10\frac{5}{16}\text{ ft} \,-\!\!-\!\!+\!\!-\!\!-\, 8\frac{3}{4}\text{ ft} \,-\!\!-\!\!\dashv$$

$$\vdash\!\!-\!\!-\!\!-\!\!-\!\!-\, x \,-\!\!-\!\!-\!\!-\!\!-\!\!\dashv$$

Translate. We write an addition sentence.

$$\underbrace{\text{First length}}_{\downarrow \; 10\frac{5}{16}} \quad \underset{+}{\overset{\text{plus}}{\downarrow}} \quad \underbrace{\text{Second length}}_{\downarrow \; 8\frac{3}{4}} \quad \underset{=}{\overset{\text{is}}{\downarrow}} \quad \underbrace{\text{Total length}}_{\downarrow \; x}$$

Solve. We carry out the addition. The LCD is 16.

$$\begin{array}{rcr} 10\frac{5}{16} & = & 10\frac{5}{16} \\ +\,8\boxed{\frac{3}{4}\cdot\frac{4}{4}} & = & +\,8\frac{12}{16} \\ \hline & & 18\frac{17}{16} = 18 + \frac{17}{16} \\ & & = 18 + 1\frac{1}{16} \\ & & = 19\frac{1}{16} \end{array}$$

Thus, $x = 19\frac{1}{16}$.

Check. We repeat the calculation. We also note that the total length is larger than either of the individual lengths, so the answer seems reasonable.

State. Janet used $19\frac{1}{16}$ ft of pipe.

47. ***Familiarize.*** We let d = the amount by which the price dropped.

Translate. We translate as follows:

$$\underbrace{\text{Opening price}}_{\downarrow \; 28\frac{7}{8}} \quad \underset{-}{\overset{-}{\downarrow}} \quad \underbrace{\text{Closing price}}_{\downarrow \; 27\frac{1}{4}} \quad \underset{=}{\overset{=}{\downarrow}} \quad \underbrace{\text{Drop in price}}_{\downarrow \; d}$$

Solve. We carry out the subtraction. The LCD is 8.

$$\begin{array}{rcrcr} 28\frac{7}{8} & = & 28\frac{7}{8} & = & 28\frac{7}{8} \\ -27\frac{1}{4} & = & -27\boxed{\frac{1}{4}\cdot\frac{2}{2}} & = & -27\frac{2}{8} \\ \hline & & & & 1\frac{5}{8} \end{array}$$

Thus, $d = 1\frac{5}{8}$.

Check. To check we add the drop in price to the closing price.

$$27\frac{1}{4} + 1\frac{5}{8} = 27\frac{2}{8} + 1\frac{5}{8} = 28\frac{7}{8}$$

This checks.

State. The stock price dropped $\$1\frac{5}{8}$.

49. ***Familiarize.*** We let h = Renée's height.

Translate. We translate as follows:

$$\underbrace{\text{Daughter's height}}_{\downarrow \; 169\frac{3}{10}} \quad \underset{+}{\overset{+}{\downarrow}} \quad \underbrace{\text{Renée's additional height}}_{\downarrow \; 4\frac{1}{2}} \quad \underset{=}{\overset{=}{\downarrow}} \quad \underbrace{\text{Renée's height}}_{\downarrow \; h}$$

Solve. We carry out the addition. The LCD is 10.

$$\begin{array}{rcrcr} 169\frac{3}{10} & = & 169\frac{3}{10} & = & 169\frac{3}{10} \\ +\,4\frac{1}{2} & = & +\,4\boxed{\frac{1}{2}\cdot\frac{5}{5}} & = & +\,4\frac{5}{10} \\ \hline & & & & 173\frac{8}{10} = 173\frac{4}{5} \end{array}$$

Thus, $h = 173\frac{4}{5}$.

Check. We can repeat the calculation. Also note that the answer is greater than the daughter's height as we would expect.

State. Renée is $173\frac{4}{5}$ cm tall.

51. ***Familiarize.*** We draw a picture.

$8\frac{1}{2}$ in.

$9\frac{3}{4}$ in. $9\frac{3}{4}$ in.

$8\frac{1}{2}$ in.

Translate. We let D = the distance around the cover of the book.

$$\underbrace{\text{Top distance}}_{8\frac{1}{2}} \ \text{plus}\ \underbrace{\text{Right-side distance}}_{9\frac{3}{4}} \ \text{plus}\ \underbrace{\text{Bottom distance}}_{8\frac{1}{2}} \ \text{plus}\ \underbrace{\text{Left-side distance}}_{9\frac{3}{4}} \ \text{is}\ \underbrace{\text{Total distance}}_{D}$$

$$8\frac{1}{2} + 9\frac{3}{4} + 8\frac{1}{2} + 9\frac{3}{4} = D$$

Solve. To solve we carry out the addition. The LCD is 4.

$$\begin{array}{rcl} 8\,\boxed{\frac{1}{2}\cdot\frac{2}{2}} & = & 8\,\frac{2}{4} \quad \text{and} \\ 9\,\frac{3}{4} & = & 9\,\frac{3}{4}. \\ 8\,\frac{1}{2} & = & 8\,\frac{2}{4} \\ +9\,\frac{3}{4} & = & +\;9\,\frac{3}{4} \\ \hline & & 34\,\frac{10}{4} = 36\frac{2}{4} = 36\frac{1}{2} \end{array}$$

Thus, $D = 36\frac{1}{2}$.

Check. We repeat the calculation.

State. The distance around the cover of the book is $36\frac{1}{2}$ in.

53. ***Familiarize.*** We draw a picture. We let d = the plane's distance from its original point of departure.

640 km

d $\quad 320\frac{3}{10}$ km

Translate. We translate as follows:

$$\underbrace{\text{Distance originally flown}}_{640} - \underbrace{\text{Distance of return flight}}_{320\frac{3}{10}} = \underbrace{\text{Distance from original departure point}}_{d}$$

$$640 - 320\frac{3}{10} = d$$

Solve. We carry out the subtraction.

$$\begin{array}{rcl} 640 & = & 639\,\frac{10}{10} \\ -\;320\,\frac{3}{10} & = & -\;320\,\frac{3}{10} \\ \hline & & 319\,\frac{7}{10} \end{array}$$

Thus, $d = 319\frac{7}{10}$.

Check. We add the distance from the original point of departure to the distance of the return flight.

$$320\frac{3}{10} + 319\frac{7}{10} = 639\frac{10}{10} = 640$$

This checks.

State. The plane is $319\frac{7}{10}$ km from its original point of departure.

55. ***Familiarize.*** We make a drawing. We let t = the number of hours the teaching assistant worked on the third day.

$2\frac{1}{2}$ hr $\quad 4\frac{1}{5}$ hr $\quad t$

$10\frac{1}{2}$ hr

Translate. We write an addition sentence.

$$2\frac{1}{2} + 4\frac{1}{5} + t = 10\frac{1}{2}$$

Solve. This is a two-step problem.

a) First we add $2\frac{1}{2} + 4\frac{1}{5}$ to find the time worked on the first two days. The LCD is 10.

$$\begin{array}{rcl} 2\,\boxed{\frac{1}{2}\cdot\frac{5}{5}} & = & 2\,\frac{5}{10} \\ +\;4\,\boxed{\frac{1}{5}\cdot\frac{2}{2}} & = & +\;4\,\frac{2}{10} \\ \hline & & 6\,\frac{7}{10} \end{array}$$

b) Then we subtract $6\frac{7}{10}$ from $10\frac{1}{2}$ to find the time worked on the third day. The LCD is 10.

$$6\frac{7}{10} + t = 10\frac{1}{2}$$

$$t = 10\frac{1}{2} - 6\frac{7}{10}$$

$$\begin{array}{rcccl} 10\,\boxed{\frac{1}{2}\cdot\frac{5}{5}} & = & 10\,\frac{5}{10} & = & 9\,\frac{15}{10} \\ -\;6\,\frac{7}{10} & = & -\;6\,\frac{7}{10} & = & -\;6\,\frac{7}{10} \\ \hline & & & & 3\,\frac{8}{10} = 3\frac{4}{5} \end{array}$$

Thus, $t = 3\frac{4}{5}$.

Check. We repeat the calculations.

State. The teaching assistant worked $3\frac{4}{5}$ hr the third day.

57. We see that d and the two smallest distances combined are the same as the largest distance. We translate and solve.

$$2\frac{3}{4} + d + 2\frac{3}{4} = 12\frac{7}{8}$$

$$d = 12\frac{7}{8} - 2\frac{3}{4} - 2\frac{3}{4}$$

$$= 10\frac{1}{8} - 2\frac{3}{4} \quad \text{Subtracting } 2\frac{3}{4} \text{ from } 12\frac{7}{8}$$

$$= 7\frac{3}{8} \quad \text{Subtracting } 2\frac{3}{4} \text{ from } 10\frac{1}{8}$$

The length of d is $7\frac{3}{8}$ ft.

59. We add the lengths of the sides. The LCD is 12.

Perimeter

$$= 5\frac{1}{4}\text{ yd} + 2\frac{1}{3}\text{ yd} + 5\frac{1}{4}\text{ yd} + 2\frac{1}{3}\text{ yd}$$

$$= \left(5\frac{1}{4} + 2\frac{1}{3} + 5\frac{1}{4} + 2\frac{1}{3}\right)\text{ yd}$$

$$= \left(5\frac{3}{12} + 2\frac{4}{12} + 5\frac{3}{12} + 2\frac{4}{12}\right)\text{ yd}$$

$$= 14\frac{14}{12}\text{ yd} = 15\frac{2}{12}\text{ yd}$$

$$= 15\frac{1}{6}\text{ yd}$$

61. $8\frac{3}{5} - 9\frac{2}{5} = 8\frac{3}{5} + \left(-9\frac{2}{5}\right)$

Since $9\frac{2}{5}$ is greater than $8\frac{3}{5}$, the answer will be negative. The difference in absolute values is

$$\begin{array}{rcr} 9\frac{2}{5} & = & 8\frac{7}{5} \\ -8\frac{3}{5} & = & -8\frac{3}{5} \\ \hline & & \frac{4}{5} \end{array}$$

so $8\frac{3}{5} - 9\frac{2}{5} = -\frac{4}{5}$.

63. $3\frac{1}{2} - 6\frac{3}{4} = 3\frac{1}{2} + \left(-6\frac{3}{4}\right)$

Since $6\frac{3}{4}$ is greater than $3\frac{1}{2}$, the answer will be negative. The difference in absolute values is

$$\begin{array}{rcrcr} 6\frac{3}{4} & = & 6\frac{3}{4} & = & 6\frac{3}{4} \\ -3\frac{1}{2} & = & -3\boxed{\frac{1}{2}\cdot\frac{2}{2}} & = & -3\frac{2}{4} \\ \hline & & & & 3\frac{1}{4} \end{array}$$

so $3\frac{1}{2} - 6\frac{3}{4} = -3\frac{1}{4}$.

65. $3\frac{4}{5} - 7\frac{2}{3} = 3\frac{4}{5} + \left(-7\frac{2}{3}\right)$

Since $7\frac{2}{3}$ is greater than $3\frac{4}{5}$, the answer will be negative. The difference in absolute values is

$$\begin{array}{rcrcrcr} 7\frac{2}{3} & = & 7\boxed{\frac{2}{3}\cdot\frac{5}{5}} & = & 7\frac{10}{15} & = & 6\frac{25}{15} \\ -3\frac{4}{5} & = & -3\boxed{\frac{4}{5}\cdot\frac{3}{3}} & = & -3\frac{12}{15} & = & -3\frac{12}{15} \\ \hline & & & & & & 3\frac{13}{15} \end{array}$$

so $3\frac{4}{5} - 7\frac{2}{3} = -3\frac{13}{15}$.

67. $-3\frac{1}{5} - 4\frac{2}{5} = -3\frac{1}{5} + \left(-4\frac{2}{5}\right)$

We add the absolute values and make the answer negative.

$$\begin{array}{r} 3\frac{1}{5} \\ +4\frac{2}{5} \\ \hline 7\frac{3}{5} \end{array}$$

Thus, $-3\frac{1}{5} - 4\frac{2}{5} = -7\frac{3}{5}$.

69. $-4\frac{2}{5} - 6\frac{3}{7} = -4\frac{2}{5} + \left(-6\frac{3}{7}\right)$

We add the absolute values and make the answer negative.

$$\begin{array}{rcrcr} 4\frac{2}{5} & = & 4\boxed{\frac{2}{5}\cdot\frac{7}{7}} & = & 4\frac{14}{35} \\ +6\frac{3}{7} & = & +6\boxed{\frac{3}{7}\cdot\frac{5}{5}} & = & +6\frac{15}{35} \\ \hline & & & & 10\frac{29}{35} \end{array}$$

Thus, $-4\frac{2}{5} - 6\frac{3}{7} = -10\frac{29}{35}$.

71. $-6\frac{1}{9} - \left(-4\frac{2}{9}\right) = -6\frac{1}{9} + 4\frac{2}{9}$

Since $-6\frac{1}{9}$ has the greater absolute value, the answer will be negative. The difference in absolute values is

$$\begin{array}{rcr} 6\frac{1}{9} & = & 5\frac{10}{9} \\ -4\frac{2}{9} & = & -4\frac{2}{9} \\ \hline & & 1\frac{8}{9} \end{array}$$

so $-6\frac{1}{9} - \left(-4\frac{2}{9}\right) = -1\frac{8}{9}$.

73. $\dfrac{12}{25} \div \dfrac{24}{5} = \dfrac{12}{25} \cdot \dfrac{5}{24} = \dfrac{12 \cdot 5}{25 \cdot 24} = \dfrac{12 \cdot 5 \cdot 1}{5 \cdot 5 \cdot 12 \cdot 2} =$

$\dfrac{12 \cdot 5}{12 \cdot 5} \cdot \dfrac{1}{5 \cdot 2} = \dfrac{1}{10}$

75.

77. Use a calculator.

$$\begin{array}{rcrcr} 3289\frac{1047}{1189} & = & 3289\ \ \frac{1047}{1189} & = & 3289\frac{1047}{1189} \\ +5278\frac{32}{41} & = & +5278\boxed{\frac{32}{41}\cdot\frac{29}{29}} & = & +5278\frac{928}{1189} \\ \hline & & & & 8567\frac{1975}{1189} = 8568\frac{786}{1189} \end{array}$$

79. ***Familiarize***. We visualize the situation.

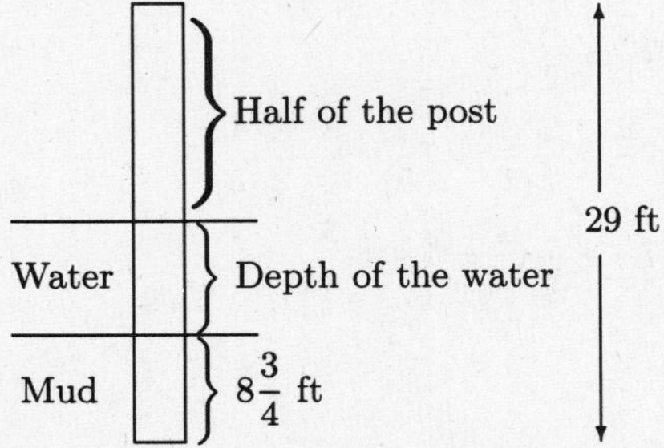

Translate and Solve. First we find the length of the post that extends above the water's surface. We let $p =$ this length.

$$\underbrace{\text{Half}}_{\frac{1}{2}}\ \underbrace{\text{of}}_{\cdot}\ \underbrace{\text{the post}}_{29}\ \underbrace{\text{is}}_{=}\ \underbrace{\text{above the water}}_{p}$$

To solve we carry out the multiplication.

$$\frac{1}{2} \cdot 29 = p$$

$$\frac{29}{2} = p$$

$$14\frac{1}{2} = p$$

Now we find the depth of the water. We let $d =$ the depth.

$$\underbrace{\text{Length of post above water}}_{14\frac{1}{2}}\ \underbrace{\text{plus}}_{+}\ \underbrace{\text{Depth of water}}_{d}\ \underbrace{\text{plus}}_{+}\ \underbrace{\text{Length of post in mud}}_{8\frac{3}{4}}\ \underbrace{\text{is}}_{=}\ \underbrace{\text{Total length of post}}_{29}$$

We solve the equation.

$$14\frac{1}{2} + d + 8\frac{3}{4} = 29$$

$23\frac{1}{4} + d = 29$ Adding on the left side

$d = 29 - 23\frac{1}{4}$ Subtracting $23\frac{1}{4}$ on both sides

$d = 5\frac{3}{4}$ Simplifying

Check. We repeat the calculations.

State. The water is $5\frac{3}{4}$ ft deep at that location.

Exercise Set 4.7

1. $8 \cdot 2\frac{5}{6}$

$= \frac{8}{1} \cdot \frac{17}{6}$ Writing fractional notation

$= \dfrac{8 \cdot 17}{1 \cdot 6} = \dfrac{2 \cdot 4 \cdot 17}{1 \cdot 2 \cdot 3} = \dfrac{2}{2} \cdot \dfrac{4 \cdot 17}{1 \cdot 3} = \dfrac{68}{3} = 22\dfrac{2}{3}$

3. $3\frac{5}{8} \cdot \frac{2}{3}$

$= \frac{29}{8} \cdot \frac{2}{3}$ Writing fractional notation

$= \dfrac{29 \cdot 2}{8 \cdot 3} = \dfrac{29 \cdot 2}{2 \cdot 4 \cdot 3} = \dfrac{2}{2} \cdot \dfrac{29}{4 \cdot 3} = \dfrac{29}{12} = 2\dfrac{5}{12}$

5. $-9 \cdot 4\frac{2}{5} = -9 \cdot \frac{22}{5} = -\frac{198}{5} = -39\frac{3}{5}$

7. $4\frac{1}{3} \cdot 6\frac{2}{5} = \frac{13}{3} \cdot \frac{32}{5} = \frac{416}{15} = 27\frac{11}{15}$

9. $3\frac{1}{2} \cdot 2\frac{1}{3} = \frac{7}{2} \cdot \frac{7}{3} = \frac{49}{6} = 8\frac{1}{6}$

11. $-3\frac{2}{5} \cdot 2\frac{7}{8} = -\frac{17}{5} \cdot \frac{23}{8} = -\frac{391}{40} = -9\frac{31}{40}$

13. $4\frac{7}{10} \cdot 5\frac{3}{10} = \frac{47}{10} \cdot \frac{53}{10} = \frac{2491}{100} = 24\frac{91}{100}$

15. $\left(-20\frac{1}{2}\right)\left(-10\frac{1}{5}\right) = -\frac{41}{2}\left(-\frac{51}{5}\right) = \frac{2091}{10} = 209\frac{1}{10}$

17. $20 \div 3\frac{1}{5}$

$= 20 \div \frac{16}{5}$ Writing fractional notation

$= 20 \cdot \frac{5}{16}$ Multiplying by the reciprocal

$= \dfrac{20 \cdot 5}{16} = \dfrac{4 \cdot 5 \cdot 5}{4 \cdot 4} = \dfrac{4}{4} \cdot \dfrac{5 \cdot 5}{4} = \dfrac{25}{4} = 6\dfrac{1}{4}$

19. $8\frac{2}{5} \div 7$

$= \frac{42}{5} \div 7$ Writing fractional notation

$= \frac{42}{5} \cdot \frac{1}{7}$ Multiplying by the reciprocal

$= \frac{42 \cdot 1}{5 \cdot 7} = \frac{6 \cdot 7}{5 \cdot 7} = \frac{7}{7} \cdot \frac{6}{5} = \frac{6}{5} = 1\frac{1}{5}$

21. $4\frac{3}{4} \div 1\frac{1}{3} = \frac{19}{4} \div \frac{4}{3} = \frac{19}{4} \cdot \frac{3}{4} = \frac{19 \cdot 3}{4 \cdot 4} = \frac{57}{16} = 3\frac{9}{16}$

23. $-1\frac{7}{8} \div 1\frac{2}{3} = -\frac{15}{8} \div \frac{5}{3} = -\frac{15}{8} \cdot \frac{3}{5} = -\frac{15 \cdot 3}{8 \cdot 5} = -\frac{5 \cdot 3 \cdot 3}{8 \cdot 5}$

$= \frac{5}{5}\left(-\frac{3 \cdot 3}{8}\right) = -\frac{3 \cdot 3}{8} = -\frac{9}{8} = -1\frac{1}{8}$

25. $5\frac{1}{10} \div 4\frac{3}{10} = \frac{51}{10} \div \frac{43}{10} = \frac{51}{10} \cdot \frac{10}{43} = \frac{51 \cdot 10}{10 \cdot 43}$

$= \frac{10}{10} \cdot \frac{51}{43} = \frac{51}{43} = 1\frac{8}{43}$

27. $20\frac{1}{4} \div (-90) = \frac{81}{4} \div (-90) = \frac{81}{4}\left(-\frac{1}{90}\right) = -\frac{81 \cdot 1}{4 \cdot 90} =$

$-\frac{9 \cdot 9 \cdot 1}{4 \cdot 9 \cdot 10} = \frac{9}{9} \cdot \left(-\frac{9 \cdot 1}{4 \cdot 10}\right) = -\frac{9}{40}$

29. $mv = 7 \cdot 3\frac{2}{5}$

$= 7 \cdot \frac{17}{5}$

$= \frac{119}{5} = 23\frac{4}{5}$

31. $vt = 5\frac{2}{3}\left(-2\frac{3}{8}\right)$

$= \frac{17}{3}\left(-\frac{19}{8}\right) = -\frac{17 \cdot 19}{3 \cdot 8}$

$= -\frac{323}{24} = -13\frac{11}{24}$

33. $M \div NP = 2\frac{1}{4} \div (-5)\left(2\frac{1}{3}\right)$

$= \frac{9}{4} \div (-5)\left(\frac{7}{3}\right)$

$= \frac{9}{4} \cdot \left(-\frac{1}{5}\right)\left(\frac{7}{3}\right)$

$= -\frac{9 \cdot 1}{4 \cdot 5} \cdot \frac{7}{3} = -\frac{9 \cdot 1 \cdot 7}{4 \cdot 5 \cdot 3}$

$= -\frac{3 \cdot 3 \cdot 1 \cdot 7}{4 \cdot 5 \cdot 3} = -\frac{3 \cdot 1 \cdot 7}{4 \cdot 5} \cdot \frac{3}{3}$

$= -\frac{21}{20} = -1\frac{1}{20}$

35. $a - bc = 18 - 2\frac{1}{5} \cdot 3\frac{3}{4}$

$= 18 - \frac{11}{5} \cdot \frac{15}{4}$

$= 18 - \frac{11 \cdot 15}{5 \cdot 4}$

$= 18 - \frac{11 \cdot 3 \cdot 5}{5 \cdot 4} = 18 - \frac{11 \cdot 3}{4} \cdot \frac{5}{5}$

$= 18 - \frac{33}{4} = \frac{72}{4} - \frac{33}{4}$

$= \frac{39}{4} = 9\frac{3}{4}$

37. $m + n \div p = 7\frac{2}{5} + 4\frac{1}{2} \div 6$

$= 7\frac{2}{5} + \frac{9}{2} \div 6$

$= 7\frac{2}{5} + \frac{9}{2} \cdot \frac{1}{6}$

$= 7\frac{2}{5} + \frac{9 \cdot 1}{2 \cdot 6} = 7\frac{2}{5} + \frac{3 \cdot 3 \cdot 1}{2 \cdot 2 \cdot 3}$

$= 7\frac{2}{5} + \frac{3 \cdot 1}{2 \cdot 2} = 7\frac{2}{5} + \frac{3}{4}$

$= 7\frac{8}{20} + \frac{15}{20} = 7\frac{23}{20}$

$= 8\frac{3}{20}$

39. ***Familiarize***. We draw a picture.

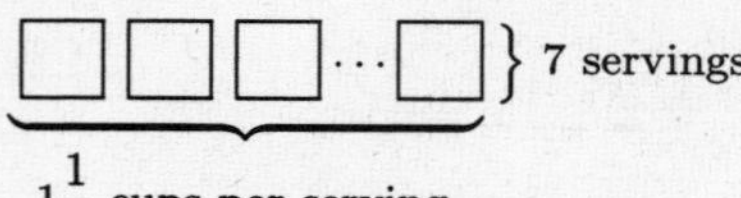

We let s = the number of cups of spaghetti needed to serve 7 people.

Translate. The situation corresponds to a multiplication sentence.

$s = 7 \cdot \frac{1}{12}$

Solve. We carry out the multiplication.

$s = 7 \cdot 1\frac{1}{2} = 7 \cdot \frac{3}{2} = \frac{7 \cdot 3}{2}$

$= \frac{21}{2} = 10\frac{1}{2}$

Check. We repeat the calculation.

State. $10\frac{1}{2}$ cups of spaghetti is needed to serve 7 people.

41. ***Familiarize***. We draw a picture.

$\frac{1}{3}$ lb	$\frac{1}{3}$ lb	....	$\frac{1}{3}$ lb

⟵ $5\frac{1}{2}$ lb ⟶

We let s = the number of servings that can be prepared from $5\frac{1}{2}$ lb of flounder fillet.

Translate. The situation corresponds to a division sentence.

$$s = 5\frac{1}{2} \div \frac{1}{3}$$

Solve. We carry out the division.

$$s = 5\frac{1}{2} \div \frac{1}{3} = \frac{11}{2} \div \frac{1}{3}$$
$$= \frac{11}{2} \cdot \frac{3}{1} = \frac{33}{2}$$
$$= 16\frac{1}{2}$$

Check. We check by multiplying. If $16\frac{1}{2}$ servings are prepared, then

$$16\frac{1}{2} \cdot \frac{1}{3} = \frac{33}{2} \cdot \frac{1}{3} = \frac{3 \cdot 11 \cdot 1}{2 \cdot 3} = \frac{3}{3} \cdot \frac{11 \cdot 1}{2} = \frac{11}{2} = 5\frac{1}{2} \text{ lb}$$

of flounder is used. Our answer checks.

State. $16\frac{1}{2}$ servings can be prepared from $5\frac{1}{2}$ lb of flounder fillet.

43. ***Familiarize.*** We let w = the weight of $5\frac{1}{2}$ cubic feet of water.

Translate. We write an equation.

$$\underbrace{\text{Weight per cubic foot}} \cdot \underbrace{\text{Number of cubic feet}} = \underbrace{\text{Total weight}}$$
$$62\frac{1}{2} \cdot 5\frac{1}{2} = w$$

Solve. To solve the equation we carry out the multiplication.

$$w = 62\frac{1}{2} \cdot 5\frac{1}{2}$$
$$= \frac{125}{2} \cdot \frac{11}{2} = \frac{125 \cdot 11}{2 \cdot 2}$$
$$= \frac{1375}{4} = 343\frac{3}{4}$$

Check. We repeat the calculation. We also note that $62\frac{1}{2} \approx 60$ and $5\frac{1}{2} \approx 5$. Then the product is about 300. Our answer seems reasonable.

State. The weight of $5\frac{1}{2}$ cubic feet of water is $343\frac{3}{4}$ lb.

45. ***Familiarize.*** We let t = the number of inches of tape used in 60 sec of recording.

Translate. We write an equation.

$$\underbrace{\text{Inches per second}} \cdot \underbrace{\text{Number of seconds}} = \underbrace{\text{Tape used}}$$
$$1\frac{3}{8} \cdot 60 = t$$

Solve. We carry out the multiplication.

$$t = 1\frac{3}{8} \cdot 60 = \frac{11}{8} \cdot 60$$
$$= \frac{11 \cdot 4 \cdot 15}{2 \cdot 4} = \frac{11 \cdot 15}{2} \cdot \frac{4}{4}$$
$$= \frac{165}{2} = 82\frac{1}{2}$$

Check. We repeat the calculation.

State. $82\frac{1}{2}$ in. of tape are used in 60 sec of recording in short-play mode.

47. ***Familiarize.*** We let t = the Fahrenheit temperature.

Translate.

$$\underbrace{\text{Celsius temperature}} \text{ times } 1\frac{4}{5} \text{ plus } 32° \text{ is } \underbrace{\text{Fahrenheit temperature}}$$
$$20 \cdot 1\frac{4}{5} + 32 = t$$

Solve. We multiply and then add, according to the rules for order of operations.

$$t = 20 \cdot 1\frac{4}{5} + 32 = \frac{20}{1} \cdot \frac{9}{5} + 32 = \frac{20 \cdot 9}{1 \cdot 5} + 32 =$$
$$\frac{4 \cdot 5 \cdot 9}{1 \cdot 5} + 32 = \frac{5}{5} \cdot \frac{4 \cdot 9}{1} + 32 = 36 + 32 = 68$$

Check. We repeat the calculation.

State. 68° Fahrenheit corresponds to 20° Celsius.

49. ***Familiarize, Translate, and Solve.*** To find the ingredients for $\frac{1}{2}$ recipe, we multiply each ingredient by $\frac{1}{2}$.

$$2 \cdot \frac{1}{2} = \frac{2 \cdot 1}{2} = \frac{2}{2} = 1$$
$$1\frac{1}{2} \cdot \frac{1}{2} = \frac{3}{2} \cdot \frac{1}{2} = \frac{3 \cdot 1}{2 \cdot 2} = \frac{3}{4}$$
$$3 \cdot \frac{1}{2} = \frac{3 \cdot 1}{2} = \frac{3}{2} = 1\frac{1}{2}$$
$$2\frac{1}{2} \cdot \frac{1}{2} = \frac{5}{2} \cdot \frac{1}{2} = \frac{5 \cdot 1}{2 \cdot 2} = \frac{5}{4} = 1\frac{1}{4}$$
$$1 \cdot \frac{1}{2} = \frac{1 \cdot 1}{2} = \frac{1}{2}$$
$$\frac{1}{3} \cdot \frac{1}{2} = \frac{1 \cdot 1}{3 \cdot 2} = \frac{1}{6}$$
$$\frac{1}{4} \cdot \frac{1}{2} = \frac{1 \cdot 1}{4 \cdot 2} = \frac{1}{8}$$

Check. We repeat the calculation.

State. The ingredients for $\frac{1}{2}$ recipe are 1 chicken bouillon cube, $\frac{3}{4}$ cup hot water, $1\frac{1}{2}$ tablespoons margarine, $1\frac{1}{2}$ tablespoons flour, $1\frac{1}{4}$ cups diced cooked chicken, $\frac{1}{2}$ cup cooked peas, $\frac{1}{2}$ 4-oz can sliced mushrooms (drained), $\frac{1}{6}$ cup sliced cooked carrots, $\frac{1}{8}$ cup chopped onions, 1 tablespoon chopped pimiento, $\frac{1}{2}$ teaspoon salt.

Familiarize, Translate, and Solve. To find the ingredients for 3 recipes, we multiply each ingredient by 3.

$$2 \cdot 3 = 6$$

$$1\frac{1}{2} \cdot 3 = \frac{3}{2} \cdot \frac{3}{1} = \frac{3 \cdot 3}{2 \cdot 1} = \frac{9}{2} = 4\frac{1}{2}$$

$$3 \cdot 3 = 9$$

$$2\frac{1}{2} \cdot 3 = \frac{5}{2} \cdot \frac{3}{1} = \frac{5 \cdot 3}{2 \cdot 1} = \frac{15}{2} = 7\frac{1}{2}$$

$$1 \cdot 3 = 3$$

$$\frac{1}{3} \cdot 3 = \frac{1 \cdot 3}{3} = \frac{3}{3} = 1$$

$$\frac{1}{4} \cdot 3 = \frac{1 \cdot 3}{4} = \frac{3}{4}$$

Check. We repeat the calculation.

State. The ingredients for 3 recipes are 6 chicken bouillon cubes, $4\frac{1}{2}$ cups hot water, 9 tablespoons margarine, 9 tablespoons flour, $7\frac{1}{2}$ cups diced cooked chicken, 3 cups cooked peas, 3 4-oz cans sliced mushrooms (drained), 1 cup sliced cooked carrots, $\frac{3}{4}$ cup chopped onions, 6 tablespoons chopped pimiento, 3 teaspoons salt.

51. ***Familiarize.*** Visualize the situation as a rectangular array containing 24 hours with $1\frac{1}{2}$ hours in each row. We must determine how many rows the array has. (The last row may be incomplete.) We let n = the number of orbits made every 24 hours.

Translate. The division that corresponds to this situation is

$$24 \div 1\frac{1}{2} = n.$$

Solve. We carry out the division.

$$n = 24 \div 1\frac{1}{2} = 24 \div \frac{3}{2} = 24 \cdot \frac{2}{3} = \frac{24 \cdot 2}{3} = \frac{3 \cdot 8 \cdot 2}{3 \cdot 1} =$$

$$\frac{3}{3} \cdot \frac{8 \cdot 2}{1} = 16$$

Check. We check by multiplying the number of orbits by the time it takes to make one orbit.

$$16 \cdot 1\frac{1}{2} = 16 \cdot \frac{3}{2} = \frac{16 \cdot 3}{2} = \frac{2 \cdot 8 \cdot 3}{2 \cdot 1} = \frac{2}{2} \cdot \frac{8 \cdot 3}{1} = 24$$

The answer checks.

State. The shuttle makes 16 orbits every 24 hr.

53. ***Familiarize.*** The question can be regarded as asking how many times 213 can be divided by $14\frac{2}{10}$. We let m = the number of miles per gallon.

Translate. The situation corresponds to a division sentence.

$$m = 213 \div 14\frac{2}{10}$$

Solve. We carry out the division.

$$m = 213 \div 14\frac{2}{10} = 213 \div \frac{142}{10}$$

$$= 213 \cdot \frac{10}{142} = \frac{3 \cdot 71 \cdot 2 \cdot 5}{2 \cdot 71 \cdot 1}$$

$$= \frac{3 \cdot 5}{1} \cdot \frac{71 \cdot 2}{71 \cdot 2} = 15$$

Check. We check by multiplying.

$$15 \cdot 14\frac{2}{10} = 15 \cdot \frac{142}{10} = \frac{3 \cdot 5 \cdot 2 \cdot 71}{2 \cdot 5 \cdot 1}$$

$$\frac{3 \cdot 71}{1} = 213$$

Our answer checks.

State. Toni's taxi got 15 miles per gallon.

55. ***Familiarize.*** The figure is composed of two rectangles. One has dimensions s by $\frac{1}{2} \cdot s$, or $6\frac{7}{8}$ in. by $\frac{1}{2} \cdot 6\frac{7}{8}$ in. The other has dimensions $\frac{1}{2} \cdot s$ by $\frac{1}{2} \cdot s$, or $\frac{1}{2} \cdot 6\frac{7}{8}$ in. by $\frac{1}{2} \cdot 6\frac{7}{8}$ in. The total area is the sum of the areas of these two rectangles. We let A = the total area.

Translate. We write an equation.

$$A = \left(6\frac{7}{8}\right) \cdot \left(\frac{1}{2} \cdot 6\frac{7}{8}\right) + \left(\frac{1}{2} \cdot 6\frac{7}{8}\right) \cdot \left(\frac{1}{2} \cdot 6\frac{7}{8}\right)$$

Solve. We carry out each multiplication and then add.

$$A = \left(6\frac{7}{8}\right) \cdot \left(\frac{1}{2} \cdot 6\frac{7}{8}\right) + \left(\frac{1}{2} \cdot 6\frac{7}{8}\right) \cdot \left(\frac{1}{2} \cdot 6\frac{7}{8}\right)$$

$$= \frac{55}{8} \cdot \left(\frac{1}{2} \cdot \frac{55}{8}\right) + \left(\frac{1}{2} \cdot \frac{55}{8}\right) \cdot \left(\frac{1}{2} \cdot \frac{55}{8}\right)$$

$$= \frac{55}{8} \cdot \frac{55}{16} + \frac{55}{16} \cdot \frac{55}{16}$$

$$= \frac{3025}{128} + \frac{3025}{256} = \frac{3025}{128} \cdot \frac{2}{2} + \frac{3025}{256}$$

$$= \frac{6050}{256} + \frac{3025}{256} = \frac{9075}{256}$$

$$= 35\frac{115}{256}$$

Check. We repeat the calculation.

State. The area is $35\frac{115}{256}$ in^2.

57. ***Familiarize.*** We make a drawing.

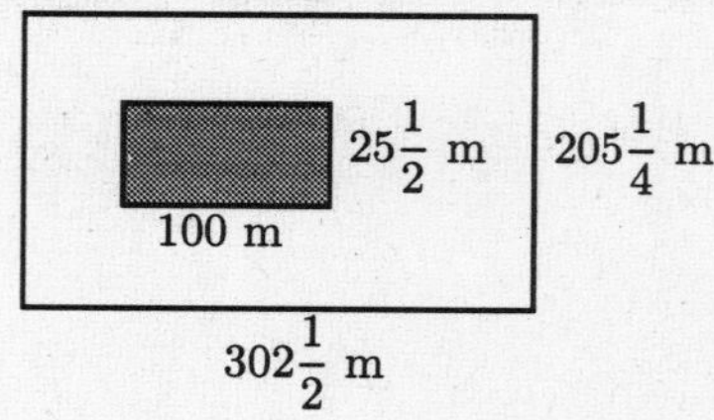

Translate. We let $A =$ the area of the lot not covered by the building.

$$\underbrace{\text{Area left over}}_{\downarrow} \quad \underset{\downarrow}{\text{is}} \quad \underbrace{\text{Area of lot}}_{\downarrow} \quad \underset{\downarrow}{\text{minus}} \quad \underbrace{\text{Area of building}}_{\downarrow}$$

$$A \quad = \left(302\frac{1}{2}\right) \cdot \left(205\frac{1}{4}\right) \quad - \quad (100) \cdot \left(25\frac{1}{2}\right)$$

Solve. We do each multiplication and then find the difference.

$$A = \left(302\frac{1}{2}\right) \cdot \left(205\frac{1}{4}\right) - (100) \cdot \left(25\frac{1}{2}\right)$$
$$= \frac{605}{2} \cdot \frac{821}{4} - \frac{100}{1} \cdot \frac{51}{2}$$
$$= \frac{605 \cdot 821}{2 \cdot 4} - \frac{100 \cdot 51}{1 \cdot 2}$$
$$= \frac{605 \cdot 821}{2 \cdot 4} - \frac{2 \cdot 50 \cdot 51}{1 \cdot 2} = \frac{605 \cdot 821}{2 \cdot 4} - \frac{2}{2} \cdot \frac{50 \cdot 51}{1}$$
$$= \frac{496,705}{8} - 2550 = 62,088\frac{1}{8} - 2550$$
$$= 59,538\frac{1}{8}$$

Check. We repeat the calculation.

State. The area left over is $59,538\frac{1}{8}$ m^2.

59. $-8(x-3) = -8 \cdot x - (-8)(3) = -8x - (-24) = -8x + 24$

61. $\left(-\frac{1}{29}\right)(-29) = \frac{-1(-29)}{29} = \frac{29}{29} = 1$

63. $-198 \div (-6) = \frac{-198}{-6} = 33$

65. 

67. Use a calculator.

$15\frac{2}{11} \cdot 23\frac{31}{43} = \frac{167}{11} \cdot \frac{1020}{43} = \frac{167 \cdot 1020}{11 \cdot 43} = \frac{170,340}{473} = 360\frac{60}{473}$

69. $-8 \div \frac{1}{2} + \frac{3}{4} + \left(-5 - \frac{5}{8}\right)^2 = -8 \div \frac{1}{2} + \frac{3}{4} + \left(-\frac{40}{8} - \frac{5}{8}\right)^2 = -8 \div \frac{1}{2} + \frac{3}{4} + \left(-\frac{45}{8}\right)^2 = -8 \div \frac{1}{2} + \frac{3}{4} + \frac{2025}{64} = -8 \cdot 2 + \frac{3}{4} + \frac{2025}{64} = -16 + \frac{3}{4} + \frac{2025}{64} = -\frac{1024}{64} + \frac{48}{64} + \frac{2025}{64} = \frac{1049}{64} = 16\frac{25}{64}$

71. $\frac{1}{3} \div \left(\frac{1}{2} - \frac{1}{5}\right) \times \frac{1}{4} + \frac{1}{6}$

$$= \frac{1}{3} \div \left(\frac{5}{10} - \frac{2}{10}\right) \times \frac{1}{4} + \frac{1}{6}$$
$$= \frac{1}{3} \div \frac{3}{10} \times \frac{1}{4} + \frac{1}{6}$$
$$= \frac{1}{3} \times \frac{10}{3} \times \frac{1}{4} + \frac{1}{6}$$
$$= \frac{10}{9} \times \frac{1}{4} + \frac{1}{6}$$
$$= \frac{2 \times 5 \times 1}{9 \times 2 \times 2} + \frac{1}{6} = \frac{2}{2} \times \frac{5 \times 1}{9 \times 2} + \frac{1}{6}$$
$$= \frac{5}{18} + \frac{1}{6} = \frac{5}{18} + \frac{3}{18} = \frac{8}{18} = \frac{4}{9}$$

73. $ab + ac = 3\frac{1}{4} \cdot 5\frac{1}{3} + 3\frac{1}{4} \cdot 4\frac{5}{8}$

$$= \frac{13}{4} \cdot \frac{16}{3} + \frac{13}{4} \cdot \frac{37}{8}$$
$$= \frac{13 \cdot 4 \cdot 4}{4 \cdot 3} + \frac{13 \cdot 37}{4 \cdot 8}$$
$$= \frac{4}{4} \cdot \frac{13 \cdot 4}{3} + \frac{13 \cdot 37}{4 \cdot 8}$$
$$= \frac{52}{3} + \frac{481}{32}$$
$$= \frac{52}{3} \cdot \frac{32}{32} + \frac{481}{32} \cdot \frac{3}{3}$$
$$= \frac{1664}{3 \cdot 32} + \frac{1443}{32 \cdot 3}$$
$$= \frac{3107}{96} = 32\frac{35}{96}$$

$a(b + c) = 3\frac{1}{4}\left(5\frac{1}{3} + 4\frac{5}{8}\right)$

$$= 3\frac{1}{4}\left(5\frac{8}{24} + 4\frac{15}{24}\right)$$
$$= 3\frac{1}{4} \cdot 9\frac{23}{24}$$
$$= \frac{13}{4} \cdot \frac{239}{24} = \frac{13 \cdot 239}{4 \cdot 24}$$
$$= \frac{3107}{96} = 32\frac{35}{96}$$

75. $a^3 + a^2 = \left(-3\frac{1}{2}\right)^3 + \left(-3\frac{1}{2}\right)^2$

$$= \left(-\frac{7}{2}\right)^3 + \left(-\frac{7}{2}\right)^2$$
$$= -\frac{343}{8} + \frac{49}{4}$$
$$= -\frac{343}{8} + \frac{98}{8}$$
$$= -\frac{245}{8} = -30\frac{5}{8}$$

$$\begin{aligned} a^2(a+1) &= \left(-3\frac{1}{2}\right)^2\left(-3\frac{1}{2}+1\right) \\ &= \left(-\frac{7}{2}\right)^2\left(-3\frac{1}{2}+1\right) \\ &= \frac{49}{4}\left(-3\frac{1}{2}+1\right) \\ &= \frac{49}{4}\left(-2\frac{1}{2}\right) = \frac{49}{4}\left(-\frac{5}{2}\right) \\ &= -\frac{245}{8} = -30\frac{5}{8} \end{aligned}$$

Chapter 5

Decimal Notation

Exercise Set 5.1

1. a) Write a word name for the integer. [Twenty-three]

b) Write "and" for the decimal point — Twenty-three [and]

c) Write a word name for the number to the right of the decimal point, followed by the place value of the last digit. — Twenty-three and [two-tenths]

A word name for 23.2 is twenty-three and two-tenths.

3. Seven → 7, and → ., eighty-nine hundredths → 89

7 . 89

A word name for 7.89 is seven and eighty-nine hundredths.

5. Forty-two → 42, and → ., three hundred fifty-nine thousandths → 359

42 . 359

A word name for 42.359 is forty-two and three hundred fifty-nine thousandths.

7. Negative five hundred twenty-eight and thirty-seven hundredths

9. Write "and 99 cents" as "and $\frac{99}{100}$ dollars." A word name for \$216.99 is two hundred sixteen and $\frac{99}{100}$ dollars.

11. Write "and 2 cents" as "and $\frac{2}{100}$ dollars." A word name for \$47.02 is forty-seven and $\frac{2}{100}$ dollars.

13. 9.$\underline{2}$ 9.2. $\frac{92}{1\underline{0}}$

1 place Move 1 place. 1 zero

$9.2 = \frac{92}{10}$

15. 0.$\underline{17}$ 0.17. $\frac{17}{1\underline{00}}$

2 places Move 2 places. 2 zeros

$0.17 = \frac{17}{100}$

17. 1.$\underline{46}$ 1.46. $\frac{146}{1\underline{00}}$

2 places Move 2 places. 2 zeros

$1.46 = \frac{146}{100}$

19. −304.$\underline{7}$ -304.7. $-\frac{3047}{1\underline{0}}$

1 place Move 1 place. 1 zero

$-304.7 = -\frac{3047}{10}$

21. 3.$\underline{142}$ 3.142. $\frac{3142}{1\underline{000}}$

3 places Move 3 places. 3 zeros

$3.142 = \frac{3142}{1000}$

23. 46.$\underline{03}$ 46.03. $\frac{4603}{1\underline{00}}$

2 places Move 2 places. 2 zeros

$46.03 = \frac{4603}{100}$

25. −0.$\underline{00013}$ -0.00013. $\frac{-13}{1\underline{00,000}}$

5 places Move 5 places. 5 zeros

$-0.00013 = \frac{-13}{100,000}$, or $-\frac{13}{100,000}$

27. 20.$\underline{003}$ 20.003. $\frac{20,003}{1\underline{000}}$

3 places Move 3 places. 3 zeros

$20.003 = \frac{20,003}{1000}$

29. 2.$\underline{0114}$ 2.0114. $\frac{20,114}{1\underline{0,000}}$

4 places Move 4 places. 4 zeros

$2.0114 = \frac{20,114}{10,000}$

31. $-4567.\underline{2}$ -4567.2. $\dfrac{-45,672}{1\underline{0}}$

1 place   Move 1 place.   1 zero

$-4567.2 = \dfrac{-45,672}{10}$, or $-\dfrac{45,672}{10}$

33. $\dfrac{8}{1\underline{0}}$ 0.8.

1 zero   Move 1 place.

$\dfrac{8}{10} = 0.8$

35. $-\dfrac{29}{1\underline{00}}$ −0.29.

2 zeros   Move 2 places.

$-\dfrac{29}{100} = -0.29$

37. $\dfrac{97}{1\underline{0}}$ 9.7.

1 zero   Move 1 place.

$\dfrac{97}{10} = 9.7$

39. $\dfrac{889}{1\underline{00}}$ 8.89.

2 zeros   Move 2 places.

$\dfrac{889}{100} = 8.89$

41. $\dfrac{-2508}{1\underline{0}}$ −250.8.

1 zero   Move 1 place.

$\dfrac{-2508}{10} = -250.8$

43. $-\dfrac{3798}{1\underline{000}}$ −3.798.

3 zeros   Move 3 places.

$-\dfrac{3798}{1000} = -3.798$

45. $\dfrac{78}{10,000}$ 0.0078.

4 zeros   Move 4 places.

$\dfrac{78}{10,000} = 0.0078$

47. $\dfrac{56,788}{100,000}$ 0.56788.

5 zeros   Move 5 places.

$\dfrac{56,788}{100,000} = 0.56788$

49. $\dfrac{2173}{1\underline{00}}$ 21.73.

2 zeros   Move 2 places.

$\dfrac{2173}{100} = 21.73$

51. $-\dfrac{66}{1\underline{00}}$ −0.66.

2 zeros   Move 2 places.

$-\dfrac{66}{100} = -0.66$

53. To compare two numbers in decimal notation, start at the left and compare corresponding digits moving from left to right. When two digits differ, the number with the larger digit is the larger of the two numbers.

0.42

Starting at the left, these digits are the first to differ; 2 is larger than 1.

0.418

Thus, 0.42 is larger. In symbols, 0.42 > 0.418.

55. 0.10

Starting at the left, these digits are the first to differ; 1 is larger than 0.

0.111

Thus, 0.111 is larger. In symbols, 0.111 > 0.1.

57. 0.0009

Starting at the left, these digits are the first to differ, and 1 is larger than 0.

0.001

Thus, 0.001 is larger. In symbols, 0.001 > 0.0009.

59. −234.07

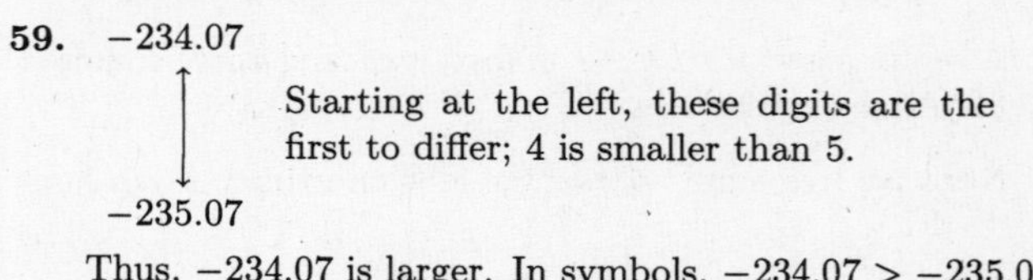

Starting at the left, these digits are the first to differ; 4 is smaller than 5.

−235.07

Thus, −234.07 is larger. In symbols, −234.07 > −235.07.

61. 0.4545

These digits differ; 4 is larger than 0.

0.05454

Thus, 0.4545 is larger. In symbols, 0.4545 > 0.05454.

63. −0.54

These digits differ; 5 is smaller than 7.

−0.78

Thus, −0.54 is larger. In symbols, −0.54 > −0.78.

65. −0.8437

Starting at the left, these digits are the first to differ; 7 is smaller than 8.

−0.84384

Thus, −0.8437 is larger. In symbols, −0.8437 > −0.84384.

67.

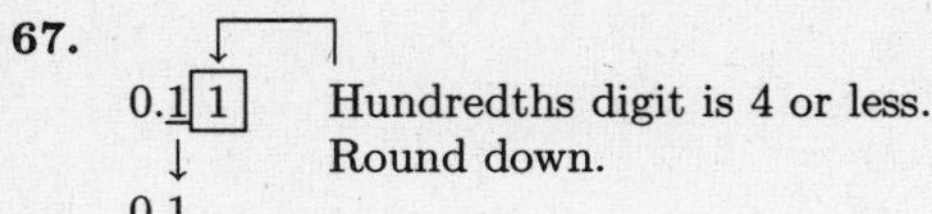

The answer is 0.1.

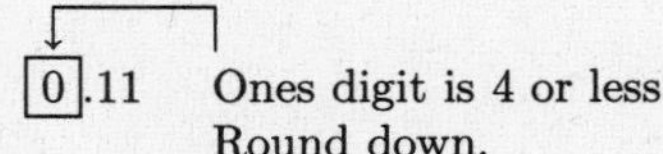

The answer is 0.

69.

Round 0.16 up to 0.2.

The answer is −0.2. Since $-0.2 < -0.16$, we have actually rounded down.

−_0.16 Ones digit is 4 or less.

Round 0.16 down to 0.

The answer is 0. Since $0 > -0.16$, we have actually rounded up.

71.

The answer is 0.6.

_0.5794 Ones digit is 4 or less.
Round down.

The answer is 0.

73.

−2.7449 Hundredths digit is 4 or less.

Round 2.7449 down to 2.7.

The answer is −2.7. Since $-2.7 > -2.7449$, we have actually rounded up.

−_2.7449 Ones digit is 4 or less.

Round 2.7449 down to 0.

The answer is 0. Since $0 > -2.7449$, we have actually rounded up.

75.

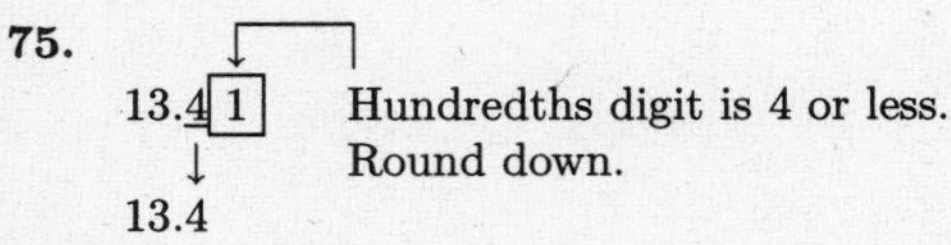

The answer is 13.4.

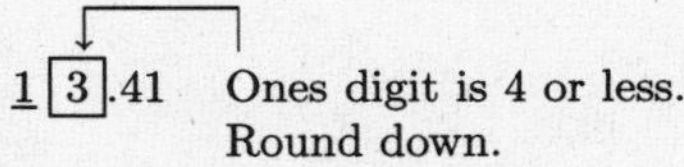

The answer is 10.

77.

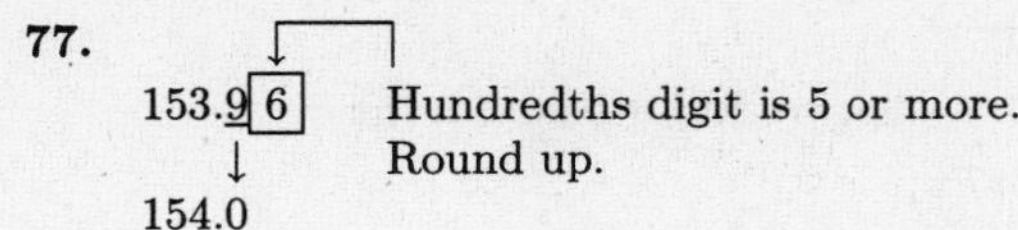

Since the tenths digit becomes 10, we carry 1 to the ones place. The answer is 154.0.

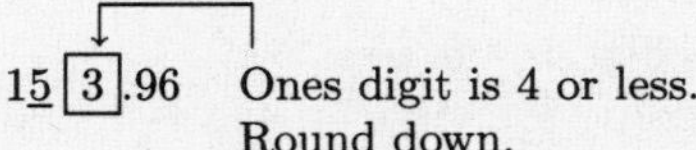

The answer is 150.

79.

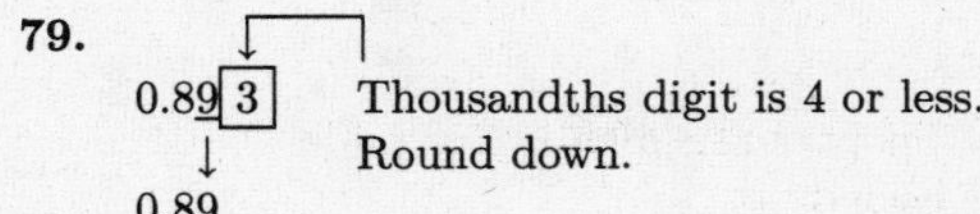

The answer is 0.89.

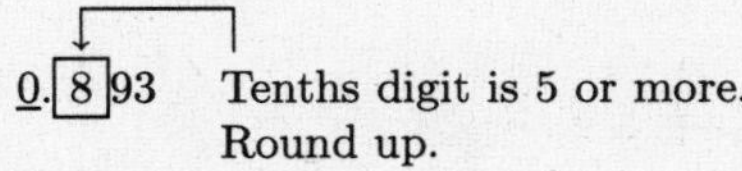

The answer is 1.

81.

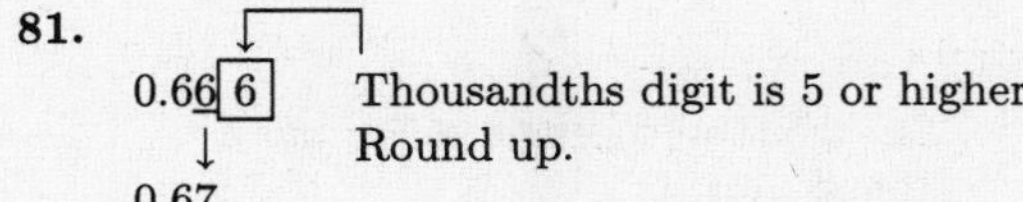

The answer is 0.67.

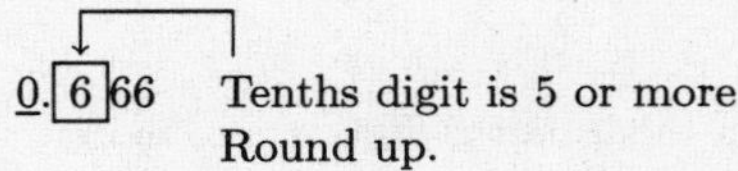

The answer is 1.

83.

−0.4246 Thousandths digit is 4 or less.

Round 0.4246 down to 0.42.

The answer is −0.42. Since $-0.42 > -0.4246$, we have actually rounded up.

−0.4246 Tenths digit is 4 or less.

Round 0.4246 down to 0.

The answer is 0. Since $0 > -0.4246$, we have actually rounded up.

85.

Since the hundredths digit becomes 10, we carry 1 to the tenths place. The answer is 2.40.

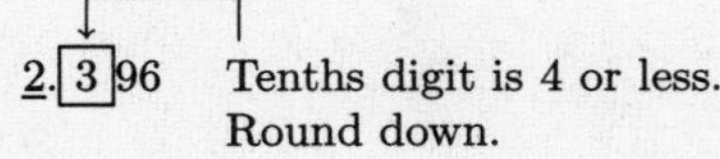

The answer is 2.

87. $-0.00\boxed{7}$ Thousandths digit is 5 or more.

Round 0.007 up to 0.01.

The answer is -0.01. Since $-0.01 < -0.007$, we have actually rounded down.

$-\underline{0}.\boxed{0}07$ Tenths digit is 4 or less.

Round 0.007 down to 0.

The answer is 0. Since $0 > -0.007$, we have actually rounded up.

89. $1.4\underline{2}\boxed{5}1$ Thousandths digit is 5 or more. Round up.

$\downarrow$

1.43

The answer is 1.43.

$\underline{1}.\boxed{4}251$ Ones digit is 4 or less. Round down.

The answer is 1.

91. $0.32\underline{4}\boxed{6}$ Ten-thousandths digit is 5 or more. Round up.

$\downarrow$

0.325

The answer is 0.325.

93. $-0.66\underline{6}\boxed{6}$ Ten-thousandths digit is 5 or more.

Round 0.6666 up to 0.667.

The answer is -0.667. Since $-0.667 < -0.6666$, we have actually rounded down.

95. $17.00\underline{1}\boxed{5}3$ Ten-thousandths digit is 5 or more. Round up.

$\downarrow$

17.002

The answer is 17.002.

97. $0.00\underline{0}\boxed{9}$ Ten-thousandths digit is 5 or more. Round up.

$\downarrow$

0.001

The answer is 0.001.

99. $\frac{0}{n} = 0$, for any integer n that is not 0.

Thus, $\frac{0}{-19} = 0$.

101.
$$\frac{4}{9} - \frac{2}{3} = \frac{4}{9} - \frac{2}{3} \cdot \frac{3}{3}$$
$$= \frac{4}{9} - \frac{6}{9}$$
$$= \frac{-2}{9}, \text{ or } -\frac{2}{9}$$

103.
$$3x - 8 = 21$$
$$3x - 8 + 8 = 21 + 8$$
$$3x = 29$$
$$\frac{3x}{3} = \frac{29}{3}$$
$$x = \frac{29}{3}$$

The solution is $\frac{29}{3}$.

105. ◈

107. $6.7834\underline{6}\boxed{123}$ ←Drop all decimal places past the fifth place.

$\downarrow$

6.78346

The answer is 6.78346.

109. $99.9999\underline{9}\boxed{9999}$ ←Drop all decimal places past the fifth place.

$\downarrow$

99.99999

The answer is 99.99999.

111. $\frac{7}{50} = \frac{7}{50} \cdot \frac{2}{2} = \frac{14}{100}$

$\frac{14}{100} = 0.14$, so $\frac{7}{50} = 0.14$.

Exercise Set 5.2

1.
$$\begin{array}{r} \scriptstyle 1 \\ 3\,1\,6.2\,5 \\ +\ 1\,8.1\,2 \\ \hline 3\,3\,4.3\,7 \end{array}$$

Add hundredths.
Add tenths.
Write a decimal point in the answer.
Add ones.
Add tens.
Add hundreds.

3.
$$\begin{array}{r} \scriptstyle 1\ 1 \\ 6\,5\,9.4\,0\,3 \\ +\ 9\,1\,6.8\,1\,2 \\ \hline 1\,5\,7\,6.2\,1\,5 \end{array}$$

Add thousandths.
Add hundredths.
Add tenths.
Write a decimal point in the answer.
Add ones.
Add tens.
Add hundreds.

5.
$$\begin{array}{r} \scriptstyle 1\ \ \ 1 \\ 9.1\,0\,4 \\ +\,1\,2\,3.4\,5\,6 \\ \hline 1\,3\,2.5\,6\,0 \end{array}$$

7.
$$\begin{array}{r} \scriptstyle 1 \\ 6\,1.0\,0\,6 \\ +\ \ 3.4\,0\,7 \\ \hline 6\,4.4\,1\,3 \end{array}$$

9. Line up the decimal points.

$$\begin{array}{r} 2\,0.0\,1\,2\,4 \\ +\,3\,0.0\,1\,2\,4 \\ \hline 5\,0.0\,2\,4\,8 \end{array}$$

11. Line up the decimal points.

```
  0.8 3 0    Writing an extra zero
+ 0.0 0 5
---------
  0.8 3 5
```

13. Line up the decimal points.

```
      1
      0.3 4 0    Writing an extra zero
      3.5 0 0    Writing 2 extra zeros
      0.1 2 7
+ 7 6 8.0 0 0    Writing in the decimal point
-------------    and 3 extra zeros
  7 7 1.9 6 7    Adding
```

15.

```
  1   1 1
  1 7.0 0 0 0    Writing in the decimal point.
    3.2 4 0 0    You may find it helpful to
    0.2 5 6 0    write extra zeros.
+   0.3 6 8 9
-------------
  2 0.8 6 4 9
```

17.

```
  1 2 1     1
      2.7 0 3 0
    7 8.3 3 0 0
    2 8.0 0 0 9
+ 1 1 8.4 3 4 1
---------------
  2 2 7.4 6 8 0
```

19.

```
  1 2 1     1
      9 9.6 0 0 1
  7 2 8 5.1 8 0 0
    5 0 0.0 4 2 0
+   8 7 0.0 0 0 0
-----------------
  8 7 5 4.8 2 2 1
```

21.

```
  4 12
  5̸.2̸      Borrow ones to subtract tenths.
- 3.9      Subtract tenths.
-----
  1.3      Write a decimal point in the answer.
           Subtract ones.
```

23.

```
  4 11 2 11    Borrow tenths to subtract hundredths.
  5̸ 1̸.3̸ 1̸     Subtract hundredths.
-   2.2 9      Subtract tenths.
---------
  4 9.0 2      Write a decimal point in the answer.
               Borrow tens to subtract ones.
               Subtract ones.
               Subtract tens.
```

25.

```
    4 9 9 10
  2.5̸ 0̸ 0̸ 0̸    Writing 3 extra zeros
- 0.0 0 2 5
-----------
  2.4 9 7 5
```

27.

```
      11
    8 1̸ 13
    9̸ 2̸.3̸ 4 1
-     6.4 2
-----------
    8 5.9 2 1
```

29.

```
  2 9  10 6 14
  3̸.0̸  0̸ 7̸ 4̸
- 1. 3  4 0 8
-------------
  1. 6  6 6 6
```

31.

```
        6 9 10
  6. 0 7̸ 0̸ 0̸    Writing 2 extra zeros
- 2. 0 0 7 8
------------
  4. 0 6 2 2
```

33. Line up the decimal points. Write an extra zero if desired.

```
    17 11
  1 7  1̸ 10
  2̸ 8̸ .2̸ 0̸
- 1 9 .3 5
----------
    8 .8 5
```

35.

```
    3 10
  3 4̸.0̸ 7
- 3 0.7
--------
    3.3 7
```

37.

```
      4 10
  8.4 5̸ 0̸
- 7.4 0 5
---------
  1.0 4 5
```

39.

```
  5 10
  6̸.0̸ 0 3
- 2.3
--------
  3.7 0 3
```

41.

```
     9 9 9 10
  1 0̸ 0̸ 0̸ 0̸     Writing in the decimal point
- 0. 0 0 9 8    and 4 extra zeros
------------
  0. 9 9 0 2    Subtracting
```

43.

```
    9 9  9 10
  1 0̸ 0̸. 0̸ 0̸
-      0. 3 4
-------------
     9 9. 6 6
```

45.

```
  6 14
  7̸.4̸8
- 2.6
-----
  4.88
```

47.

```
    4 10
  2 5̸.0̸ 0 8
- 1 2.4
----------
  1 2.6 0 8
```

49.

```
              7 10
  2 5 4 8. 9 8̸ 0̸
-          2. 0 0 7
-------------------
  2 5 4 6. 9 7 3
```

51.

```
    4  9 9 10
  4 5̸. 0̸ 0̸ 0̸
-   0. 9 9 9
------------
  4 4. 0 0 1
```

53. $-8.02 + 9.73$ A positive and a negative number

a) $|-8.02| = 8.02$, $|9.73| = 9.73$, and $|9.73| > |-8.02|$, so the answer is positive.

b)

```
  9 .73    Finding the difference in
- 8 .02    the absolute values
-------
  1 .71
```

c) $-8.02 + 9.73 = 1.71$

55. $12.9 - 15.4 = 12.9 + (-15.4)$
We add the opposite of 15.4. We have a positive and a negative number.

a) $|12.9| = 12.9$, $|-15.4| = 15.4$, and $|15.4| > |12.9|$, so the answer is negative.

b) $\begin{array}{r} \overset{4}{1}\,\overset{14}{\not 5}.\not 4 \\ -\ 1\,2.9 \\ \hline 2.5 \end{array}$ Finding the difference in the absolute values

c) $12.9 - 15.4 = -2.5$

57. $-2.9 + (-4.3)$ Two negative numbers

a) $\begin{array}{r} \overset{1}{2}.9 \\ +\ 4.3 \\ \hline 7.2 \end{array}$ Adding the absolute values

b) $-2.9 + (-4.3) = -7.2$ The sum of two negative numbers is negative.

59. $-4.301 + 7.68$ A negative and a positive number

a) $|-4.301| = 4.301$, $|7.68| = 7.68$, and $|7.68| > |-4.301|$, so the answer is positive.

b) $\begin{array}{r} 7.6\,\overset{7}{\not 8}\,\overset{10}{\not 0} \\ -\ 4.3\,0\,1 \\ \hline 3.3\,7\,9 \end{array}$ Finding the difference in the absolute values

c) $-4.301 + 7.68 = 3.379$

61. $-13.4 - 9.2$
$= -13.4 + (-9.2)$ Adding the opposite of 9.2
$= -22.6$ The sum of two negatives is negative.

63. $-2.1 - (-4.6)$
$= -2.1 + 4.6$ Adding the opposite of -4.6
$= 2.5$ Subtracting absolute values. Since 4.6 has the larger absolute value, the answer is positive.

65. $14.301 + (-17.82)$
$= -3.519$ Subtracting absolute values. Since -17.82 has the larger absolute value, the answer is negative.

67. $7.201 - (-2.4)$
$= 7.201 + 2.4$ Adding the opposite of -2.4
$= 9.601$ Adding

69. $23.9 + (-9.4)$
$= 14.5$ Subtracting absolute values. Since 23.9 has the larger absolute value, the answer is positive.

71. $-8.9 - (-12.7)$
$= -8.9 + 12.7$ Adding the opposite of -12.7
$= 3.8$ Subtracting absolute values. Since 12.7 has the larger absolute value, the answer is positive.

73. $-4.9 - 5.392$
$= -4.9 + (-5.392)$ Adding the opposite of 5.392
$= -10.292$ The sum of two negatives is negative.

75. $14.7 - 23.5$
$= 14.7 + (-23.5)$ Adding the opposite of 23.5
$= -8.8$ Subtracting absolute values. Since -23.5 has the larger absolute value, the answer is negative.

77. $5.1x + 3.6x$
$= (5.1 + 3.6)x$ Using the distributive law
$= 8.7x$ Adding

79. $17.59a - 12.73a$
$= (17.59 - 12.73)a$
$= 4.86a$

81. $15.2t + 7.9 + 5.9t$
$= 15.2t + 5.9t + 7.9$ Using the commutative law
$= (15.2 + 5.9)t + 7.9$ Using the distributive law
$= 21.1t + 7.9$

83. $9.208t - 14.519t$
$= (9.208 - 14.519)t$ Using the distributive law
$= (9.208 + (-14.519))t$ Adding the opposite of 14.519
$= -5.311t$ Subtracting absolute values. The coefficient is negative since -14.519 has the larger absolute value.

85. $-4.2a + 7.9a$
$= (-4.2 + 7.9)a$
$= 3.7a$ Subtracting absolute values. The coefficient is positive since 7.9 has the larger absolute value.

87. $-23.09x + 3.24x = (-23.09 + 3.24)x$
$= -19.85x$

89. $4.906y - 7.1 + 3.2y$
$= 4.906y + 3.2y - 7.1$
$= (4.906 + 3.2)y - 7.1$
$= 8.106y - 7.1$

91. $4.8x + 1.9y - 5.7x + 1.2y$
$= 4.8x + 1.9y + (-5.7x) + 1.2y$ Rewriting as addition
$= 4.8x + (-5.7x) + 1.9y + 1.2y$ Using the commutative law
$= (4.8 + (-5.7))x + (1.9 + 1.2)y$
$= -0.9x + 3.1y$

93. $4.9 - 3.9t + 2.3 - 4.5t$
$= 4.9 + (-3.9t) + 2.3 + (-4.5t)$
$= 4.9 + 2.3 + (-3.9t) + (-4.5t)$
$= (4.9 + 2.3) + (-3.9 + (-4.5))t$
$= 7.2 + (-8.4t)$
$= 7.2 - 8.4t$

95. $\frac{0}{n} = 0$, for any integer n that is not 0.

Thus, $\frac{0}{-92} = 0$.

97. $\frac{3}{5} - \frac{7}{10} = \frac{3}{5} \cdot \frac{2}{2} - \frac{7}{10}$ The LCM is 10.

$= \frac{6}{10} - \frac{7}{10}$

$= \frac{-1}{10}$, or $-\frac{1}{10}$

99.
$$\begin{aligned} 7x + 19 &= 40 \\ 7x + 19 - 19 &= 40 - 19 \\ 7x &= 21 \\ \frac{7x}{7} &= \frac{21}{7} \\ x &= 3 \end{aligned}$$

The solution is 3.

101. 

103. We regroup and simplify.

$$\begin{aligned} &-3.928 - 4.39a + 7.4b - 8.073 + 2.0001a - \\ &\quad 9.931b - 9.8799a + 12.897b \\ &= -3.928 - 8.073 - 4.39a + 2.0001a - 9.8799a + \\ &\quad 7.4b - 9.931b + 12.897b \\ &= (-3.928 - 8.073) + (-4.39 + 2.0001 - 9.8799)a + \\ &\quad (7.4 - 9.931 + 12.897)b \\ &= -12.001 - 12.2698a + 10.366b \end{aligned}$$

105. First, "undo" the incorrect addition by subtracting 235.7 from the incorrect answer:

$$\begin{array}{r} 817.2 \\ -\ 235.7 \\ \hline 581.5 \end{array}$$

The original minuend was 581.5. Now subtract 235.7 from this as the student originally intended:

$$\begin{array}{r} 581.5 \\ -\ 235.7 \\ \hline 345.8 \end{array}$$

The correct answer is 345.8.

Exercise Set 5.3

1.
$$\begin{array}{rl} 6.8 & \text{(1 decimal place)} \\ \times\ \ 7 & \text{(0 decimal places)} \\ \hline 47.6 & \text{(1 decimal place)} \end{array}$$

3.
$$\begin{array}{rl} 0.84 & \text{(2 decimal places)} \\ \times\ \ 8 & \text{(0 decimal places)} \\ \hline 6.72 & \text{(2 decimal places)} \end{array}$$

5.
$$\begin{array}{rl} 6.3 & \text{(1 decimal place)} \\ \times\ 0.04 & \text{(2 decimal places)} \\ \hline 0.252 & \text{(3 decimal places)} \end{array}$$

7.
$$\begin{array}{rl} 17.2 & \text{(1 decimal place)} \\ \times\ 0.006 & \text{(3 decimal places)} \\ \hline 0.1032 & \text{(4 decimal places)} \end{array}$$

9. $1\underline{0} \times 23.76$ 23.7.6

1 zero Move 1 place to the right.

$10 \times 23.76 = 237.6$

11. $-1000 \times 783.686852 = -(1000 \times 783.686852)$

$-(1\underline{000} \times 783.686852)$ −783.686.852

3 zeros Move 3 places to the right.

$-1000 \times 783.686852 = -783,686.852$

13. $-7.8 \times 1\underline{00}$ −7.80.

2 zeros Move 2 places to the right.

$-7.8 \times 100 = -780$

15. $0.\underline{1} \times 89.23$ 8.9.23

1 decimal place Move 1 place to the left.

$0.1 \times 89.23 = 8.923$

17. $0.\underline{001} \times 97.68$ 0.097.68

3 decimal places Move 3 places to the left.

$0.001 \times 97.68 = 0.09768$

19. $28.7 \times (-0.01) = -(28.7 \times 0.01)$

$-(28.7 \times 0.\underline{01})$ −0.28.7

2 decimal places Move 2 places to the left.

$28.7 \times (-0.01) = -0.287$

21.
$$\begin{array}{rl} 2.73 & \text{(2 decimal places)} \\ \times\ \ 16 & \text{(0 decimal places)} \\ \hline 1638 & \\ 2730 & \\ \hline 43.68 & \text{(2 decimal places)} \end{array}$$

23.
$$\begin{array}{rl} 0.984 & \text{(3 decimal places)} \\ \times\ \ 3.3 & \text{(1 decimal place)} \\ \hline 2952 & \\ 29520 & \\ \hline 3.2472 & \text{(4 decimal places)} \end{array}$$

25. We multiply the absolute values.

$$\begin{array}{rl} 37.4 & \text{(1 decimal place)} \\ \times\ \ 2.4 & \text{(1 decimal place)} \\ \hline 1496 & \\ 7480 & \\ \hline 89.76 & \text{(2 decimal places)} \end{array}$$

Since the product of two negative numbers is positive, the answer is 89.76.

27. We multiply the absolute values.

$$\begin{array}{rl} 749 & \text{(0 decimal places)} \\ \times\ 0.43 & \text{(2 decimal places)} \\ \hline 2247 & \\ 29960 & \\ \hline 322.07 & \text{(2 decimal places)} \end{array}$$

Since the product of a positive number and a negative number is negative, the answer is −322.07.

29.
```
      0. 8 7    (2 decimal places)
  ×     6 4     (0 decimal places)
  ---------
      3 4 8
    5 2 2 0
  ---------
    5 5. 6 8    (2 decimal places)
```

31.
```
        4 6. 5 0    (2 decimal places)
  ×         7 5     (0 decimal places)
  -------------
      2 3 2 5 0
    3 2 5 5 0 0
  -------------
    3 4 8 7. 5 0    (2 decimal places)
```

Since the last decimal place is 0, we could also write this answer as 3487.5.

33. We multiply the absolute values.
```
    0. 2 3 1    (3 decimal places)
  ×     0. 5    (1 decimal place)
  ----------
  0. 1 1 5 5    (4 decimal places)
```

Since the product of two negative numbers is positive, the answer is 0.1155.

35. $9.42 \times (-1000) = -(9.42 \times 1000)$

$-(9.42 \times 1\underline{000})$ $\quad -9.420.$

3 zeros $\quad$ Move 3 places to the right.

$9.42 \times (-1000) = -9420$

37. $-95.3 \times (-0.0001) = 95.3 \times 0.0001$

$95.3 \times 0.\underline{0001}$ $\quad 0.0095.3$

4 decimal places $\quad$ Move 4 places to the left.

$-95.3 \times (-0.0001) = 0.00953$

39. Move 2 places to the right.

$28.88.¢

Change from $ sign in front to ¢ sign at end.

$28.88 = 2888¢

41. Move 2 places to the right.

$0.66.¢

Change from $ sign in front to ¢ sign at end.

$0.66 = 66¢

43. Move 2 places to the left.

$0.34.¢

Change from ¢ sign at end to $ sign in front.

34¢ = $0.34

45. Move 2 places to the left.

$34.45.¢

Change from ¢ sign at end to $ sign in front.

3345¢ = $34.45

47. $\$4.03 \text{ trillion} = \$4.03 \times 1,\underbrace{000,000,000,000}_{12 \text{ zeros}}$

$4.030000000000.

Move 12 places to the right.

$\$4.03 \text{ trillion} = \$4,030,000,000,000$

49. $4.3 \text{ billion} = 4.3 \times 1,\underbrace{000,000,000}_{9 \text{ zeros}}$

4.300000000.

Move 9 places to the right.

$4.3 \text{ billion} = 4,300,000,000$

51.
$$\begin{aligned} & P + Prt \\ &= 10,000 + 10,000(0.04)(2.5) && \text{Substituting} \\ &= 10,000 + 400(2.5) && \text{Multiplying and dividing} \\ &= 10,000 + 1000 && \text{in order from left to right} \\ &= 11,000 && \text{Adding} \end{aligned}$$

53.
$$\begin{aligned} & vt + at^2 \\ &= 10(1.5) + 4.9(1.5)^2 \\ &= 10(1.5) + 4.9(1.5)(1.5) \\ &= 10(1.5) + 4.9(2.25) && \text{Squaring first} \\ &= 15 + 11.025 \\ &= 26.025 \end{aligned}$$

55. a)
$$\begin{aligned} P &= 2l + 2w \\ &= 2(12.5) + 2(9.5) \\ &= 25 + 19 \\ &= 44 \end{aligned}$$

The perimeter is 44 ft.

b)
$$\begin{aligned} A &= l \cdot w \\ &= (12.5)(9.5) \\ &= 118.75 \end{aligned}$$

The area is 118.75 ft^2.

57. $\frac{n}{n} = 1$, for any integer n that is not 0.

Thus, $\frac{-109}{-109} = 1$.

59.
$$\begin{aligned} \frac{2}{9} - \frac{5}{18} &= \frac{2}{9} \cdot \frac{2}{2} - \frac{5}{18} && \text{The LCM is 18.} \\ &= \frac{4}{18} - \frac{5}{18} \\ &= \frac{-1}{18}, \text{ or } -\frac{1}{18} \end{aligned}$$

61. $-\frac{3}{20}+\frac{3}{4}=-\frac{3}{20}+\frac{3}{4}\cdot\frac{5}{5}$ The LCM is 20.

$= -\frac{3}{20}+\frac{15}{20}$

$= \frac{12}{20}$

$= \frac{4\cdot 3}{4\cdot 5} = \frac{4}{4}\cdot\frac{3}{5}$

$= \frac{3}{5}$

63. 

65. Use a calculator.

$d+vt+at^2$

$= 79.2+3.029(7.355)+9.8(7.355)^2$

$= 79.2+3.029(7.355)+9.8(54.096025)$

$= 79.2+22.278295+530.141045$

$= 101.478295+530.141045$

$= 631.61934$

67. (1 trillion) · (1 billion)

$= 1,\underbrace{000,000,000,000}_{12 \text{ zeros}}\times 1,\underbrace{000,000,000}_{9 \text{ zeros}}$

$= 1,\underbrace{000,000,000,000,000,000,000}_{21 \text{ zeros}}$

$= 10^{21}$

Exercise Set 5.4

1.

```
    14.6
 5 )73.0
    50
    23
    20
     30    Write an extra 0.
     30
      0
```

3.

```
    2 3. 7 8
 4 )9 5. 1 2
    8 0 0 0
    1 5 1 2
    1 2 0 0
      3 1 2
      2 8 0
        3 2
        3 2
          0
```

Divide as though dividing whole numbers. Place the decimal point directly above the decimal point in the dividend.

5.

```
        7. 4 8
 1 2 )8 9. 7 6
      8 4 0 0
        5 7 6
        4 8 0
          9 6
          9 6
            0
```

7.

```
        7. 2
 3 3 )2 3 7. 6
      2 3 1 0
          6 6
          6 6
            0
```

9. We first consider $9.144 \div 8$.

```
    1. 1 4 3
 8 )9. 1 4 4
    8 0 0 0
    1 1 4 4
      8 0 0
      3 4 4
      3 2 0
        2 4
        2 4
          0
```

Since a positive number divided by a negative number is negative, the answer is -1.143.

11.

```
     4. 0 4 1
 3 )1 2. 1 2 3
    1 2 0 0 0
        1 2 3
        1 2 0
            3
            3
            0
```

13. We first consider $0.35 \div 5$.

```
    0. 0 7
 5 )0. 3 5
       3 5
         0
```

Since a negative number divided by a positive number is negative, the answer is -0.07.

15.

```
              7 0.
 0.1 2^ )8.4 0^
          8 4 0
              0
```

Multiply the divisor by 100 (move the decimal point 2 places). Multiply the same way in the dividend (move 2 places). Then divide.

17.

```
                4 0.
 3.2^ )1 2 8.0^
        1 2 8 0
              0
```

Put a decimal point at the end of the whole number. Multiply the divisor by 10 (move the decimal point 1 place). Multiply the same way in the dividend (move 1 place), adding an extra 0. Then divide.

19. We first consider $6 \div 15$.

```
       0.4
 1 5 )6.0
      6 0
        0
```

Put a decimal point at the end of the whole number. Write an extra 0 to the right of the decimal point. Then divide.

Since a positive number divided by a negative number is negative, the answer is -0.4.

21.
```
       0.4 1
  3 6 ⟌1 4.7 6
       1 4 4 0
           3 6
           3 6
             0
```

23.
```
          7.5
  6.2^ ⟌4 6.5^0
        4 3 4
          3 1 0   Write an extra 0.
          3 1 0
              0
```

25. We first consider $39.06 \div 4.2$.
```
          9.3
  4.2^ ⟌3 9.0^6
        3 7 8 0
          1 2 6
          1 2 6
              0
```

Since a positive number divided by a negative number is negative, the answer is -9.3.

27. We first consider $5 \div 8$.
```
     0 .6 2 5
  8 ⟌5 .0 0 0
     4 8
       2 0     Write an extra 0.
       1 6
         4 0   Write an extra 0.
         4 0
           0
```

Since a negative number divided by a negative number is positive, the answer is 0.625.

29.
```
              0.2 6
  0.4 7^ ⟌0. 1 2^2 2
                9 4 0
                2 8 2
                2 8 2
                    0
```

31.
```
          1 5.6 2 5
  4.8^ ⟌7 5.0^0 0 0
        4 8 0
        2 7 0
        2 4 0
          3 0 0
          2 8 8
            1 2 0
              9 6
              2 4 0
              2 4 0
                  0
```

33.
```
                  2.3 4
  0.0 3 2^ ⟌0. 0 7 4^8 8
                6 4 0 0
                1 0 8 8
                  9 6 0
                  1 2 8
                  1 2 8
                      0
```

35. We first consider $24.969 \div 82$.
```
       0.3 0 4 5
  8 2 ⟌2 4.9 6 9 0
       2 4 6 0 0
           3 6 9
           3 2 8
             4 1 0   Write an extra 0.
             4 1 0
                 0
```

Since a negative number divided by a positive number is negative, the answer is -0.3045.

37. $\dfrac{213.4567}{1000}$ 0.213.4567 (Move 3 places to the left.)

3 zeros Move 3 places to the left.

$$\frac{213.4567}{1000} = 0.2134567$$

39. $\dfrac{-213.4567}{10}$ $-21.3.4567$

1 zero Move 1 place to the left.

$$\frac{-213.4567}{10} = -21.34567$$

41. $\dfrac{1.0237}{0.001}$ 1.023.7

3 decimal places Move 3 places to the right.

$$\frac{1.0237}{0.001} = 1023.7$$

43. $\dfrac{56.78}{-0.001} = -\dfrac{56.78}{0.001}$

$-\dfrac{56.78}{0.001}$ $-56.780.$

3 decimal places Move 3 places to the right.

$$\frac{56.78}{-0.001} = -56,780$$

45. $\dfrac{743.92}{-100} = -\dfrac{743.92}{100}$

$-\dfrac{743.92}{100}$ $-7.43.92$

2 zeros Move 2 places to the left.

$$\frac{743.92}{-100} = -7.4392$$

47. $\dfrac{0.97}{0.\underline{1}}$ $\quad 0.9.7$ (Move 1 place to the right.)

1 decimal place Move 1 place to the right.

$$\frac{0.97}{0.1} = 9.7$$

49. $\dfrac{75.3}{-0.001} = -\dfrac{75.3}{0.001}$

$-\dfrac{75.3}{0.\underline{001}}$ $\quad -75.300.$

3 decimal places Move 3 places to the right.

$$\frac{75.3}{-0.001} = -75,300$$

51. $\dfrac{23,001}{1\underline{00}}$ $\quad 230.01.$

2 zeros Move 2 places to the left.

$$\frac{23,001}{100} = 230.01$$

53. $\dfrac{-57.281}{-10} = \dfrac{57.281}{10}$

$\dfrac{57.281}{1\underline{0}}$ $\quad 5.7.281$

1 zero Move 1 place to the left.

$$\frac{-57.281}{-10} = 5.7281$$

55. $14 \times (82.6 + 67.9) = 14 \times (150.5)$ Doing the calculation inside the parentheses

$= 2107$ Multiplying

57. $0.003 + 3.03 \div (-0.01) = 0.003 - 303$ Dividing first

$= -302.997$ Subtracting

59. $42 \times (10.6 + 0.024)$

$= 42 \times 10.624$ Doing the calculation inside the parentheses

$= 446.208$ Multiplying

61. $4.2 \times 5.7 + 0.7 \div 3.5$

$= 23.94 + 0.2$ Doing the multiplications and divisions in order from left to right

$= 24.14$ Adding

63. $-9.0072 + 0.04 \div 0.1^2$

$= -9.0072 + 0.04 \div 0.01$ Evaluating the exponential expression

$= -9.0072 + 4$ Dividing

$= -5.0072$ Adding

65. $(5 - 0.04)^2 \div 4 + 8.7 \times 0.4$

$= (4.96)^2 \div 4 + 8.7 \times 0.4$ Doing the calculation inside the parentheses

$= 24.6016 \div 4 + 8.7 \times 0.4$ Evaluating the exponential expression

$= 6.1504 + 3.48$ Doing the multiplications and divisions in order from left to right

$= 9.6304$ Adding

67. $4 \div 0.4 - 0.1 \times 5 + 0.1^2$

$= 4 \div 0.4 - 0.1 \times 5 + 0.01$ Evaluating the exponential expression

$= 10 - 0.5 + 0.01$ Doing the multiplications and divisions in order from left to right

$= 9.51$ Adding and subtracting in order from left to right

69. $5.5^2 \times [(6 - 7.8) \div 0.06 + 0.12]$

$= 5.5^2 \times [-1.8 \div 0.06 + 0.12]$ Doing the calculation in the innermost parentheses first

$= 5.5^2 \times [-30 + 0.12]$ Doing the calculation inside the parentheses

$= 5.5^2 \times (-29.88)$

$= 30.25 \times (-29.88)$ Evaluating the exponential expression

$= -903.87$ Multiplying

71. $0.01 \times \{[(4 - 0.25) \div 2.5] - (4.5 - 4.025)\}$

$= 0.01 \times \{[3.75 \div 2.5] - 0.475\}$ Doing the calculations in the innermost parentheses first

$= 0.01 \times \{1.5 - 0.475\}$ Again, doing the calculations in the innermost parentheses

$= 0.01 \times 1.025$ Subtracting inside the parentheses

$= 0.01025$ Multiplying

73. We add the temperatures for the years 1990 through 1994 and then divide by the number of addends, 5.

$(15.47 + 15.41 + 15.13 + 15.20 + 15.31) \div 5 =$
$76.52 \div 5 = 15.304$

The average temperature for the years 1990 through 1994 was 15.304° C.

75. From Exercise 73, we know that the average temperature for the years 1990 through 1994 was 15.304° C. To find the average temperature for the years 1984 through 1988, we add the temperatures for those years and then divide by the number of addends, 5.

$(15.11 + 15.11 + 15.16 + 15.32 + 15.35) \div 5 =$
$76.05 \div 5 = 15.21$

We subtract to find the difference in the averages.

$$15.304 - 15.21 = 0.094$$

The average temperature for 1990 through 1994 exceeds the average for 1984 through 1988 by 0.094° C.

77. First we find the eleven-year-average.

$(15.11 + 15.11 + 15.16 + 15.32 + 15.35 + 15.25 +$
$15.47 + 15.41 + 15.13 + 15.20 + 15.31) \div 11 =$
$167.82 \div 11 \approx 15.256$

From Exercise 75, we know that the average of the first five years was 15.21° C. We subtract to find the difference.

$15.256 - 15.21 = 0.046$

The eleven-year-average exceeds the average of the first five years by about 0.046° C.

79. $10\frac{1}{2} + 4\frac{5}{8} = 10\frac{4}{8} + 4\frac{5}{8}$
$= 14\frac{9}{8} = 15\frac{1}{8}$

81. $\frac{36}{42} = \frac{6 \cdot 6}{6 \cdot 7} = \frac{6}{6} \cdot \frac{6}{7} = \frac{6}{7}$

83. $7 - 8\frac{2}{3} = 7 + \left(-8\frac{2}{3}\right)$

Since $-8\frac{2}{3}$ has the greater absolute value, the answer will be negative. We find the difference in absolute values.

$$\begin{array}{r} 8\,\frac{2}{3} \\ -\ 7 \\ \hline 1\,\frac{2}{3} \end{array}$$

Thus, $7 - 8\frac{2}{3} = -1\frac{2}{3}$.

85. 

87. Use a calculator.

$9.0534 - 2.041^2 \times 0.731 \div 1.043^2$
$= 9.0534 - 4.165681 \times 0.731 \div 1.087849$ Evaluating the exponential expressions
$= 9.0534 - 3.045112811 \div 1.087849$ Multiplying and dividing in order from left to right
$= 9.0534 - 2.799205415$
$= 6.254194585$

89. $439.57 \times 0.01 \div 1000 \times \_\_\_ = 4.3957$
$4.3957 \div 1000 \times \_\_\_ = 4.3957$
$0.0043957 \times \_\_\_ = 4.3957$

We need to multiply 0.0043957 by a number that moves the decimal point 3 places to the right. Thus, we need to multiply by 1000. This is the missing value.

91. $0.0329 \div 0.001 \times 10^4 \div \_\_\_ = 3290$
$0.0329 \div 0.001 \times 10{,}000 \div \_\_\_ = 3290$
$32.9 \times 10{,}000 \div \_\_\_ = 3290$
$329{,}000 \div \_\_\_ = 3290$

We need to divide 329,000 by a number that moves the decimal point 2 places to the left. Thus, we need to divide by 100. This is the missing value.

Exercise Set 5.5

1. We are estimating the sum

$$\$109.95 + \$249.95.$$

We round both numbers to the nearest ten. The estimate is

$$\$110 + \$250 = \$360.$$

Answer (d) is correct.

3. We are estimating the difference

$$\$299 - \$249.95.$$

We round both numbers to the nearest ten. The estimate is

$$\$300 - \$250 = \$50.$$

Answer (c) is correct.

5. We are estimating the product

$$9 \times \$299.$$

We round \$299 to the nearest ten. The estimate is

$$9 \times \$300 = \$2700.$$

Answer (a) is correct.

7. We are estimating the quotient

$$\$1700 \div \$299.$$

Rounding \$299, we get \$300. Since \$1700 is close to \$1800, which is a multiple of \$300, we estimate

$$\$1800 \div \$300,$$

so the answer is about 6.

Answer (c) is correct.

9. This is about $0.0 + 1.3 + 0.3$, so the answer is about 1.6.

11. This is about $6 + 0 + 0$, so the answer is about 6.

13. This is about $52 + 1 + 7$, so the answer is about 60.

15. This is about $2.7 - 0.4$, so the answer is about 2.3.

17. This is about $200 - 20$, so the answer is about 180.

19. This is about 50×8, rounding 49 to the nearest ten and 7.89 to the nearest one, so the answer is about 400. Answer (a) is correct.

21. This is about 100×0.08, rounding 98.4 to the nearest ten and 0.083 to the nearest hundredth, so the answer is about 8. Answer (c) is correct.

23. This is about $4 \div 4$, so the answer is about 1. Answer (b) is correct.

25. This is about $75 \div 25$, so the answer is about 3. Answer (b) is correct.

27.
$$\begin{array}{r}
3 \\
3\,\overline{)\;9} \\
3\,\overline{)\;27} \\
2\,\overline{)\;54} \\
2\,\overline{)\;108}
\end{array}$$

$108 = 2 \cdot 2 \cdot 3 \cdot 3 \cdot 3$

29.
$$\begin{array}{ccccc}
 & & 325 & & \\
 & \swarrow & & \searrow & \\
25 & & \cdot & & 13 \\
\swarrow\;\searrow & & & & | \\
5 \quad \cdot \quad 5 & & \cdot & & 13
\end{array}$$

$325 = 5 \cdot 5 \cdot 13$

31. $\dfrac{125}{400} = \dfrac{25 \cdot 5}{25 \cdot 16} = \dfrac{25}{25} \cdot \dfrac{5}{16} = \dfrac{5}{16}$

33. $\dfrac{72}{81} = \dfrac{9 \cdot 8}{9 \cdot 9} = \dfrac{9}{9} \cdot \dfrac{8}{9} = \dfrac{8}{9}$

35. 

37. We round each factor to the nearest ten. The estimate is $180 \times 60 = 10,800$. The estimate is close to the result given, so the decimal point was placed correctly.

39. We round each number on the left to the nearest one. The estimate is $19 - 1 \times 4 = 19 - 4 = 15$. The estimate is not close to the result given, so the decimal point was not placed correctly.

Exercise Set 5.6

1. $\dfrac{7}{16} = 7 \div 16$

$$\begin{array}{r}
0.4375 \\
16\,\overline{)\,7.0000} \\
\underline{6\,4} \\
6\,0 \\
\underline{4\,8} \\
1\,2\,0 \\
\underline{1\,1\,2} \\
8\,0 \\
\underline{8\,0} \\
0
\end{array}$$

$\dfrac{7}{16} = 0.4375$

(Note that we could also have multiplied $\dfrac{7}{16}$ by $\dfrac{625}{625}$ to get a denominator of 10,000.)

3. $\dfrac{13}{40} = \dfrac{13}{40} \cdot \dfrac{25}{25}$ We use $\dfrac{25}{25}$ for 1 to get a denominator of 1000.

$= \dfrac{325}{1000} = 0.325$

(Note that we could also have performed the division $13 \div 40$.)

5. $\dfrac{1}{5} = \dfrac{1}{5} \cdot \dfrac{2}{2} = \dfrac{2}{10} = 0.2$

7. $\dfrac{17}{20} = \dfrac{17}{20} \cdot \dfrac{5}{5} = \dfrac{85}{100} = 0.85$

9. $\dfrac{21}{40} = \dfrac{21}{40} \cdot \dfrac{25}{25} = \dfrac{525}{1000} = 0.525$

11. $-\dfrac{51}{40} = -\dfrac{51}{40} \cdot \dfrac{25}{25} = -\dfrac{1275}{1000} = -1.275$

13. $\dfrac{13}{25} = \dfrac{13}{25} \cdot \dfrac{4}{4} = \dfrac{52}{100} = 0.52$

15. $\dfrac{2502}{125} = \dfrac{2502}{125} \cdot \dfrac{8}{8} = \dfrac{20,016}{1000} = 20.016$

17. $\dfrac{-1}{4} = \dfrac{-1}{4} \cdot \dfrac{25}{25} = \dfrac{-25}{100} = -0.25$

19. $\dfrac{23}{40} = \dfrac{23}{40} \cdot \dfrac{25}{25} = \dfrac{575}{1000} = 0.575$

21. $-\dfrac{5}{8} = -\dfrac{5}{8} \cdot \dfrac{125}{125} = -\dfrac{625}{1000} = -0.625$

23. $\dfrac{37}{25} = \dfrac{37}{25} \cdot \dfrac{4}{4} = \dfrac{148}{100} = 1.48$

25. $\dfrac{4}{15} = 4 \div 15$

$$\begin{array}{r}
0.\,2\,6\,6 \\
15\,\overline{)\,4.\,0\,0\,0} \\
\underline{3\,0} \\
1\,0\,0 \\
\underline{9\,0} \\
1\,0\,0 \\
\underline{9\,0} \\
1\,0
\end{array}$$

Since 10 keeps reappearing as a remainder, the digits repeat and

$\dfrac{4}{15} = 0.2666\ldots$ or $0.2\overline{6}$.

27. $\dfrac{1}{3} = 1 \div 3$

$$\begin{array}{r}
0.\,3\,3\,3 \\
3\,\overline{)\,1.\,0\,0\,0} \\
\underline{9} \\
1\,0 \\
\underline{9} \\
1\,0 \\
\underline{9} \\
1
\end{array}$$

Since 1 keeps reappearing as a remainder, the digits repeat and

$\dfrac{1}{3} = 0.333\ldots$ or $0.\overline{3}$.

29. First consider $\frac{4}{3}$.

$\frac{4}{3} = 4 \div 3$

$$\begin{array}{r} 1.33 \\ 3\,\overline{)4.00} \\ \underline{3} \\ 1\,0 \\ \underline{9} \\ 1\,0 \\ \underline{9} \\ 1 \end{array}$$

Since 1 keeps reappearing as a remainder, the digits repeat and

$\frac{4}{3} = 1.333\ldots$ or $1.\overline{3}$.

Thus, $\frac{-4}{3} = -1.\overline{3}$.

31. $\frac{7}{6} = 7 \div 6$

$$\begin{array}{r} 1.166 \\ 6\,\overline{)7.000} \\ \underline{6} \\ 1\,0 \\ \underline{6} \\ 4\,0 \\ \underline{3\,6} \\ 4\,0 \\ \underline{3\,6} \\ 4 \end{array}$$

Since 4 keeps reappearing as a remainder, the digits repeat and

$\frac{7}{6} = 1.166\ldots$ or $1.1\overline{6}$.

33. $\frac{4}{7} = 4 \div 7$

$$\begin{array}{r} 0.5714285 \\ 7\,\overline{)4.0000000} \\ \underline{3\,5} \\ 5\,0 \\ \underline{4\,9} \\ 1\,0 \\ \underline{7} \\ 3\,0 \\ \underline{2\,8} \\ 2\,0 \\ \underline{1\,4} \\ 6\,0 \\ \underline{5\,6} \\ 4\,0 \\ \underline{3\,5} \\ 5 \end{array}$$

Since 5 reappears as a remainder, the sequence repeats and

$\frac{4}{7} = 0.571428571428\ldots$ or $0.\overline{571428}$.

35. First consider $\frac{11}{12}$.

$\frac{11}{12} = 11 \div 12$

$$\begin{array}{r} 0.9166 \\ 12\,\overline{)11.0000} \\ \underline{10\,8} \\ 2\,0 \\ \underline{1\,2} \\ 8\,0 \\ \underline{7\,2} \\ 8\,0 \\ \underline{7\,2} \\ 8 \end{array}$$

Since 8 keeps reappearing as a remainder, the digits repeat and $\frac{11}{12} = 0.91666\ldots$ or $0.91\overline{6}$.

Thus, $-\frac{11}{12} = -0.91\overline{6}$.

37. Round 0.2 [6] 6 6 ... to the nearest tenth. (Hundredths digit is 5 or more.)

0.3 Round up.

Round 0.2 6 [6] 6 ... to the nearest hundredth. (Thousandths digit is 5 or more.)

0.2 7 Round up.

Round 0.2 6 6 [6] ... to the nearest thousandth. (Ten-thousandths digit is 5 or more.)

0.2 6 7 Round up.

39. Round 0.3 [3] 3 3 ... to the nearest tenth. (Hundredths digit is 4 or less.)

0.3 Round down.

Round 0.3 3 [3] 3 ... to the nearest hundredth. (Thousandths digit is 4 or less.)

0.3 3 Round down.

Round 0.3 3 3 [3] ... to the nearest thousandth. (Ten-thousandths digit is 4 or less.)

0.3 3 3 Round down.

41. Round −1.3 [3] 3 3 ... to the nearest tenth. (Hundredths digit is 4 or less.)

Round 1.3333... down to 1.3.

The answer is −1.3. Since $-1.3 > -1.3333\ldots$, we have actually rounded up.

Round −1.3 3 [3] 3 ... to the nearest hundredth. (Thousandths digit is 4 or less.)

Round 1.3333... down to 1.33.

The answer is −1.33. Since $-1.33 > -1.3333\cdots$, we have actually rounded up.

Round −1.3 3 3 [3] ... to the nearest thousandth. (Ten-thousandths digit is 4 or less.)

Round $1.3333\cdots$ down to 1.333.

The answer is -1.333. Since $-1.333 > -1.3333\cdots$, we have actually rounded up.

43. Round 1. 1 6 6 6 ... to the nearest tenth.

Hundredths digit is 5 or more.

1. 2 Round up.

Round 1. 1 6 6 6 ... to the nearest hundredth.

Thousandths digit is 5 or more.

1. 1 7 Round up.

Round 1. 1 6 6 6 ... to the nearest thousandth.

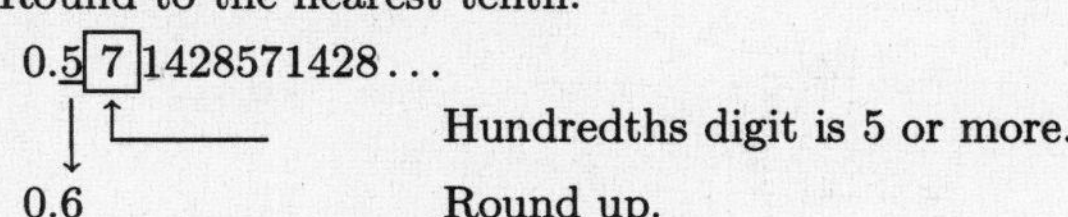

Ten-thousandths digit is 5 or more.

1. 1 6 7 Round up.

45. $0.\overline{571428}$

Round to the nearest tenth.

0.5 7 1428571428...

Hundredths digit is 5 or more.

0.6 Round up.

Round to the nearest hundredth.

0.57 1 428571428...

Thousandths digit is 4 or less.

0.57 Round down.

Round to the nearest thousandth.

0.571 4 28571428...

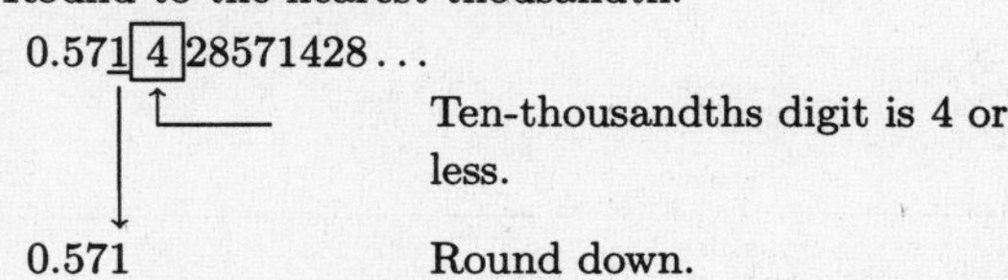

Ten-thousandths digit is 4 or less.

0.571 Round down.

47. Round $-0.$ 9 1 6 6 ... to the nearest tenth.

Hundredths digit is 4 or less.

Round $0.9166\cdots$ down to 0.9.

The answer is -0.9. Since $-0.9 > -0.9166$, we have actually rounded up.

Round $-0.$ 9 1 6 6 ... to the nearest hundredth.

Thousandths digit is 5 or more.

Round $0.9166\cdots$ up to 0.92.

The answer is -0.92. Since $-0.92 < -0.9166\cdots$, we have actually rounded down.

Round $-0.$ 9 1 6 6 ... to the nearest thousandth.

Ten-thousandths digit is 5 or more.

Round $0.9166\cdots$ up to 0.917.

The answer is -0.917. Since $-0.917 < 0.9166\cdots$, we have actually rounded down.

49. We will use the first method discussed in the text.

$$\frac{7}{8} \times 12.64 = \frac{7}{8} \times \frac{12.64}{1} = \frac{7 \times 12.64}{8} = \frac{88.48}{8} = 11.06$$

51. We will use the second method discussed in the text.

$$\frac{47}{9} \times 79.95 = 5.\overline{2} \times 79.95$$
$$\approx 5.222 \times 79.95 = 417.4989$$

Note that this answer is not as accurate as those found using either of the other methods, due to rounding.

53. We will use the first method discussed in the text.

$$\frac{5}{6} \times 0.0765 + \frac{5}{4} \times 0.1124 = \frac{5}{6} \times \frac{0.0765}{1} + \frac{5}{4} \times \frac{0.1124}{1}$$
$$= \frac{5 \times 0.0765}{6 \times 1} + \frac{5 \times 0.1124}{4 \times 1}$$
$$= \frac{0.3825}{6} + \frac{0.562}{4}$$
$$= 0.06375 + 0.1405$$
$$= 0.20425$$

55. We will use the third method discussed in the text.

$$\frac{3}{4} \times 2.56 - \frac{7}{8} \times 3.94$$
$$= \frac{3}{4} \times \frac{256}{100} - \frac{7}{8} \times \frac{394}{100}$$
$$= \frac{768}{400} - \frac{2758}{800}$$
$$= \frac{768}{400} \cdot \frac{2}{2} - \frac{2758}{800}$$
$$= \frac{1536}{800} - \frac{2758}{800}$$
$$= \frac{-1222}{800} = -\frac{1222}{800}$$
$$= -\frac{2 \cdot 611}{2 \cdot 400} = \frac{2}{2} \cdot \left(-\frac{611}{400}\right)$$
$$= -\frac{611}{400}, \text{ or } -1.5275$$

57. ***Familiarize***. We draw a picture and recall that the formula for the area A of a triangle with base b and height h is $A = \frac{1}{2} \times b \times h$.

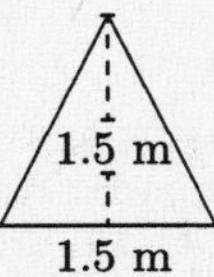

Translate. We substitute 1.5 for b and 1.5 for h.

$$A = \frac{1}{2} \times b \times h = \frac{1}{2} \times 1.5 \times 1.5$$

Solve. We carry out the computation.

$$A = \frac{1}{2} \times 1.5 \times 1.5$$
$$= \frac{1.5}{2} \times 1.5 \quad \text{Multiplying } \frac{1}{2} \text{ and } 1.5$$
$$= 0.75 \times 1.5 \quad \text{Dividing}$$
$$= 1.125 \quad \text{Multiplying}$$

Check. We repeat the calculations using a different method.

$$\frac{1}{2} \times 1.5 \times 1.5 = 0.5 \times (1.5 \times 1.5) = 0.5 \times 2.25 = 1.125$$

Our answer checks.

State. The area of the sign is 1.125 m^2.

59. ***Familiarize.*** We draw a picture and recall that the formula for the area A of a triangle with base b and height h is $A = \frac{1}{2} \times b \times h$.

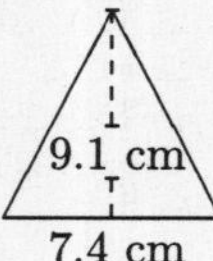

Translate. We substitute 7.4 for b and 9.1 for h.

$$A = \frac{1}{2} \times b \times h = \frac{1}{2} \times 7.4 \times 9.1$$

Solve. We carry out the computation.

$$\begin{aligned} A &= \frac{1}{2} \times 7.4 \times 9.1 \\ &= \frac{7.4}{2} \times 9.1 && \text{Multiplying } \frac{1}{2} \text{ and } 7.4 \\ &= 3.7 \times 9.1 && \text{Dividing} \\ &= 33.67 && \text{Multiplying} \end{aligned}$$

Check. We repeat the calculations using a different method.

$\frac{1}{2} \times 7.4 \times 9.1 = 0.5 \times (7.4 \times 9.1) = 0.5 \times 67.34 = 33.67$

Our answer checks.

State. The area of the reflector is 33.67 cm^2.

61.
$$\begin{array}{rcr} 20 & = & 19\frac{5}{5} \\ -16\frac{3}{5} & = & -16\frac{3}{5} \\ \hline & & 3\frac{2}{5} \end{array}$$

63. $\frac{n}{1} = n$, for any integer n.

Thus, $\frac{95}{-1} = \frac{-95}{1} = -95$.

65.
$$\begin{aligned} & 9 - 4 + 2 \div (-1) \cdot 6 \\ &= 9 - 4 - 2 \cdot 6 && \text{Multiplying and dividing in} \\ &= 9 - 4 - 12 && \text{order from left to right} \\ &= 5 - 12 && \text{Adding and subtracting in} \\ &= -7 && \text{order from left to right} \end{aligned}$$

67. 

69. Using a calculator we find that $\frac{1}{7} = 1 \div 7 = 0.\overline{142857}$.

71. Using a calculator we find that $\frac{3}{7} = 3 \div 7 = 0.\overline{428571}$.

73. Using a calculator we find that $\frac{5}{7} = 5 \div 7 = 0.\overline{714285}$.

75. Using a calculator we find that $\frac{1}{9} = 1 \div 9 = 0.\overline{1}$.

77. Using a calculator we find that $\frac{1}{999} = 0.\overline{001}$.

79. We substitute $\frac{22}{7}$ for π and 2.1 for r.

$$\begin{aligned} A &= \pi r^2 \\ &= \frac{22}{7}(2.1)^2 \\ &= \frac{22}{7}(4.41) \\ &= \frac{22 \times 4.41}{7} \\ &= \frac{97.02}{7} \\ &= 13.86 \text{ cm}^2 \end{aligned}$$

81. We substitute 3.14 for π and $\frac{3}{4}$ for r.

$$\begin{aligned} A &= \pi r^2 \\ &= 3.14\left(\frac{3}{4}\right)^2 \\ &= 3.14\left(\frac{9}{16}\right) \\ &= \frac{3.14 \times 9}{16} \\ &= \frac{28.26}{16} \\ &= 1.76625 \text{ ft}^2 \end{aligned}$$

83. 

Exercise Set 5.7

1.
$$\begin{aligned} 4.2 \cdot x &= 39.06 \\ \frac{4.2 \cdot x}{4.2} &= \frac{39.06}{4.2} && \text{Dividing on both sides by 4.2} \\ x &= 9.3 \end{aligned}$$

$$\begin{array}{r} 0\,9.3 \\ 4.2_\wedge \overline{)\,3\,9.0_\wedge 6} \\ 3\,7\,8\,0 \\ \hline 1\,2\,6 \\ 1\,2\,6 \\ \hline 0 \end{array}$$

To do a partial check, we can approximate. Note that $4.2 \approx 4$ and $9.3 \approx 9.5$. Since $4 \cdot 9.5 = 38 \approx 39.06$, the answer checks. The solution is 9.3.

3.
$$\begin{aligned} 1000 \cdot y &= 9.0678 \\ \frac{1000 \cdot y}{1000} &= \frac{9.0678}{1000} && \text{Dividing on both sides by 1000} \\ y &= 0.0090678 && \text{Moving the decimal point 3 places to the left} \end{aligned}$$

To do a partial check, we can approximate. Note that $0.0090678 \approx 0.01$. Since $1000 \cdot 0.01 = 10 \approx 9.0678$, the answer checks. The solution is 0.0090678.

5. $-23.4 = 5.2a$

$\frac{-23.4}{5.2} = \frac{5.2a}{5.2}$ Dividing by 5.2 on both sides

$-4.5 = a$

The solution is -4.5.

7. $-9.2x = -94.76$

$\frac{-9.2x}{-9.2} = \frac{-94.76}{-9.2}$

$x = 10.3$

The solution is 10.3.

9. $t - 19.27 = 24.51$

$t - 19.27 + 19.27 = 24.51 + 19.27$ Adding 19.27 on both sides

$t = 43.78$

To check, note that $43.78 - 19.27 = 24.51$. The solution is 43.78.

11. $-5.9 + m = 8.42$

$-5.9 + m + 5.9 = 8.42 + 5.9$ Adding 5.9 on both sides

$m = 14.32$

To check, note that $-5.9 + 14.32 = 8.42$. The solution is 14.32.

13. $x + 13.9 = 4.2$

$x + 13.9 - 13.9 = 4.2 - 13.9$ Adding -13.9 (or subtracting 13.9) on both sides

$x = -9.7$

The solution is -9.7.

15. $23.1 + y = 12.06$

$23.1 + y - 23.1 = 12.06 - 23.1$

$y = -11.04$

The solution is -11.04.

17. $7.1x - 9.3 = 8.45$

$7.1x - 9.3 + 9.3 = 8.45 + 9.3$ Adding 9.3 on both sides

$7.1x = 17.75$

$\frac{7.1x}{7.1} = \frac{17.75}{7.1}$ Dividing by 7.1 on both sides

$x = 2.5$

Check: $7.1x - 9.3 = 8.45$

$7.1(2.5) - 9.3$	? 8.45	
$17.75 - 9.3$		
8.45	8.45	TRUE

The solution is 2.5.

19. $12.4 + 3.7t = 2.04$

$12.4 + 3.7t - 12.4 = 2.04 - 12.4$ Subtracting 12.04 on both sides

$3.7t = -10.36$

$\frac{3.7t}{3.7} = \frac{-10.36}{3.7}$ Dividing by 3.7 on both sides

$t = -2.8$

Check: $12.4 + 3.7t = 2.04$

$12.4 + 3.7(-2.8)$	? 2.04	
$12.4 - 10.36$		
2.04	2.04	TRUE

The solution is -2.8.

21. $-26.05 = 7.5x + 9.2$

$-26.05 - 9.2 = 7.5x + 9.2 - 9.2$

$-35.25 = 7.5x$

$-4.7 = x$

The solution is -4.7.

23. $1.2a + 7.48 = 3.4$

$1.2a + 7.48 - 7.48 = 3.4 - 7.48$

$1.2a = -4.08$

$\frac{1.2a}{1.2} = \frac{-4.08}{1.2}$

$a = -3.4$

The solution is -3.4.

25. $-4.2x + 3.04 = -4.1$

$-4.2x + 3.04 - 3.04 = -4.1 - 3.04$

$-4.2x = -7.14$

$\frac{-4.2x}{-4.2} = \frac{-7.14}{-4.2}$

$x = 1.7$

The solution is 1.7.

27. $7x + 3 = 2x + 19$

$7x + 3 - 2x = 2x + 19 - 2x$ Subtracting $2x$ on both sides

$5x + 3 = 19$

$5x + 3 - 3 = 19 - 3$ Subtracting 3 on both sides

$5x = 16$

$\frac{5x}{5} = \frac{16}{5}$ Dividing by 5 on both sides

$x = 3.2$

Check: $7x + 3 = 2x + 19$

$7 \cdot 3.2 + 3$	? $2 \cdot 3.2 + 19$	
$22.4 + 3$	$6.4 + 19$	
25.4	25.4	TRUE

The solution is 3.2.

29. $5.2x - 17 = 2.1x + 45$

$5.2x - 17 - 2.1x = 2.1x + 45 - 2.1x$ Subtracting $2.1x$ on both sides

$3.1x - 17 = 45$

$3.1x - 17 + 17 = 45 + 17$ Adding 17 on both sides

$3.1x = 62$

$\frac{3.1x}{3.1} = \frac{62}{3.1}$ Dividing by 3.1 on both sides

$x = 20$

Check: $5.2x - 17 = 2.1x + 45$

$5.2(20) - 17$	? $2.1(20) + 45$	
$104 - 17$	$42 + 45$	
87	87	TRUE

The solution is 20.

31.
$$\begin{aligned} 6y-5 &= 8+10y \\ 6y-5-6y &= 8+10y-6y \quad \text{Subtracting } 6y \text{ on both sides} \\ -5 &= 8+4y \\ -5-8 &= 8+4y-8 \quad \text{Subtracting 8 on both sides} \\ -13 &= 4y \\ \frac{-13}{4} &= \frac{4y}{4} \quad \text{Dividing by 4 on both sides} \\ -3.25 &= y \end{aligned}$$

The solution is -3.25.

33.
$$\begin{aligned} 3.1a+12 &= 7.5a-5.16 \\ 3.1a+12-3.1a &= 7.5a-5.16-3.1a \\ 12 &= 4.4a-5.16 \\ 12+5.16 &= 4.4a-5.16+5.16 \\ 17.16 &= 4.4a \\ \frac{17.16}{4.4} &= \frac{4.4a}{4.4} \\ 3.9 &= a \end{aligned}$$

The solution is 3.9.

35.
$$\begin{aligned} 22.3+52.3t &= 21.2t-49.23 \\ 22.3+52.3t-21.2t &= 21.2t-49.23-21.2t \\ 22.3+31.1t &= -49.23 \\ 22.3+31.1t-22.3 &= -49.23-22.3 \\ 31.1t &= -71.53 \\ \frac{31.1t}{31.1} &= \frac{-71.53}{31.1} \\ t &= -2.3 \end{aligned}$$

The solution is -2.3.

37.
$$\begin{aligned} 31.79-4.1a &= 10.7-9.8a \\ 31.79-4.1a+9.8a &= 10.7-9.8a+9.8a \\ 31.79+5.7a &= 10.7 \\ 31.79+5.7a-31.79 &= 10.7-31.79 \\ 5.7a &= -21.09 \\ \frac{5.7a}{5.7} &= \frac{-21.09}{5.7} \\ a &= -3.7 \end{aligned}$$

The solution is -3.7.

39.
$$\begin{aligned} 2(x+3) &= 4x-11 \\ 2x+6 &= 4x-11 \quad \text{Using the distributive law} \\ 2x+6-2x &= 4x-11-2x \\ 6 &= 2x-11 \\ 6+11 &= 2x-11+11 \\ 17 &= 2x \\ \frac{17}{2} &= \frac{2x}{2} \\ 8.5 &= x \end{aligned}$$

Check: $2(x+3) = 4x-11$

$2(8.5+3)$	? $4\cdot 8.5-11$	
$2(11.5)$	$34-11$	
23	23	TRUE

The solution is 8.5.

41.
$$\begin{aligned} 2a+17 &= 12(a-1) \\ 2a+17 &= 12a-12 \quad \text{Using the distributive law} \\ 2a+17-2a &= 12a-12-2a \\ 17 &= 10a-12 \\ 17+12 &= 10a-12+12 \\ 29 &= 10a \\ \frac{29}{10} &= \frac{10a}{10} \\ 2.9 &= a \end{aligned}$$

Check: $2a+17 = 12(a-1)$

$2\cdot 2.9+17$	? $12(2.9-1)$	
$5.8+17$	$12(1.9)$	
22.8	22.8	TRUE

The solution is 2.9.

43.
$$\begin{aligned} 3.4(x+7.3) &= 9.1x-0.83 \\ 3.4x+24.82 &= 9.1x-0.83 \quad \text{Using the distributive law} \\ 3.4x+24.82-3.4x &= 9.1x-0.83-3.4x \\ 24.82 &= 5.7x-0.83 \\ 24.82+0.83 &= 5.7x-0.83+0.83 \\ 25.65 &= 5.7x \\ \frac{25.65}{5.7} &= \frac{5.7x}{5.7} \\ 4.5 &= x \end{aligned}$$

The solution is 4.5.

45.
$$\begin{aligned} -7.37-3.2t &= 4.9(t+6.1) \\ -7.37-3.2t &= 4.9t+29.89 \\ -7.37-3.2t+3.2t &= 4.9t+29.89+3.2t \\ -7.37 &= 8.1t+29.89 \\ -7.37-29.89 &= 8.1t+29.89-29.89 \\ -37.26 &= 8.1t \\ -4.6 &= t \end{aligned}$$

The solution is -4.6.

47.
$$\begin{aligned} 9.7(x-4.2)+3x &= 4.1x+7.42 \\ 9.7x-40.74+3x &= 4.1x+7.42 \\ 12.7x-40.74 &= 4.1x+7.42 \quad \text{Collecting like terms} \\ 12.7x-40.74-4.1x &= 4.1x+7.42-4.1x \\ 8.6x-40.74 &= 7.42 \\ 8.6x-40.74+40.74 &= 7.42+40.74 \\ 8.6x &= 48.16 \\ \frac{8.6x}{8.6} &= \frac{48.16}{8.6} \\ x &= 5.6 \end{aligned}$$

The solution is 5.6.

49.
$$\begin{aligned}4.3(7-2x)+3.4 &= 5(x-4.1)+74.4\\ 30.1-8.6x+3.4 &= 5x-20.5+74.4\\ &\quad\text{Using the distributive law twice}\\ 33.5-8.6x &= 5x+53.9 \quad \text{Collecting like terms}\\ 33.5-8.6x+8.6x &= 5x+53.9+8.6x\\ 33.5 &= 13.6x+53.9\\ 33.5-53.9 &= 13.6x+53.9-53.9\\ -20.4 &= 13.6x\\ \frac{-20.4}{13.6} &= \frac{13.6x}{13.6}\\ -1.5 &= x\end{aligned}$$

The solution is -1.5.

51. $\frac{n}{n}=1$, for any integer n that is not 0.

Thus, $\frac{-43}{-43}=1$.

53.
$$\begin{aligned}\frac{7}{10}-\frac{3}{25} &= \frac{7}{10}\cdot\frac{5}{5}-\frac{3}{25}\cdot\frac{2}{2} \quad \text{The LCM is 50.}\\ &= \frac{35}{50}-\frac{6}{50}\\ &= \frac{29}{50}\end{aligned}$$

55. We add in order from left to right.

$-17+24+(-9)=7+(-9)=-2$

57.

59.
$$\begin{aligned}7.035(4.91x-8.21)+17.401 &=\\ &\quad 23.902x-7.372815\\ 34.54185x-57.75735+17.401 &=\\ &\quad 23.902x-7.372815\\ 34.54185x-40.35635 &=\\ &\quad 23.902x-7.372815\\ 34.54185x-40.35635-23.902x &=\\ &\quad 23.902x-7.372815-23.902x\\ 10.63985x-40.35635 &= -7.372815\\ 10.63985x-40.35635+40.35635 &=\\ &\quad -7.372815+40.35635\\ 10.63985x &= 32.983535\\ \frac{10.63985x}{10.63985} &= \frac{32.983535}{10.63985}\\ x &= 3.1\end{aligned}$$

The solution is 3.1.

61.
$$\begin{aligned}5(x-4.2)+3[2x-5(x+7)] &=\\ &\quad 39+2(7.5-6x)+3x\\ 5(x-4.2)+3[2x-5x-35] &=\\ &\quad 39+2(7.5-6x)+3x\\ 5(x-4.2)+3[-3x-35] &= 39+2(7.5-6x)+3x\\ 5x-21-9x-105 &= 39+15-12x+3x\\ -4x-126 &= 54-9x\\ -4x-126+9x &= 54-9x+9x\\ 5x-126 &= 54\\ 5x-126+126 &= 54+126\\ 5x &= 180\\ \frac{5x}{5} &= \frac{180}{5}\\ x &= 36\end{aligned}$$

The solution is 36.

Exercise Set 5.8

1. ***Familiarize.*** Repeated addition fits this situation. We let C = the cost of 8 pairs of socks.

$$\underbrace{\boxed{\$4.95}+\boxed{\$4.95}+\cdots+\boxed{\$4.95}}_{\text{8 addends}}$$

Translate.

Price per pair	times	Number of pairs	is	Total cost
↓	↓	↓	↓	↓
4.95	×	8	=	C

Solve. We carry out the multiplication.

$$\begin{array}{r} 4.95\\ \times\quad 8\\ \hline 39.60\end{array}$$

Thus, $C=39.60$.

Check. We obtain a partial check by rounding and estimating:

$$4.95\times 8\approx 5\times 8=40\approx 39.60.$$

State. Eight pairs of socks cost \$39.60.

3. ***Familiarize.*** Repeated addition fits this situation. We let c = the cost of 17.7 gal of gasoline.

Translate.

Cost per gallon	times	Number of gallons	is	Total cost
↓	↓	↓	↓	↓
1.199	·	17.7	=	c

Solve. We carry out the multiplication.

$$\begin{array}{r} 1.199\\ \times\quad 17.7\\ \hline 8393\\ 83930\\ 119900\\ \hline 21.2223\end{array}$$

Thus, $c=21.2223$.

Check. We obtain a partial check by rounding and estimating:

$$1.199\times 17.7\approx 1\times 20=20\approx 21.2223.$$

State. We round \$21.2223 to the nearest cent and find that the cost of the gasoline is \$21.22.

5. ***Familiarize.*** We visualize the situation. We let c = the amount of change.

\$20	
\$16.99	c

Translate. This is a "take-away" situation.

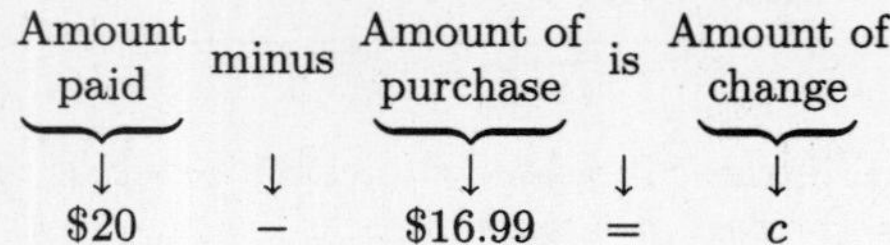

Solve. To solve the equation we carry out the subtraction.

$$\begin{array}{r} \overset{1}{\cancel{2}}\overset{9}{\cancel{0}}.\overset{9}{\cancel{0}}\overset{10}{\cancel{0}} \\ -\ 1\,6.\,9\,9 \\ \hline 3.\,0\,1 \end{array}$$

Thus, $c = \$3.01$.

Check. We check by adding 3.01 to 16.99 to get 20. This checks.

State. The change was $3.01.

7. ***Familiarize***. We visualize the situation. We let $n =$ the new temperature.

98.6°	4.2°
n	

Translate. We are combining amounts.

Normal body temperature	plus	Degrees temperature rises	is	New temperature
↓	↓	↓	↓	↓
98.6	+	4.2	=	n

Solve. To solve the equation we carry out the addition.

$$\begin{array}{r} \overset{1}{9}\,8.6 \\ +\quad 4.2 \\ \hline 1\,0\,2.8 \end{array}$$

Thus, $n = 102.8$.

Check. We can check by repeating the addition. We can also check by rounding:

$98.6 + 4.2 \approx 99 + 4 = 103 \approx 102.8$

State. The new temperature was 102.8°F.

9. ***Familiarize.*** Think of this as a rectangular array containing $47.60 arranged in 8 rows. We want to find how many dollars are in each row. We let s represent this number.

Translate. We think (Total cost) ÷ (Number of students) = (Each student's share).

Solve. To solve the equation we carry out the division.

$$\begin{array}{r} 5.9\,5 \\ 8\,\overline{)\,4\,7.6\,0} \\ 4\,0\,0\,0 \\ \hline 7\,6\,0 \\ 7\,2\,0 \\ \hline 4\,0 \\ 4\,0 \\ \hline 0 \end{array}$$

Thus, $s = 5.95$.

Check. We obtain a partial check by rounding and estimating:

$47.60 \div 8 \approx 50 \div 10 = 5 \approx 5.95$

State. Each person's share is $5.95.

11. ***Familiarize.*** We draw a picture, letting $A =$ the area.

A	312.6 ft
800.4 ft	

Translate. We use the formula $A = l \cdot w$.

$A = 800.4 \times 312.6$

Solve. We carry out the multiplication.

$$\begin{array}{r} 3\,1\,2.\,6 \\ \times\ 8\,0\,0.\,4 \\ \hline 1\,2\,5\,0\,4 \\ 2\,5\,0\,0\,8\,0\,0\,0 \\ \hline 2\,5\,0,\,2\,0\,5.\,0\,4 \end{array}$$

Thus, $A = 250,205.04$.

Check. We obtain a partial check by rounding and estimating:

$800.4 \times 312.6 \approx 800 \times 300 = 240,000 \approx 250,205.04$

State. The area is 250,205.04 ft^2.

13. ***Familiarize***. We visualize the situation. We let $m =$ the odometer reading at the end of the trip.

22,456.8 mi	234.7 mi
m	

Translate. We are combining amounts.

Reading before trip	plus	Miles driven	is	Reading at end of trip
↓	↓	↓	↓	↓
22,456.8	+	234.7	=	m

Solve. To solve the equation we carry out the addition.

$$\begin{array}{r} 2\,2{,}4\,\overset{1}{5}\,\overset{1}{6}.8 \\ +\quad 2\,3\,4.7 \\ \hline 2\,2{,}6\,9\,1.5 \end{array}$$

Thus, $m = 22,691.5$.

Check. We can check by repeating the addition. We can also check by rounding:

$22,456.8 + 234.7 \approx 22,460 + 230 = 22,690 \approx 22,691.5$

State. The odometer reading at the end of the trip was 22,691.5.

15. ***Familiarize***. We visualize the situation. We let $n =$ the number by which hamburgers exceed hot dogs, in billions.

24.8 billion	
15.9 billion	n

Translate. This is a "how-much-more" situation.

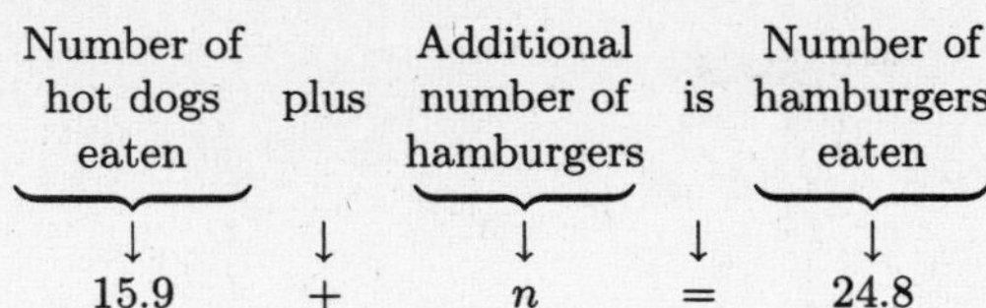

Solve. We subtract 15.9 on both sides.

$n = 24.8 - 15.9$
$n = 8.9$

$$\begin{array}{r} \overset{1}{\not 2}\,\overset{13}{\not 4}.\overset{18}{\not 8} \\ -\ 1\,5.9 \\ \hline 8.9 \end{array}$$

Check. We check by adding 8.9 to 15.9 to get 24.8. This checks.

State. Americans eat 8.9 billion more hamburgers than hot dogs.

17. ***Familiarize***. Because the question asks "how much greater?" we let d = the difference in age.

Translate. We translate to an equation.

Difference in age	is	1992 age	minus	1970 age
↓	↓	↓	↓	↓
d	$=$	24.4	$-$	20.8

Solve. We carry out the subtraction.

$d = 24.4 - 20.8$
$d = 3.6$

$$\begin{array}{r} 2\,\overset{3}{\not 4}.\overset{14}{\not 4} \\ -\ 2\,0.8 \\ \hline 3.6 \end{array}$$

Check. We can check by adding 3.6 to 20.8 to get 24.4. This checks.

State. The median age of a bride in 1992 was 3.6 years greater than in 1970.

19. ***Familiarize.*** This is a two-step problem. First, we find the number of miles that have been driven between fillups. This is a "how-much-more" situation. We let n = the number of miles driven.

Translate and Solve.

First odometer reading	plus	Number of miles driven	is	Second odometer reading
↓	↓	↓	↓	↓
26,342.8	$+$	n	$=$	26,736.7

To solve the equation we subtract 26,342.8 on both sides.

$n = 26,736.7 - 26,342.8$
$n = 393.9$

$$\begin{array}{r} 2\,6,\,7\,3\,6.7 \\ -\ 2\,6,\,3\,4\,2.8 \\ \hline 3\,9\,3.9 \end{array}$$

Second, we divide the total number of miles driven by the number of gallons. This gives us m = the number of miles per gallon.

$393.9 \div 19.5 = m$

To find the number m, we divide.

```
          2 0.2
1 9.5∧)3 9 3. 9∧0
       3 9 0 0
       -------
         3 9 0
         3 9 0
         -----
             0
```

Thus, $m = 20.2$.

Check. To check, we first multiply the number of miles per gallon times the number of gallons:

$19.5 \times 20.2 = 393.9$

Then we add 393.9 to 26,342.8:

$26,342.8 + 393.9 = 26,736.7$

The number 20.2 checks.

State. The van gets 20.2 miles per gallon.

21. ***Familiarize.*** We visualize a rectangular array consisting of 748.45 objects with 62.5 objects in each row. We want to find n, the number of rows.

Translate. We think (Total number of pounds) ÷ (Pounds per cubic foot) = (Number of cubic feet).

$748.45 \div 62.5 = n$

Solve. We carry out the division.

```
           1 1.9 7 5 2
6 2.5∧)7 4 8. 4∧5 0 0 0
        6 2 5 0 0
        ---------
        1 2 3 4 5
          6 2 5 0
        ---------
          6 0 9 5
          5 6 2 5
          -------
            4 7 0 0
            4 3 7 5
            -------
              3 2 5 0
              3 1 2 5
              -------
                1 2 5 0
                1 2 5 0
                -------
                      0
```

Thus, $n = 11.9752$.

Check. We obtain a partial check by rounding and estimating:

$748.45 \div 62.5 \approx 700 \div 70 = 10 \approx 11.9752$

State. The tank holds 11.9752 cubic feet of water.

23. ***Familiarize.*** This is a two-step problem. First, we find the number of games that can be played in one hour. Think of an array containing 60 minutes (1 hour = 60 minutes) with 1.5 minutes in each row. We want to find how many rows there are. We let g represent this number.

Translate and Solve. We think (Number of minutes) ÷ (Number of minutes per game) = (Number of games).

$60 \div 1.5 = g$

To solve the equation we carry out the division.

```
        4 0.
1.5∧)6 0. 0∧
     6 0 0
     -----
         0
         0
         -
         0
```

Thus, $g = 40$.

Second, we find the cost t of playing 40 video games. Repeated addition fits this situation. (We express 25¢ as \$0.25.)

$$\underbrace{\text{Cost of one game}}_{0.25} \ \underbrace{\text{times}}_{\times} \ \underbrace{\text{Number of games played}}_{40} \ \underbrace{\text{is}}_{=} \ \underbrace{\text{Total cost}}_{t}$$

To solve the equation we carry out the multiplication.

$$\begin{array}{r} 0.\,2\,5 \\ \times \quad 4\,0 \\ \hline 1\,0.\,0\,0 \end{array}$$

Thus, $t = 10$.

Check. To check, we first divide the total cost by the cost per game to find the number of games played:

$10 \div 0.25 = 40$

Then we multiply 40 by 1.5 to find the total time:

$1.5 \times 40 = 60$

The number 10 checks.

State. It costs \$10 to play video games for one hour.

25. ***Familiarize***. We let $d =$ the distance around the figure.

Translate. We are combining lengths.

$$\underbrace{\text{The sum of the lengths of the 5 sides}}_{8.9 + 23.8 + 4.7 + 22.1 + 18.6} \ \underbrace{\text{is}}_{=} \ \underbrace{\text{the distance around the figure.}}_{d}$$

Solve. To solve we carry out the addition.

$$\begin{array}{r} {}^{2\ 3} \\ 8.9 \\ 2\,3.8 \\ 4.7 \\ 2\,2.1 \\ +\,1\,8.6 \\ \hline 7\,8.1 \end{array}$$

Thus, $d = 78.1$.

Check. To check we can repeat the addition. We can also check by rounding:

$8.9+23.8+4.7+22.1+18.6 \approx 9+24+5+22+19 = 79 \approx 78.1$

State. The distance around the figure is 78.1 cm.

27. ***Familiarize***. This is a multistep problem. First we find the sum s of the two 0.8 cm segments. Then we use this length to find d.

Translate and Solve.

$$\underbrace{\text{Length of one small segment}}_{0.8} \ \underbrace{\text{plus}}_{+} \ \underbrace{\text{Length of other small segment}}_{0.8} \ \underbrace{\text{is}}_{=} \ \underbrace{\text{Total length.}}_{s}$$

To solve we carry out the addition.

$$\begin{array}{r} {}^{1} \\ 0.\,8 \\ +\,0.\,8 \\ \hline 1.\,6 \end{array}$$

Thus, $s = 1.6$.

Now we find d.

$$\underbrace{\text{Total length of smaller segments}}_{1.6} \ \underbrace{\text{plus}}_{+} \ \underbrace{\text{length of } d}_{d} \ \underbrace{\text{is}}_{=} \ \underbrace{3.91\text{ cm}}_{3.91}$$

To solve we subtract 1.6 on both sides of the equation.

$d = 3.91 - 1.6$
$d = 2.31$

$$\begin{array}{r} 3.9\,1 \\ -\,1.6\,0 \\ \hline 2.3\,1 \end{array}$$

Check. We repeat the calculations.

State. The length d is 2.31 cm.

29. ***Familiarize.*** This is a two-step problem. First, we find how many minutes there are in 2 hr. We let m represent this number. Repeated addition fits this situation (Remember that 1 hr = 60 min.)

Translate and Solve.

$$\underbrace{\text{Number of minutes in 1 hour}}_{60} \ \underbrace{\text{times}}_{\cdot} \ \underbrace{\text{Number of hours}}_{2} \ \underbrace{\text{is}}_{=} \ \underbrace{\text{Total number of minutes}}_{m}$$

To solve the equation we carry out the multiplication.

$$\begin{array}{r} 6\,0 \\ \times \quad 2 \\ \hline 1\,2\,0 \end{array}$$

Thus, $m = 120$.

Next, we find how many calories are burned in 120 minutes. We let t represent this number. Repeated addition fits this situation also.

$$\underbrace{\text{Number of calories burned in 1 minute}}_{8.6} \ \underbrace{\text{times}}_{\times} \ \underbrace{\text{Number of minutes}}_{120} \ \underbrace{\text{is}}_{=} \ \underbrace{\text{Total number of calories burned}}_{t}$$

To solve the equation we carry out the multiplication.

$$\begin{array}{r} 1\,2\,0 \\ \times \quad 8.\,6 \\ \hline 7\,2\,0 \\ 9\,6\,0\,0 \\ \hline 1\,0\,3\,2.\,0 \end{array}$$

Thus, $t = 1032$.

Check. To check, we first divide the total number of calories by the number of calories burned in one minute to find the total number of minutes the person mowed:

$1032 \div 8.6 = 120$

Then we divide 120 by 60 to find the number of hours:

$120 \div 60 = 2$

The number 1032 checks.

State. In 2 hr of mowing, 1032 calories would be burned.

31. ***Familiarize.*** This is a multistep problem. We will first find the total amount of the checks. Then we will find how much is left in the account after the checks are written. Finally, we will use this amount and the amount of the deposit to find the balance in the account after all the changes. We will let $c =$ the total amount of the checks.

Translate and Solve. We are combining amounts.

$$\underbrace{\text{First check}}_{\downarrow\; 23.82} \;\; \underset{\downarrow\;+}{\text{plus}} \;\; \underbrace{\text{Second check}}_{\downarrow\; 507.88} \;\; \underset{\downarrow\;+}{\text{plus}} \;\; \underbrace{\text{Third check}}_{\downarrow\; 98.32} \;\; \underset{\downarrow\;=}{\text{is}} \;\; \underbrace{\text{Total amount of checks}}_{\downarrow\; c}$$

To solve the equation we carry out the addition.

$$\begin{array}{r} \overset{1}{2}\,\overset{2}{3}.\overset{2}{8}\,\overset{1}{2} \\ 5\,0\,7.8\,8 \\ +\;\;9\,8.3\,2 \\ \hline 6\,3\,0.0\,2 \end{array}$$

Thus, $c = 630.02$.

Now we let $a =$ the amount in the account after the checks are written.

$$\underbrace{\text{Original amount}}_{\downarrow\; 1123.56} \;\; \underset{\downarrow\;-}{\text{less}} \;\; \underbrace{\text{Check amount}}_{\downarrow\; 630.02} \;\; \underset{\downarrow\;=}{\text{is}} \;\; \underbrace{\text{New amount}}_{\downarrow\; a}$$

To solve the equation we carry out the subtraction.

$$\begin{array}{r} \overset{10}{\not{0}}\;\; \\ \not{1}\,\not{1}\,\overset{12}{\not{2}}\,3.5\,6 \\ -\;\;6\;3\;0.0\,2 \\ \hline 4\;9\;3.5\,4 \end{array}$$

Thus, $a = 493.54$.

Finally, we let $f =$ the amount in the account after the paycheck is deposited.

$$\underbrace{\text{Amount after checks}}_{\downarrow\; 493.54} \;\; \underset{\downarrow\;+}{\text{plus}} \;\; \underbrace{\text{Amount of deposit}}_{\downarrow\; 678.20} \;\; \underset{\downarrow\;=}{\text{is}} \;\; \underbrace{\text{Final amount}}_{\downarrow\; f}$$

We carry out the addition.

$$\begin{array}{r} \overset{1}{4}\,\overset{1}{9}\,3.5\,4 \\ +\,6\,7\,8.2\,0 \\ \hline 1\,1\,7\,1.7\,4 \end{array}$$

Thus, $f = 1171.74$.

Check. We repeat the calculations.

State. There is \$1171.74 in the account after the changes.

33. ***Familiarize.*** We make and label a drawing. The question deals with a rectangle and a circle, so we also list the relevant area formulas. We let $g =$ the area covered by grass.

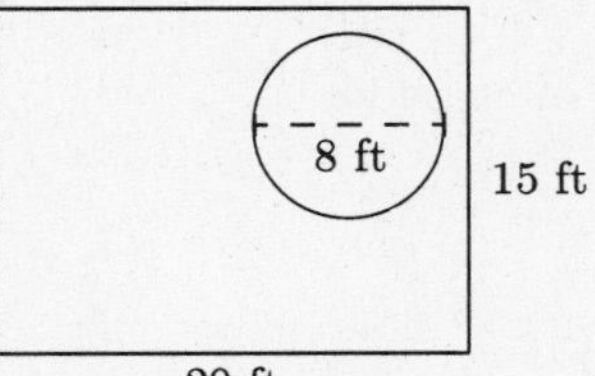

Area of a rectangle with length l and width w:
$A = l \times w$

Area of a circle with radius r: $A = \pi r^2$, where $\pi \approx 3.14$

Translate. We subtract the area of the circle from the area of the rectangle. Recall that a circle's radius is half of its diameter.

$$\underbrace{\text{Area of rectangle}}_{\downarrow\; 20\times 15} \;\; \underset{\downarrow\;-}{\text{minus}} \;\; \underbrace{\text{Area of circle}}_{\downarrow\; 3.14\left(\frac{8}{2}\right)^2} \;\; \underset{\downarrow\;=}{\text{is}} \;\; \underbrace{\text{Area covered by grass}}_{\downarrow\; g}$$

Solve. We carry out the computations.

$$\begin{aligned} 20 \times 15 - 3.14\left(\frac{8}{2}\right)^2 &= g \\ 20 \times 15 - 3.14(4)^2 &= g \\ 20 \times 15 - 3.14 \times 16 &= g \\ 300 - 50.24 &= g \\ 249.76 &= g \end{aligned}$$

Check. We can repeat the calculations. Also note that 249.76 is less than the area of the yard but more than the area of the flower garden. This agrees with the impression given by our drawing.

State. Grass covers 249.76 ft^2 of the yard.

35. ***Familiarize.*** We make and label a drawing. The question deals with a square and a circle, so we also list the relevant area formulas. We let $c =$ the amount of cloth left over.

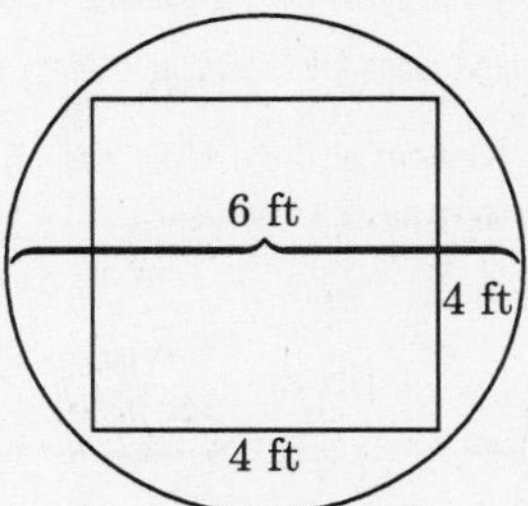

Area of square with sides of length s: $A = s^2$

Area of circle with radius r: $A = \pi r^2$, where $\pi \approx 3.14$

Translate. We subtract the area of the square from the area of the circle. Recall that a circle's radius is half of its diameter, or width.

$$\underbrace{\text{Area of circle}} \text{ minus } \underbrace{\text{Area of square}} \text{ is } \underbrace{\text{Area left over}}$$

$$3.14\left(\frac{6}{2}\right)^2 \quad - \quad 4^2 \quad = \quad c$$

Solve. We carry out the computations.

$$3.14\left(\frac{6}{2}\right)^2 - 4^2 = c$$
$$3.14(3)^2 - 4^2 = c$$
$$3.14 \times 9 - 16 = c$$
$$28.26 - 16 = c$$
$$12.26 = c$$

Check. We can repeat the calculations. Also note that 12.26 is less than the area of the square which in turn is less than the area of the circle. This agrees with the impression given by our drawing.

State. The amount of cloth left over is 12.26 ft^2.

37. ***Familiarize.*** The total cost is the sum of the daily charge and the mileage charge. The mileage charge is the product of the cost per mile and the number of miles driven. We let c = the total cost, in dollars, of driving 120 mi in 1 day. Note that we convert 27 cents to \$0.27 so that only one unit, dollars, is used.

Translate.

$$\underbrace{\text{Daily charge}} \text{ plus } \underbrace{\text{Cost per mile}} \text{ times } \underbrace{\text{Number of miles driven}} \text{ is } \underbrace{\text{Total cost}}$$

$$\$24.95 \quad + \quad \$0.27 \quad \times \quad 120 \quad = \quad c$$

Solve. We solve the equation.

$$24.95 + 0.27 \cdot 120 = c$$
$$24.95 + 32.4 = c$$
$$57.35 = c$$

Check. We repeat the calculation. Our answer checks.

State. The total cost of driving 120 mi in 1 day is \$57.35.

39. ***Familiarize.*** Repeated addition fits this situation. We let t = the total savings in 1 year.

Translate.

$$\underbrace{\text{Number of weeks}} \text{ times } \underbrace{\text{Savings per week}} \text{ is } \underbrace{\text{Total savings}}$$

$$52 \quad \cdot \quad 6.72 \quad = \quad t$$

Solve. To solve the equation we carry out the multiplication.

```
     6. 7 2
   ×    5 2
   ---------
   1 3 4 4
 3 3 6 0 0
 ---------
 3 4 9. 4 4
```

Check. We can obtain a partial check by rounding and estimating:

$$52 \cdot 6.72 \approx 50 \cdot 7 = 350 \approx 349.44.$$

State. The family would save \$349.44 in 1 year.

41. ***Familiarize.*** This is a two-step problem. First we find the number of eggs in 20 dozen (1 dozen = 12). We let n represent this number.

Translate and Solve. We think (Number of dozens) $\cdot$ (Number in a dozen) = (Number of eggs).

$$20 \cdot 12 = n$$
$$240 = n$$

Second, we find the cost c of one egg. We think (Total cost) $\div$ (Number of eggs) = (Cost of one egg).

$$\$13.80 \div 240 = c$$

We carry out the division.

```
          0.0 5 7 5
 2 4 0 ) 1 3.8 0 0 0
         1 2 0 0
           1 8 0 0
           1 6 8 0
             1 2 0 0
             1 2 0 0
                   0
```

Thus, $c = 0.0575 \approx 0.058$ (rounded to the nearest tenth of a cent).

Check. We repeat the calculations.

State. Each egg cost about \$0.058, or 5.8¢.

43. ***Familiarize.*** This is a three-step problem. We will find the area S of a standard soccer field and the area F of a standard football field using the formula Area = $l \cdot w$. Then we will find E, the amount by which the area of a soccer field exceeds the area of a football field.

Translate and Solve.

$$S = l \cdot w = 114.9 \times 74.4 = 8548.56$$

$$F = l \cdot w = 120 \times 53.3 = 6396$$

$$\underbrace{\text{Area of football field}} \text{ plus } \underbrace{\text{Excess area of soccer field}} \text{ is } \underbrace{\text{Area of soccer field}}$$

$$6396 \quad + \quad E \quad = \quad 8548.56$$

To solve the equation we subtract 6396 on both sides.

$$E = 8548.56 - 6396$$
$$E = 2152.56$$

```
      4 14
  8 5̸ 4̸ 8. 5 6
− 6 3 9 6. 0 0
--------------
  2 1 5 2. 5 6
```

Check. We can obtain a partial check by rounding and estimating:

$$114.9 \times 74.4 \approx 110 \times 75 = 8250 \approx 8548.56$$

$$120 \times 53.3 \approx 120 \times 50 = 6000 \approx 6396$$

$$8250 - 6000 = 2250 \approx 2152.56$$

State. The area of a soccer field is 2152.56 yd^2 greater than the area of a football field.

45. ***Familiarize.*** This is a three-step problem. First we find m = the number of months in 30 years.

Translate and Solve.

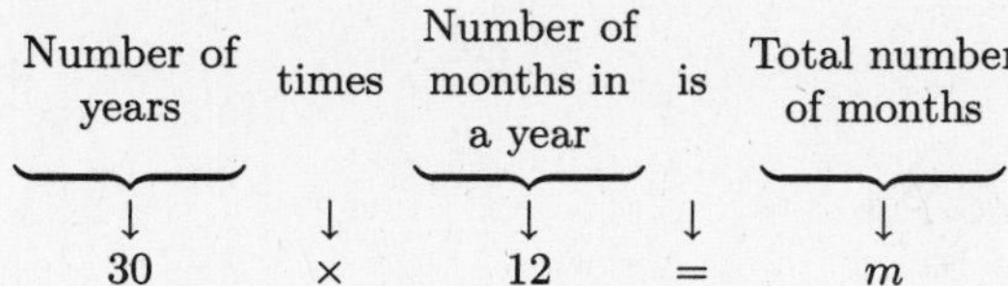

We carry out the multiplication.

$$\begin{array}{r} 1\,2 \\ \times\ 3\,0 \\ \hline 3\,6\,0 \end{array}$$

Thus, $m = 360$.

Next we find a = the amount paid back.

Monthly payment	times	Number of months	is	Amount paid back
↓	↓	↓	↓	↓
880.52	×	360	=	a

We carry out the multiplication.

$$\begin{array}{r} 8\,8\,0.\,5\,2 \\ \times\quad 3\,6\,0 \\ \hline 5\,2\,8\,3\,1\,2\,0 \\ 2\,6\,4\,1\,5\,6\,0\,0 \\ \hline 3\,1\,6,\,9\,8\,7.\,2\,0 \end{array}$$

Thus, $a = 316,987.20$.

Finally, we find p = the amount by which the amount paid back exceeds the amount of the loan. This is a "how-much-more" situation.

Amount of loan	plus	Excess amount	is	Amount paid back
↓	↓	↓	↓	↓
120,000	+	p	=	316,987.20

We subtract 120,000 on both sides.

$p = 316,987.20 - 120,000$
$p = 196,987.20$

$$\begin{array}{r} 3\,1\,6,\,9\,8\,7.\,2\,0 \\ -\ 1\,2\,0,\,0\,0\,0.\,0\,0 \\ \hline 1\,9\,6,\,9\,8\,7.\,2\,0 \end{array}$$

Check. We repeat the calculations.

State. You pay back \$316,987.20. This is \$196,987.20 more than the amount of the loan.

47. ***Familiarize.*** We let m = the number of miles that can be driven within an \$80 budget. We convert 10 cents to \$0.10 so that only one unit, dollars, is used.

Translate. We translate to an equation.

Daily rate	plus	Cost per mile	times	Number of miles driven	is	Total cost
↓	↓	↓	↓	↓	↓	↓
\$34.95	+	\$0.10	·	m	=	80

Solve. We solve the equation.

$34.95 + 0.10m = 80$

$0.10m = 45.05$ Subtracting 34.95 on both sides

$m = \frac{45.05}{0.10}$ Dividing by 0.10 on both sides

$m = 450.5$

Check. The mileage cost is found by multiplying 450.5 by \$0.10 obtaining \$45.05. Then we add \$45.05 to \$34.95, the daily rate, and get \$80. The result checks.

State. The businessperson can drive 450.5 mi on the car-rental allotment.

49. ***Familiarize.*** The total cost is the insurance charge plus the charge for the time the bike is used. The charge for the time is the cost per hour times the number of hours the bike is used. Let t = the number of hours that a person can rent a bike for \$25.00.

Translate.

Insurance charge	plus	Cost per hour	times	Number of hours used	is	Total cost
↓	↓	↓	↓	↓	↓	↓
5.50	+	2.40	·	t	=	25.00

Solve. We solve the equation.

$5.50 + 2.40t = 25.00$

$2.40t = 19.50$ Subtracting 5.50 on both sides

$\frac{2.40t}{2.40} = \frac{19.50}{2.40}$ Dividing by 2.40 on both sides

$t = 8.125$

Check. The cost for the time is found by multiplying 8.125 by \$2.40 obtaining \$19.50. Then we add \$19.50 to \$5.50, the insurance charge, and get \$25.00. The result checks.

State. With \$25.00, a person can rent a bike for 8.125 hours.

51. ***Familiarize.*** The total cost is the charge for a house call plus the charge for the repairperson's time. The charge for the time is the cost per hour times the number of hours the repairperson worked. Let h = the number of hours the repairperson worked.

Translate.

House call charge	plus	Cost per hour	times	Number of hours	is	Total cost
↓	↓	↓	↓	↓	↓	↓
\$20	+	\$27.50	·	h	=	\$116.25

Solve. We solve the equation.

$20 + 27.50h = 116.25$

$27.50h = 96.25$ Subtracting 20 on both sides

$\frac{27.50h}{27.50} = \frac{96.25}{27.50}$ Dividing by 27.50 on both sides

$h = 3.5$

Check. The cost for the time worked is $27.50 · 3.5, or $96.25. We add $96.25 to $20, the cost of a house call, and get $116.25. Our answer checks.

State. The repairperson worked 3.5 hr.

53. ***Familiarize***. This is a multistep problem. First, we find the cost of the cheese. We let c = the cost of the cheese.

Translate and Solve.

$$\underbrace{\text{Number of pounds}}_{\downarrow\; 6} \;\text{times}\; \cdot \;\underbrace{\text{Price per pound}}_{\downarrow\; \$4.79} \;\text{is}\; = \;\underbrace{\text{Cost of cheese}}_{\downarrow\; c}$$

To solve the equation we carry out the multiplication.

$$\begin{array}{r} \$4.\,7\,9 \\ \times \quad\;\; 6 \\ \hline \$2\,8.\,7\,4 \end{array}$$

Thus, $c = \$28.74$.

Next, we subtract to find how much money m is left to purchase seltzer.

$m = \$40 - \$28.74$

$m = \$11.26$

$$\begin{array}{r} \scriptstyle 3\;\,9\;\,9\;10 \\ \not{4}\not{0}.\not{0}\,\not{0} \\ -\,2\,8.\,7\,4 \\ \hline 1\,1.\,2\,6 \end{array}$$

Finally, we divide the amount of money left over by the cost of a bottle of seltzer to find how many bottles can be purchased. We let b = the number of bottles of seltzer that can be purchased.

$\$11.26 \div \$0.64 = b$

To find b we carry out the division.

$$\begin{array}{r} 1\,7. \\ 0.6\,4_{\wedge}\overline{)\,1\,1.2\,6_{\wedge}} \\ 6\,4\,0 \\ \hline 4\,8\,6 \\ 4\,4\,8 \\ \hline 3\,8 \end{array}$$

We stop dividing at this point, because Frank cannot purchase a fraction of a bottle. Thus, $b = 17$ (rounded to the nearest 1).

Check. The cost of the seltzer is $17 \cdot \$0.64$ or $10.88. The cost of the cheese is $6 \cdot \$4.79$, or $28.74. Frank has spent a total of $\$10.88 + \$28.74$, or $39.62. Frank has $\$40 - \$39.62$, or $0.38 left over. This is not enough to purchase another bottle of seltzer, so our answer checks.

State. Frank should buy 17 bottles of seltzer.

55. $\dfrac{0}{n} = 0$, for any integer n that is not 0.

Thus, $\dfrac{0}{-13} = 0$.

57. $\dfrac{8}{11} - \dfrac{4}{3} = \dfrac{8}{11} \cdot \dfrac{3}{3} - \dfrac{4}{3} \cdot \dfrac{11}{11}$ The LCM is 33.

$= \dfrac{24}{33} - \dfrac{44}{33}$

$= \dfrac{-20}{33}$, or $-\dfrac{20}{33}$

59.

$$\begin{array}{rcrcr} 4\dfrac{1}{3} & = & 4\dfrac{1}{3}\cdot\dfrac{2}{2} & = & 4\dfrac{2}{6} \\ +\,2\dfrac{1}{2} & = & +\,2\dfrac{1}{2}\cdot\dfrac{3}{3} & = & +\,2\dfrac{3}{6} \\ \hline & & & & 6\dfrac{5}{6} \end{array}$$

61. 

63. ***Familiarize***. This is a multistep problem. First we find the area that is to be seeded. We make a drawing.

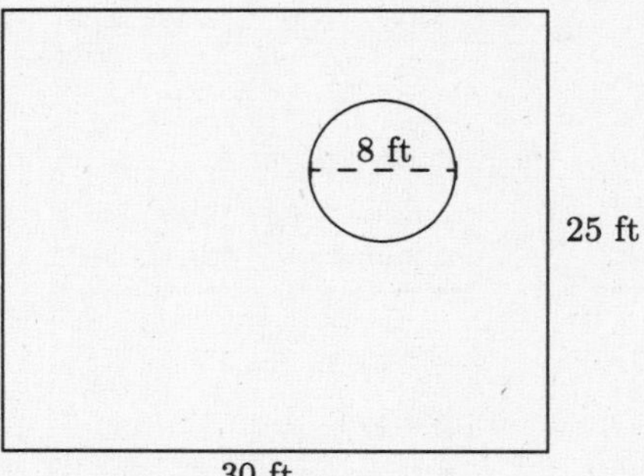

We let a = the area to be seeded. Recall that the area of a rectangle with length l and width w is $A = l \times w$ and the area of a circle with radius r is $A = \pi r^2$, where $\pi \approx 3.14$.

Translate and Solve. To find the area to be seeded we subtract the area of the base of the fountain from the area of the yard. Recall that a circle's radius is half of its diameter, or width.

$$\underbrace{\text{Area of yard}}_{\downarrow\; 30\times 25} \;\text{minus}\; - \;\underbrace{\text{Area of fountain}}_{\downarrow\; 3.14\left(\frac{8}{2}\right)^2} \;\text{is}\; = \;\underbrace{\text{Area to be seeded}}_{\downarrow\; a}$$

To solve the equation we carry out the computations.

$$\begin{aligned} 30 \times 25 - 3.14\left(\frac{8}{2}\right)^2 &= a \\ 30 \times 25 - 3.14(4)^2 &= a \\ 30 \times 25 - 3.14(16) &= a \\ 750 - 50.24 &= a \\ 699.76 &= a \end{aligned}$$

Now we find the number of pounds p of grass seed that should be purchased. We think (Number of square feet) ÷ (Number of square feet covered by 1 pound of seed) = (Number of pounds needed).

$699.76 \div 300 = p$

To solve the equation we carry out the division on a calculator. We find that $p \approx 2.3$.

Check. We recheck the calculations. Our answer checks.

State. About 2.3 lb of grass seed should be purchased.

65. ***Familiarize.*** Let $p =$ the highest price per mile Lindsey can afford.

Translate.

$$\underbrace{\text{Daily charge}}_{\$18.90} \ \underset{+}{\text{plus}} \ \underbrace{\text{Cost per miles}}_{p} \ \underset{\cdot}{\text{times}} \ \underbrace{\text{Number of miles}}_{190} \ \underset{=}{\text{is}} \ \underbrace{\text{Total cost}}_{\$55}$$

Solve. We solve the equation.

$$18.90 + 190p = 55$$

$$190p = 36.1 \quad \text{Subtracting 18.90 on both sides}$$

$$\frac{190p}{190} = \frac{36.1}{190} \quad \text{Dividing by 190 on both sides}$$

$$p = 0.19$$

Check. When the cost per mile is \$0.19 the total cost of the rental is \$18.90 + \$0.19 · 190, or \$55. Clearly, if the cost per mile is greater than \$0.19, the rental cost will be greater than \$55. Our answer checks.

State. The highest price per mile that Lindsey can afford is \$0.19.

Chapter 6

Introduction to Graphing and Statistics

Exercise Set 6.1

1. Go down the left column (headed "Penny Number") to 40. Then go across to the column headed "Length" (immediately to the right of the Penny Number column) and read the entry, 5. The Length column gives the length in inches, so a 40-penny nail is 5 inches long.

3. Go down the left column (headed "Penny Number") to 10. Then go across to the columns headed "Approx. Number Per Pound," and find the column headed "Box Nails." The entry is 94, so there are approximately 94 10-penny nails in a pound.

5. First, go down the "Penny Number" column to 16, and then go across to the "Finishing Nails" column to find how many 16-penny finishing nails there are in a pound. There are approximately 90. Then go down the "Penny Number" column to 10 and across to the "Common Nails" column to find how many 10-penny nails there are in a pound. There are approximately 69. The difference, 90 − 69, or 21, tells us that there are approximately 21 more 16-penny finishing nails in a pound.

7. Go down the "Penny Number" column to 20, and then go across to the "Box Nails" column. The entry is 52, so there are approximately 52 box nails in a pound.

9. Go down the "Penny Number" column to 4, and then go across to the "Finishing Nails" column. The entry is 548, so there are approximately 548 4-penny finishing nails in one pound. Then there are approximately 5×548, or 2740 nails in 5 pounds.

11. Go down the "Activity" column to Racquetball, and then go across to the "154 lb" column under the heading "Calories Burned in 30 Minutes." The entry is 294, so 294 calories have been burned by a 154-pound person after 30 minutes of racquetball.

13. Go down the "110 lb" column to 216, and then go across to the "Activity" column. The entry is Calisthenics, so calisthenics burns 216 calories in 30 minutes for a 110-pound person.

15. Going down the "Activity" column to Aerobic Dance and then across to the "154 lb" column, we find the entry 282. Next, going down the "Activity" column to Tennis and across to the "154 lb" column, we find the entry 222. Since $282 > 222$, aerobic dance burns more calories in 30 minutes for a 154-pound person.

17. Going down the "Activity" column to Tennis and then across to the "110 lb" column, we see that 165 calories are burned after 30 minutes of tennis for a 110-pound person. Now, 2 hours, or 120 minutes, is 4×30 minutes, so we multiply to find the answer: 4×165 calories $= 660$ calories.

19. Go down the "132 lb" column and find the smallest entry. It is 132 in the bottom row. Go across the bottom row to the "Activity" column, and read the entry, Moderate Walking.

21. We observe that 120 pounds is approximately half-way between 110 pounds and 132 pounds (or is approximately the average of 110 pounds and 132 pounds). Therefore, we would expect the number of calories burned by a 120-pound person during 30 minutes of moderate walking to be approximately half-way between (or the average of) the numbers of calories burned by a 110-pound person and a 132-pound person during 30 minutes of moderate walking. Going down the "Activity" column to Moderate Walking and then across, first to the "110 lb" column, and then to the "132 lb" column, we find the entries 111 and 132, respectively. The number half-way between these two (or their average) is $\dfrac{111+132}{2}$, or 121.5, or approximately 121. (We round down since 120 is closer to 110 than to 132.) Therefore, you would expect to burn about 121 calories during 30 minutes of moderate walking.

23. There are more bike symbols beside 1995 than any other year. Therefore, the year in which the greatest number of bikes was sold is 1995.

25. There was positive growth between every pair of consecutive years except 1993 and 1994. The pair of years for which the amount of positive growth was the least was 1991 and 1992 (represented by an increase of 1 bike symbol as opposed to 2 or 3 symbols for each of the other pairs).

27. Sales for 1992 are represented by 5 bike symbols. Since each bike symbol stands for 1000 bikes sold, we multiply to find 5×1000, or 5000, bikes were sold in 1992.

29. We look for a row of the chart containing fewer symbols than the one immediately below it. The only such row is the one showing sales in 1994. Therefore, in 1994 there was a decline in the number of bikes sold.

31. Singles are represented by 5 whole symbols (5×3) and $1/3$ of another symbol $(1/3 \times 3)$ for a total of $15 + 1$, or 16.

33. Triples are represented by 1 whole symbol, so there were 1×3, or 3, triples. In Exercise 33 we calculated that there were 16 singles. Then there were $16 - 3$, or 13 fewer triples than singles.

35. $12 \div 3 = 4$, so we are looking for a row containing exactly 4 whole symbols. Only the row representing outs contains exactly 4 whole symbols, so the player made an out exactly 12 times.

37. Since each flame symbol will represent 10 calories, we divide the number of calories for each tablespread by 10 to determine the number of symbols to use.

Jam: $54 \div 10 = 5.4$

Mayonnaise: $51 \div 10 = 5.1$

Peanut Butter: $94 \div 10 = 9.4$

Honey: $64 \div 10 = 6.4$

Syrup: $60 \div 10 = 6$

We fill in the pictograph with the appropriate number of symbols beside each tablespread.

Tablespread

Jam

Mayonnaise

Peanut butter

Honey

Syrup

= 10 calories

39. $-49 \div (-49) = \dfrac{-49}{-49} = 1$

$\left(\dfrac{n}{n} = 1, \text{ for any integer } n \text{ that is not } 0.\right)$

41. 

43. First we consider home runs: Home runs are represented by 1 whole symbol ($1 \times 3 = 3$) and 2/3 of another symbol ($2/3 \times 3 = 2$) for a total of $3 + 2$, or 5 home runs. Each home run accounts for 4 bases, so home runs accounted for 4×5, or 20 bases.

Next we consider triples: From Exercise 33 we know there were 3 triples. Each triple accounts for 3 bases, so triples accounted for 3×3, or 9 bases.

Now we consider doubles: Doubles are represented by 3 whole symbols, so there were 3×3, or 9 doubles. Each double accounts for 2 bases, so doubles accounted for 2×9, or 18 bases.

Finally, we consider singles: From Exercise 31 we know that there were 16 singles and each accounts for 1 base. Thus, singles accounted for 16 bases.

We add the number of bases for each type of hit: $20 + 9 + 18 + 16 = 63$

The player's hits accounted for 63 bases.

45. Brazil's population of 150 million will be represented by 150 million $\div$ 75 million, or 2 symbols.

Germany's population of 75 million will be represented by 75 million $\div$ 75 million, or 1 symbol.

India's population of 825 million will be represented by 825 million $\div$ 75 million, or 11 symbols.

Japan's population of 125 million will be represented by 125 million $\div$ 75 million, or $1\frac{2}{3}$ symbols.

The United States' population of 260 million will be represented by 260 million $\div$ 75 million, or $3\frac{7}{15}$, or about $3\frac{1}{2}$ symbols.

We redraw the pictograph.

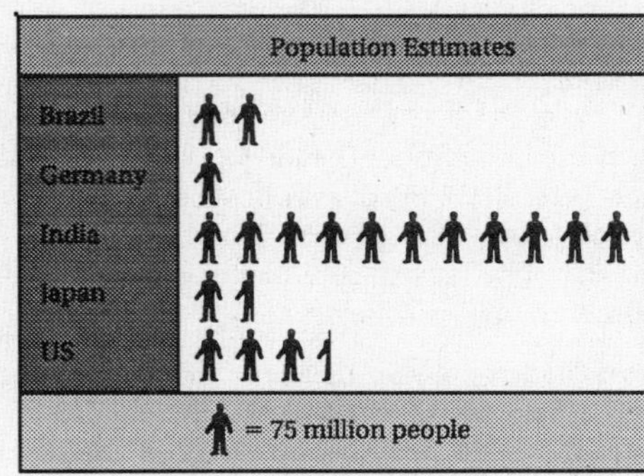

Exercise Set 6.2

1. We look for the longest bar and find that it represents Los Angeles, CA.

3. Go to the right end of the bar representing Pittsburgh, PA, and then go down to the "Weeks of Growing Season" scale. We can read, fairly accurately, that the growing season is 22 weeks long.

5. Going to the right end of the bar representing Fresno, CA, and then down to the scale, we find that the growing season in Fresno is 40 weeks long. Doing the same for the bar representing Ogden, UT, we find that the growing season in Ogden is 20 weeks long. Since $40 \div 20 = 2$, the growing season in Fresno, CA, is 2 times as long as in Ogden, UT.

7. Going to the right end of the bar representing Los Angeles, CA, and then down to the scale, we find that the growing season in Los Angeles is 52 weeks long. Now $\frac{1}{2} \times 52 = 26$, so we go to 26 on the "Weeks of Growing Season" scale and then go up, looking for the bar that ends closest to 26 weeks. It is the top bar. We then go across to the left and read the name of the city, Syracuse, NY.

9. We go to the right across the bottom of the graph to the white bar above the years 1971-74. Next we go to the top of that bar and, from there, back to the left to read 215 on the vertical scale. The average cholesterol level for women during the years 1971-1974 was approximately 215.

11. We read up the vertical scale to 217 (2/5 of the distance from 215 to 220). From there we move to the right until we come to the top of a blue bar. Moving down that bar we find that in 1960-1962 the average cholesterol level for men was 217.

13. Following the procedure described in Exercise 9, we find that the average cholesterol level for women in 1960 was approximately 222, and in 1991 it was approximately 205. We subtract to find how much the level dropped:

$222 - 205 = 17$

Thus, the average cholesterol level for women dropped approximately 17 mg/100 ml from 1960 to 1991.

15. Label the marks on the vertical scale with the names of the activities and title it "Activity." Label the horizontal scale appropriately by 100's and title it "Calories burned per hour." Then draw horizontal bars to show the calories burned by each activity.

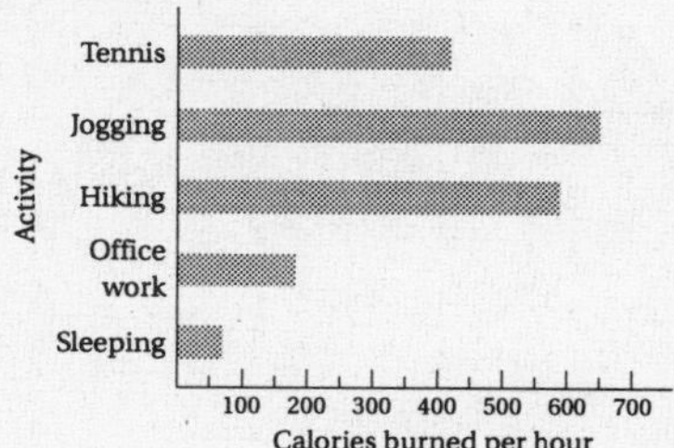

17. Disregarding jogging, the activity in which the most calories per hour are burned is hiking. Thus, if you were trying to lose weight by exercising and could not jog, hiking would be the most beneficial exercise.

19. Find the highest point on the line, and go down to the "Time" scale. The temperature was highest at 3 P.M.

21. Finding the highest point on the line and going straight across to the "Temperature" scale, we see that the highest temperature was about 51°F. Doing the same for the lowest point on the line, we see that the lowest temperature was about 36°F. The difference is 51°F − 36°F, or approximately 15°F.

23. Reading the graph from left to right, we see that the line went down the most between 5 P.M. and 6 P.M.

25. Find the highest point on the line, and go down to the "Year" scale. Estimated sales are the greatest in 1997.

27. Find 1995 on the bottom scale and go up to the point on the line directly above it. Then go straight across to the "Estimated Sales" scale. We are at about 17.0, so estimated sales in 1995 are approximately $17.0 million.

29. The graph shows estimated sales of about $19.5 million in 1997 and about $18.0 million in 1999. Now, $19.5 million − $18.0 million = $1.5 million, so estimated sales in 1997 are approximately $1.5 million greater than in 1999.

31. First, label the horizontal scale with the times, and title it "Time (P.M.)." Then label the vertical scale by 10's, and title it "Number of cars." Next, mark points above the times at appropriate levels, and then draw line segments connecting them.

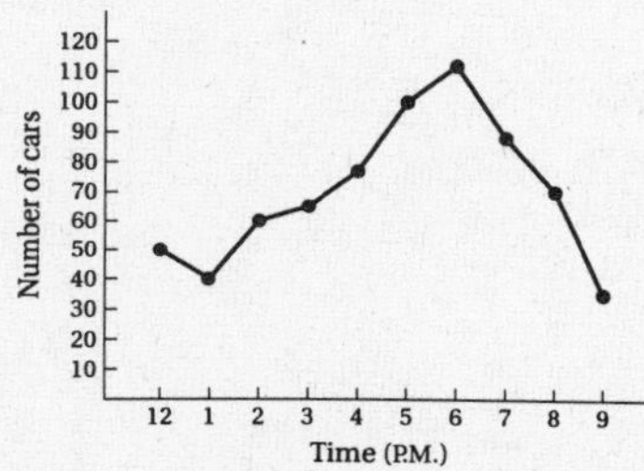

33. Reading the graph from left to right, we see that the line went down the most between 8 P.M. and 9 P.M.

35. ***Familiarize.*** We draw a picture. We let n = the number of 12-oz bottles that can be filled.

12 oz in each row

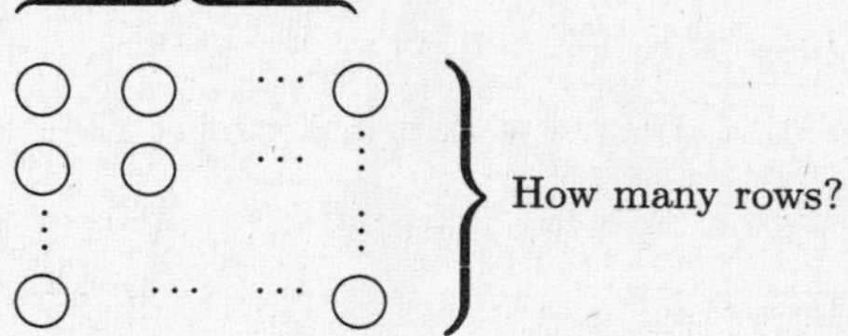

Translate and Solve. We translate to an equation and solve as follows:

$408 \div 12 = n$

$$\begin{array}{r} 34 \\ 12\overline{)408} \\ \underline{360} \\ 48 \\ \underline{48} \\ 0 \end{array}$$

Check. We check by multiplying the number of bottles by 12:

$12 \times 34 = 408$

State. 34 twelve-oz bottles can be filled.

37. ***Familiarize.*** We let n = the number.

Translate.

$$\begin{array}{ccccc} 24 & \text{is} & \frac{3}{4} & \text{of} & \underbrace{\text{what number?}} \\ \downarrow & \downarrow & \downarrow & \downarrow & \downarrow \\ 24 & = & \frac{3}{4} & \cdot & n \end{array}$$

Solve. We multiply by $\frac{4}{3}$ on boths sides.

$$24 = \frac{3}{4}n$$
$$\frac{4}{3} \cdot 24 = \frac{4}{3} \cdot \frac{3}{4}n$$
$$\frac{4 \cdot 24}{3} = n$$
$$\frac{4 \cdot 3 \cdot 8}{3 \cdot 1} = n$$
$$\frac{3}{3} \cdot \frac{4 \cdot 8}{1} = n$$
$$32 = n$$

Check. We repeat the calculations.

State. 24 is $\frac{3}{4}$ of 32.

39.

41. First we find the number of calories Bonnie consumed during the day.

Breakfast	700 calories
Lunch	615 calories
Snack	235 calories
Total	1550 calories

Next we find the number of calories she burned during the day. Jogging for 45 min, or $\frac{3}{4}$ hr, burned $\frac{3}{4} \cdot 650$, or $487\frac{1}{2}$ calories. Bonnie worked in her office for a total of $2\frac{1}{2}$ hr $+ 5\frac{1}{2}$ hr, or 8 hr. This burned $8 \cdot 180$, or 1440 calories. Playing tennis for 40 min, or $\frac{2}{3}$ hr, burned $\frac{2}{3} \cdot 420$, or 280 calories.

All together, Bonnie burned $487\frac{1}{2}$ calories + 1440 calories + 280 calories, or $2207\frac{1}{2}$ calories.

We subtract the number of calories burned from the number consumed to find how many calories were lost or gained:

1550 calories $- 2207\frac{1}{2}$ calories $= -657\frac{1}{2}$ calories.

Thus, Bonnie lost $657\frac{1}{2}$ calories in the course of the day.

Exercise Set 6.3

1. To plot $(2,5)$, we locate 2 on the first, or horizontal, axis. Then we go up 5 units and make a dot.

To plot $(-1,3)$, we locate -1 on the first, or horizontal, axis. Then we go up 3 units and make a dot.

To plot $(3,-2)$, we locate 3 on the first, or horizontal, axis. Then we go down 2 units and make a dot.

To plot $(-2,-4)$, we locate -2 on the first, on the horizontal, axis. Then we go down 4 units and make a dot.

To plot $(0,4)$, we locate 0 on the first, or horizontal, axis. Then we go up 4 units and make a dot.

To plot $(0,-5)$, we locate 0 on the first, or horizontal, axis. Then we go down 5 units and make a dot.

To plot $(5,0)$, we locate 5 on the first, or horizontal, axis. Since the second coordinate is 0, we do not move up or down. We make a dot at the point we located on the first axis.

To plot $(-5,0)$, we locate -5 on the first, or horizontal, axis. Since the second coordinate is 0, we do not move up or down. We make a dot at the point we located on the first axis.

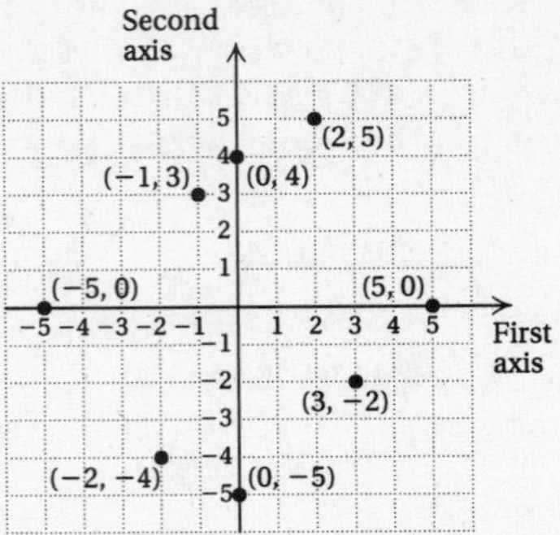

3. To plot $(-3,-1)$, we locate -3 on the first, or horizontal, axis. Then we go down 1 unit and make a dot.

To plot $(5,1)$, we locate 5 on the first, or horizontal, axis. Then we go up 1 unit and make a dot.

To plot $(-1,-5)$, we locate -1 on the first, or horizontal, axis. Then we go down 5 units and make a dot.

To plot $(0,0)$, we locate 0 on the first, or horizontal, axis. Since the second coordinate is 0, we do not move up or down. We make a dot at the point we located on the first axis.

To plot $(0,1)$, we locate 0 on the first, or horizontal, axis. Then we go up 1 unit and make a dot.

To plot $(-4,0)$, we locate -4 on the first, or horizontal, axis. Since the second coordinate is 0, we do not move up or down. We make a dot at the point we located on the first axis.

To plot $\left(2,3\frac{1}{2}\right)$, we locate 2 on the first, or horizontal, axis. Then we go up $3\frac{1}{2}$ units and make a dot.

To plot $\left(4\frac{1}{2},-2\right)$, we locate $4\frac{1}{2}$ on the first, or horizontal, axis. Then we go down 2 units and make a dot.

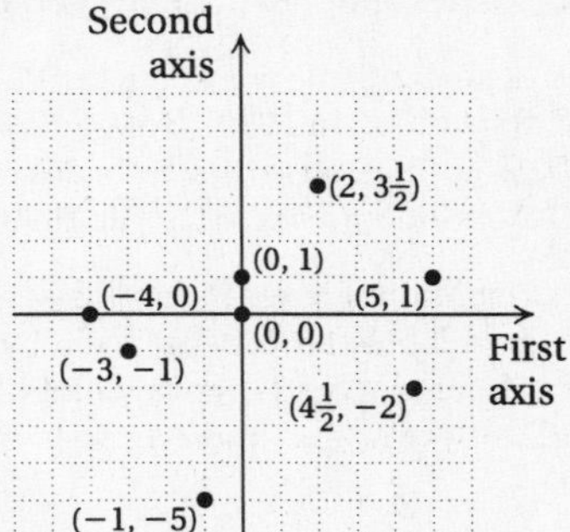

5.

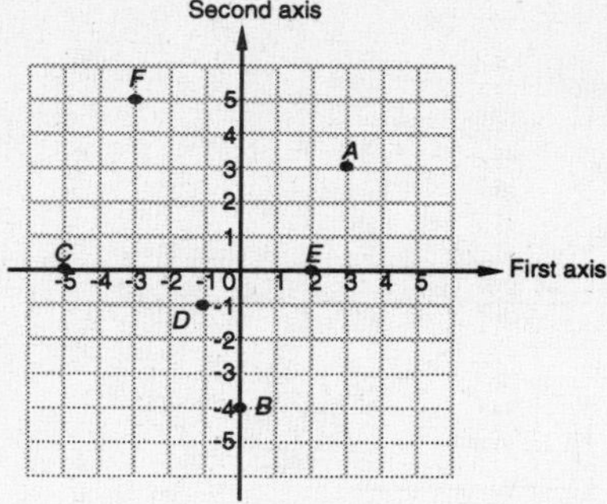

We look below point A to see that its first coordinate is 3. Looking to the left of point A, we find that its second coordinate is also 3. Thus, the coordinates of point A are $(3, 3)$.

We look above point B to see that its first coordinate is 0. Looking at the location of point B on the second, or vertical, axis, we find that its second coordinate is -4. Thus, the coordinates of point B are $(0, -4)$.

Looking at the location of point C on the first, or horizontal, axis, we see that the first coordinate of point C is -5. We look to the right of point C to see that its second coordinate is 0. Thus, the coordinates of point C are $(-5, 0)$.

We look above point D to see that its first coordinate is -1. Looking to the right of point D, we find that its second coordinate is also -1. Thus, the coordinates of point D are $(-1, -1)$.

Looking at the location of point E on the first, or horizontal, axis, we see that the first coordinate of point E is 2. We look to the left of point E to see that its second coordinate is 0. Thus, the coordinates of point E are $(2, 0)$.

We look below point F to see that its first coordinate is -3. Looking to the right of point F, we find that its second coordinate is 5. Thus, the coordinates of point F are $(-3, 5)$.

7.

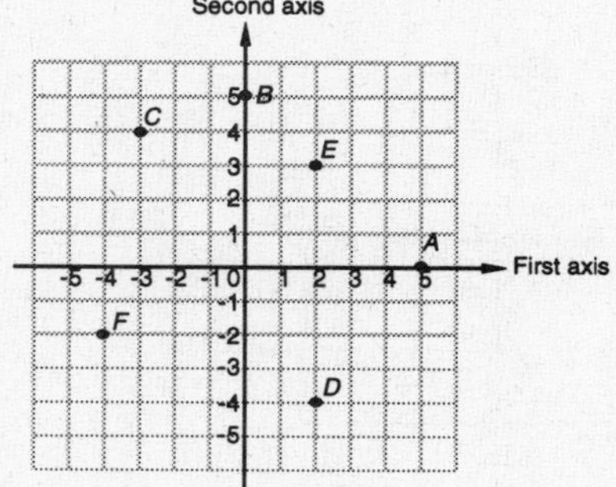

Looking at the location of point A on the first, or horizontal, axis, we see that the first coordinate of point A is 5. We look to the left of point A to see that its second coordinate is 0. Thus, the coordinates of point A are $(5, 0)$.

We look below point B to see that its first coordinate is 0. Looking at the location of point B on the second, or vertical, axis, we find that its second coordinate is 5. Thus, the coordinates of point B are $(0, 5)$.

We look below point C to see that its first coordinate is -3. Looking to the right of point C, we find that its second coordinate is 4. Thus, the coordinates of point C are $(-3, 4)$.

We look above point D to see that its first coordinate is 2. Looking to the left of point D, we find that its second coordinate is -4. Thus, the coordinates of point D are $(2, -4)$.

We look below point E to see that its first coordinate is 2. Looking to the left of point E, we find that its second coordinate is 3. Thus the coordinates of point E are $(2, 3)$.

We look above point F to see that its first coordinate is -4. Looking to the right of point F, we find that its second coordinate is -2. Thus, the coordinates of point F are $(-4, -2)$.

9. Since the first coordinate is negative and the second coordinate positive, the point $(-5, 3)$ is located in quadrant II.

11. Since the first coordinate is positive and the second coordinate negative, the point $(100, -1)$ is in quadrant IV.

13. Since both coordinates are negative, the point $(-6, -29)$ is in quadrant III.

15. Since both coordinates are positive, the point $\left(3\frac{7}{10}, 9\frac{1}{11}\right)$ is in quadrant I.

17. In quadrant III, first coordinates are always <u>negative</u> and second coordinates are always <u>negative</u>.

19.

$y = 3x - 1$		
$5 \; ? \; 3 \cdot 2 - 1$		Substituting 2 for x and 5 for y
	$6 - 1$	(alphabetical order of variables)
5	5	TRUE

The equation becomes true: $(2, 5)$ is a solution.

21.

$3x - y = 4$		
$3 \cdot 2 - (-3) \; ?$	4	Substituting 2 for x and -3 for y
$6 + 3$		
9	4	FALSE

The equation becomes false; $(2, -3)$ is not a solution.

23.

$2c + 3d = -7$		
$2(-2) + 3(-1) \; ?$	-7	Substituting -2 for c and -1 for d
$-4 - 3$		
-7	-7	TRUE

The equation becomes true; $(-2, -1)$ is a solution.

25.

$3x + y = 19$		
$3 \cdot 5 + (-4) \; ?$	19	Substituting 5 for x and -4 for y
$15 - 4$		
11	19	FALSE

The equation becomes false; $(5, -4)$ is not a solution.

27.

$2q - 3p = 3$		
$2 \cdot 5 - 3\left(2\frac{1}{3}\right)$	? 3	Substituting $2\frac{1}{3}$ for p and 5 for q
$10 - 3 \cdot \frac{7}{3}$		
$10 - 7$		
3	3	TRUE

The equation becomes true; $\left(2\frac{1}{3}, 5\right)$ is a solution.

29. $3x - 4 = 17$

$3x - 4 + 4 = 17 + 4$ Adding 4 on both sides

$3x = 21$

$\frac{3x}{3} = \frac{21}{3}$ Dividing by 3 on both sides

$x = 7$

The solution is 7.

31. $\frac{90}{51} = \frac{3 \cdot 30}{3 \cdot 17} = \frac{3}{3} \cdot \frac{30}{17} = 1 \cdot \frac{30}{17} = \frac{30}{17}$

33.

35.

$5.2x + 6.1y = -4.821$		
$5.2(-2.37) + 6.1(1.23)$	? -4.821	
$-12.324 + 7.503$		
-4.821	-4.821	TRUE

The equation becomes true; $(-2.37, 1.23)$ is a solution.

37.

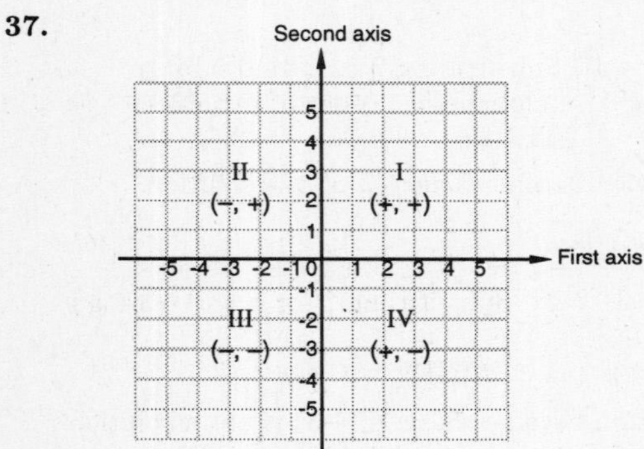

If the first coordinate is positive, then the point must be in either quadrant I or quadrant IV.

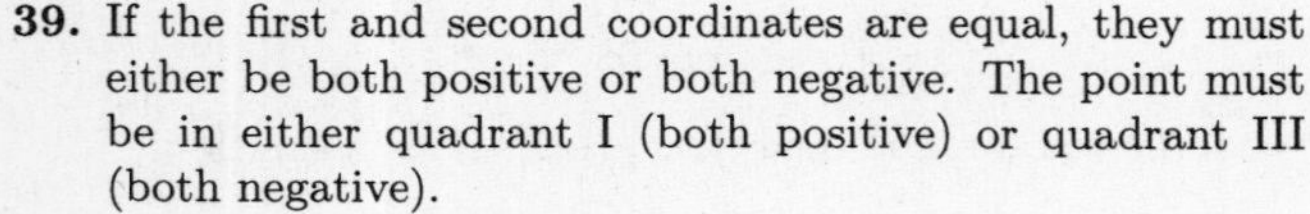

39. If the first and second coordinates are equal, they must either be both positive or both negative. The point must be in either quadrant I (both positive) or quadrant III (both negative).

41.

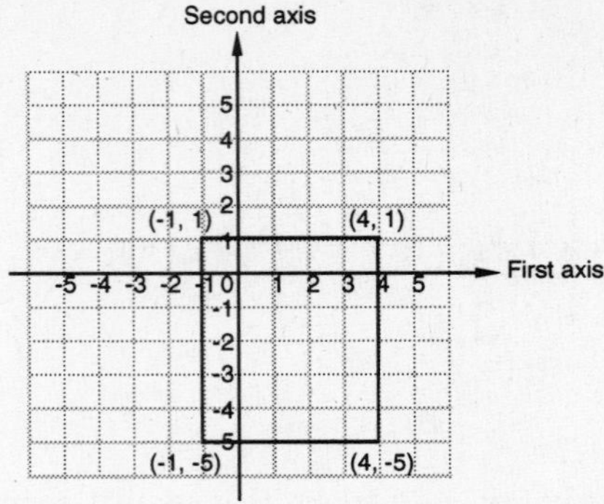

The coordinates of the fourth vertex are $(-1, -5)$.

43. Answers may vary.

We select eight points such that the sum of the coordinates for each point is 6.

$(-1, 7)$	$-1 + 7 = 6$
$(0, 6)$	$0 + 6 = 6$
$(1, 5)$	$1 + 5 = 6$
$(2, 4)$	$2 + 4 = 6$
$(3, 3)$	$3 + 3 = 6$
$(4, 2)$	$4 + 2 = 6$
$(5, 1)$	$5 + 1 = 6$
$(6, 0)$	$6 + 0 = 6$

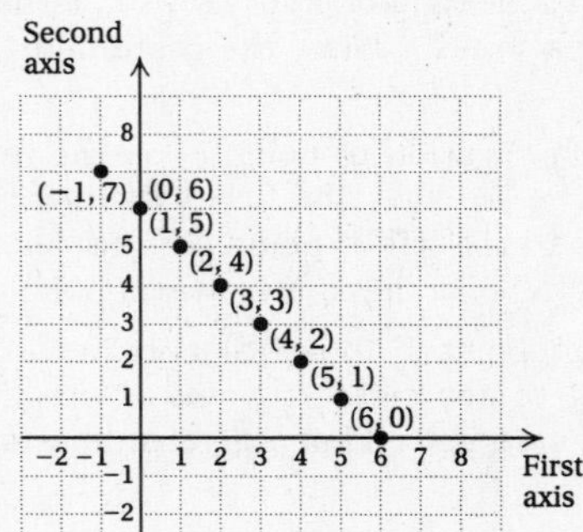

45.

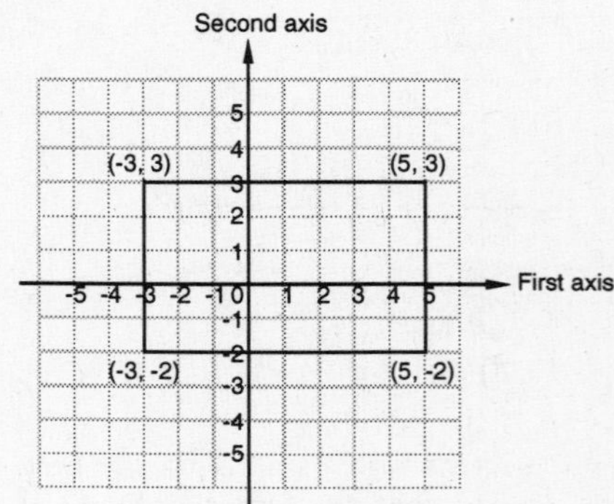

The length is 8, and the width is 5.

$P = 2l + 2w$

$P = 2 \cdot 8 + 2 \cdot 5 = 16 + 10 = 26$

Exercise Set 6.4

1.
$$\begin{aligned} x-y&=4 \\ 9-y&=4 && \text{Substituting 9 for } x \\ -y&=-5 && \text{Subtracting 9 on both sides} \\ -1\cdot y&=-5 && \text{Recall that } -a=-1\cdot a. \\ y&=5 && \text{Dividing by } -1 \text{ on both sides} \end{aligned}$$

The pair $(9,5)$ is a solution of $x-y=4$.

3.
$$\begin{aligned} 2x+y&=7 \\ 2\cdot 3+y&=7 && \text{Substituting 3 for } x \\ 6+y&=7 \\ y&=1 && \text{Subtracting 6 on both sides} \end{aligned}$$

The pair $(3,1)$ is a solution of $2x+y=7$.

5.
$$\begin{aligned} 3x-y&=7 \\ 3x-5&=7 && \text{Substituting 5 for } y \\ 3x&=12 && \text{Adding 5 on both sides} \\ x&=4 && \text{Dividing by 3 on both sides} \end{aligned}$$

The pair $(4,5)$ is a solution of $3x-y=7$.

7.
$$\begin{aligned} x+3y&=1 \\ 7+3y&=1 && \text{Substituting 7 for } x \\ 3y&=-6 && \text{Subtracting 7 on both sides} \\ y&=-2 && \text{Dividing by 3 on both sides} \end{aligned}$$

The pair $(7,-2)$ is a solution of $x+3y=1$.

9.
$$\begin{aligned} 2x+5y&=17 \\ 2\cdot 1+5y&=17 && \text{Substituting 1 for } x \\ 2+5y&=17 \\ 5y&=15 && \text{Subtracting 2 on both sides} \\ y&=3 && \text{Dividing by 5 on both sides} \end{aligned}$$

The pair $(1,3)$ is a solution of $2x+5y=17$.

11.
$$\begin{aligned} 3x-2y&=8 \\ 3x-2(-1)&=8 && \text{Substituting } -1 \text{ for } y \\ 3x+2&=8 \\ 3x&=6 && \text{Subtracting 2 on both sides} \\ x&=2 && \text{Dividing by 3 on both sides} \end{aligned}$$

The pair $(2,-1)$ is a solution of $3x-2y=8$.

13. To complete the pair $(\ \ ,7)$, we replace y with 7 and solve for x.

$$\begin{aligned} x+y&=9 \\ x+7&=9 && \text{Substituting 7 for } y \\ x&=2 && \text{Subtracting 7 on both sides} \end{aligned}$$

Thus, $(2,7)$ is a solution of $x+y=9$.

To complete the pair $(4,\ \ )$, we replace x with 4 and solve for y.

$$\begin{aligned} x+y&=9 \\ 4+y&=9 && \text{Substituting 4 for } x \\ y&=5 && \text{Subtracting 4 on both sides} \end{aligned}$$

Thus, $(4,5)$ is a solution of $x+y=9$.

15. To complete the pair $(\ \ ,3)$, we replace y with 3 and solve for x.

$$\begin{aligned} x-y&=4 \\ x-3&=4 && \text{Substituting 3 for } y \\ x&=7 && \text{Adding 3 on both sides} \end{aligned}$$

Thus, $(7,3)$ is a solution of $x-y=4$.

To complete the pair $(10,\ \ )$, we replace x with 10 and solve for y.

$$\begin{aligned} x-y&=4 \\ 10-y&=4 && \text{Substituting 10 for } x \\ -y&=-6 && \text{Subtracting 10 on both sides} \\ -1\cdot y&=-6 && \text{Recall that } -a=-1\cdot a. \\ y&=6 && \text{Dividing by } -1 \text{ on both sides} \end{aligned}$$

Thus, $(10,6)$ is a solution of $x-y=4$.

17. To complete the pair $(3,\ \ )$, we replace x with 3 and solve for y.

$$\begin{aligned} 2x+3y&=15 \\ 2\cdot 3+3y&=15 && \text{Substituting 3 for } x \\ 6+3y&=15 \\ 3y&=9 && \text{Subtracting 6 on both sides} \\ y&=3 && \text{Dividing by 3 on both sides} \end{aligned}$$

Thus, $(3,3)$ is a solution of $2x+3y=15$.

To complete the pair $(\ \ ,1)$, we replace y with 1 and solve for x.

$$\begin{aligned} 2x+3y&=15 \\ 2x+3\cdot 1&=15 && \text{Substituting 1 for } y \\ 2x+3&=15 \\ 2x&=12 && \text{Subtracting 3 on both sides} \\ x&=6 && \text{Dividing by 2 on both sides} \end{aligned}$$

Thus, $(6,1)$ is a solution of $2x+3y=15$.

19. To complete the pair $(3,\ \ )$, we replace x with 3 and solve for y.

$$\begin{aligned} 5x+2y&=11 \\ 5\cdot 3+2y&=11 && \text{Substituting 3 for } x \\ 15+2y&=11 \\ 2y&=-4 && \text{Subtracting 15 on both sides} \\ y&=-2 && \text{Dividing by 2 on both sides} \end{aligned}$$

Thus, $(3,-2)$ is a solution of $5x+2y=11$.

To complete the pair $(\ \ ,3)$, we replace y with 3 and solve for x.

$$\begin{aligned} 5x+2y&=11 \\ 5x+2\cdot 3&=11 && \text{Substituting 3 for } y \\ 5x+6&=11 \\ 5x&=5 && \text{Subtracting 6 on both sides} \\ x&=1 && \text{Dividing by 5 on both sides} \end{aligned}$$

Thus, $(1,3)$ is a solution of $5x+2y=11$.

21. To complete the pair $(\ \ ,2)$, we replace y with 2 and solve for x.

$$\begin{aligned} 3x-7y&=1 \\ 3x-7\cdot 2&=1 && \text{Substituting 2 for } y \\ 3x-14&=1 \\ 3x&=15 && \text{Adding 14 on both sides} \\ x&=5 && \text{Dividing by 3 on both sides} \end{aligned}$$

Thus, $(5,2)$ is a solution of $3x-7y=1$.

To complete the pair $(-2, \quad)$, we replace x with -2 and solve for y.

$$\begin{aligned} 3x - 7y &= 1 \\ 3(-2) - 7y &= 1 \quad \text{Substituting } -2 \text{ for } x \\ -6 - 7y &= 1 \\ -7y &= 7 \quad \text{Adding 6 on both sides} \\ y &= -1 \quad \text{Dividing by } -7 \text{ on both sides} \end{aligned}$$

Thus, $(-2, -1)$ is a solution of $3x - 7y = 1$.

23. To complete the pair $(0, \quad)$, we replace x with 0 and solve for y.

$$\begin{aligned} 2x + 5y &= 3 \\ 2 \cdot 0 + 5y &= 3 \quad \text{Substituting 0 for } x \\ 5y &= 3 \\ y &= \frac{3}{5} \quad \text{Dividing by 5 on both sides} \end{aligned}$$

Thus, $\left(0, \frac{3}{5}\right)$ is a solution of $2x + 5y = 3$.

To complete the pair $(\quad, 0)$, we replace y with 0 and solve for x.

$$\begin{aligned} 2x + 5y &= 3 \\ 2x + 5 \cdot 0 &= 3 \quad \text{Substituting 0 for } y \\ 2x &= 3 \\ x &= \frac{3}{2} \quad \text{Dividing by 2 on both sides} \end{aligned}$$

Thus, $\left(\frac{3}{2}, 0\right)$ is a solution of $2x + 5y = 3$.

25. We are free to choose any number as a replacement for x or y. To find one solution we choose to replace x with 0.

$$\begin{aligned} x + y &= 15 \\ 0 + y &= 15 \quad \text{Substituting 0 for } x \\ y &= 15 \end{aligned}$$

Thus, $(0, 15)$ is one solution of $x + y = 15$.

To find a second solution we can replace y with 5.

$$\begin{aligned} x + y &= 15 \\ x + 5 &= 15 \quad \text{Substituting 5 for } y \\ x &= 10 \quad \text{Subtracting 5 on both sides} \end{aligned}$$

Thus, $(10, 5)$ is a second solution of $x + y = 15$.

To find a third solution we can replace x with -3.

$$\begin{aligned} x + y &= 15 \\ -3 + y &= 15 \quad \text{Substituting } -3 \text{ for } x \\ y &= 18 \quad \text{Adding 3 on both sides} \end{aligned}$$

Thus, $(-3, 18)$ is a third solution of $x + y = 15$.

27. We are free to choose any number as a replacement for x or y. To find one solution we choose to replace x with 3.

$$\begin{aligned} x + 3y &= 9 \\ 3 + 3y &= 9 \quad \text{Substituting 3 for } x \\ 3y &= 6 \quad \text{Subtracting 3 on both sides} \\ y &= 2 \quad \text{Dividing by 3 on both sides} \end{aligned}$$

Thus, $(3, 2)$ is one solution of $x + 3y = 9$.

To find a second solution we can replace y with -1.

$$\begin{aligned} x + 3y &= 9 \\ x + 3(-1) &= 9 \quad \text{Substituting } -1 \text{ for } y \\ x - 3 &= 9 \\ x &= 12 \quad \text{Adding 3 on both sides} \end{aligned}$$

Thus, $(12, -1)$ is a second solution of $x + 3y = 9$.

To find a third solution we can replace y with 5.

$$\begin{aligned} x + 3y &= 9 \\ x + 3 \cdot 5 &= 9 \quad \text{Substituting 5 for } y \\ x + 15 &= 9 \\ x &= -6 \quad \text{Subtracting 15 on both sides} \end{aligned}$$

Thus, $(-6, 5)$ is a third solution of $x + 3y = 9$.

29. We are free to choose any number as a replacement for x or y. To find one solution we choose to replace y with 1.

$$\begin{aligned} 4x + y &= 9 \\ 4x + 1 &= 9 \quad \text{Substituting 1 for } y \\ 4x &= 8 \quad \text{Subtracting 1 on both sides} \\ x &= 2 \quad \text{Dividing by 4 on both sides} \end{aligned}$$

Thus, $(2, 1)$ is one solution of $4x + y = 9$.

To find a second solution we can replace x with 0.

$$\begin{aligned} 4x + y &= 9 \\ 4 \cdot 0 + y &= 9 \quad \text{Substituting 0 for } x \\ y &= 9 \end{aligned}$$

Thus, $(0, 9)$ is a second solution of $4x + y = 9$.

To find a third solution we can replace x with -2.

$$\begin{aligned} 4x + y &= 9 \\ 4(-2) + y &= 9 \quad \text{Substituting } -2 \text{ for } x \\ -8 + y &= 9 \\ y &= 17 \quad \text{Adding 8 on both sides} \end{aligned}$$

Thus, $(-2, 17)$ is a third solution of $4x + y = 9$.

31. We are free to choose any number as a replacement for x or y. Since y is isolated it is generally easiest to substitute for x and then calculate y. To find one solution we choose to replace x with -1.

$$\begin{aligned} y &= 3x - 1 \\ y &= 3(-1) - 1 \quad \text{Substituting } -1 \text{ for } x \\ y &= -3 - 1 \\ y &= -4 \end{aligned}$$

Thus, $(-1, -4)$ is one solution of $y = 3x - 1$.

To find a second solution we can replace x with 0.

$$\begin{aligned} y &= 3x - 1 \\ y &= 3 \cdot 0 - 1 \quad \text{Substituting 0 for } x \\ y &= -1 \end{aligned}$$

Thus, $(0, -1)$ is a second solution of $y = 3x - 1$.

To find a third solution we can replace x with 2.

$$\begin{aligned} y &= 3x - 1 \\ y &= 3 \cdot 2 - 1 \quad \text{Substituting 2 for } x \\ y &= 6 - 1 \\ y &= 5 \end{aligned}$$

Thus, $(2, 5)$ is a third solution of $y = 3x - 1$.

33. We are free to choose any number as a replacement for x or y. Since y is isolated it is generally easiest to substitute for x and then calculate y. To find one solution we choose to replace x with -2.

$$\begin{aligned} y &= -2x + 5 \\ y &= -2(-2) + 5 \quad \text{Substituting } -2 \text{ for } x \\ y &= 4 + 5 \\ y &= 9 \end{aligned}$$

Thus, $(-2, 9)$ is one solution of $y = -2x + 5$.

To find a second solution we can replace x with 0.

$$\begin{aligned} y &= -2x + 5 \\ y &= -2 \cdot 0 + 5 \quad \text{Substituting 0 for } x \\ y &= 5 \end{aligned}$$

Thus, $(0, 5)$ is a second solution of $y = -2x + 5$.

To find a third solution we can replace x with 4.

$$\begin{aligned} y &= -2x + 5 \\ y &= -2 \cdot 4 + 5 \quad \text{Substituting 4 for } x \\ y &= -8 + 5 \\ y &= -3 \end{aligned}$$

Thus, $(4, -3)$ is a third solution of $y = -2x + 5$.

35. We are free to choose any number as a replacement for x or y. To find one solution we choose to replace x with 6.

$$\begin{aligned} 2x + 3y &= 18 \\ 2 \cdot 6 + 3y &= 18 \quad \text{Substituting 6 for } x \\ 12 + 3y &= 18 \\ 3y &= 6 \quad \text{Subtracting 12 on both sides} \\ y &= 2 \quad \text{Dividing by 3 on both sides} \end{aligned}$$

Thus, $(6, 2)$ is one solution of $2x + 3y = 18$.

To find a second solution we can replace y with -2.

$$\begin{aligned} 2x + 3y &= 18 \\ 2x + 3(-2) &= 18 \quad \text{Substituting } -2 \text{ for } y \\ 2x - 6 &= 18 \\ 2x &= 24 \quad \text{Adding 6 on both sides} \\ x &= 12 \quad \text{Dividing by 2 on both sides} \end{aligned}$$

Thus, $(12, -2)$ is a second solution of $2x + 3y = 18$.

To find a third solution we can replace x with 9.

$$\begin{aligned} 2x + 3y &= 18 \\ 2 \cdot 9 + 3y &= 18 \quad \text{Substituting 9 for } x \\ 18 + 3y &= 18 \\ 3y &= 0 \quad \text{Subtracting 18 on both sides} \\ y &= 0 \quad \text{Dividing by 3 on both sides} \end{aligned}$$

Thus, $(9, 0)$ is a third solution of $2x + 3y = 18$.

37. We are free to choose any number as a replacement for x or y. Since x is isolated it is generally easiest to substitute for y and then calculate to find x. To find one solution we choose to replace y with -5.

$$\begin{aligned} 2 + y &= x \\ 2 + (-5) &= x \quad \text{Substituting } -5 \text{ for } y \\ -3 &= x \end{aligned}$$

Thus, $(-3, 5)$ is one solution of $2 + y = x$.

To find a second solution we can replace y with 0.

$$\begin{aligned} 2 + y &= x \\ 2 + 0 &= x \quad \text{Substituting 0 for } y \\ 2 &= x \end{aligned}$$

Thus, $(2, 0)$ is a second solution of $2 + y = x$.

To find a third solution we can replace y with 3.

$$\begin{aligned} 2 + y &= x \\ 2 + 3 &= x \quad \text{Substituting 3 for } y \\ 5 &= x \end{aligned}$$

Thus, $(5, 3)$ is a third solution of $2 + y = x$.

39. We are free to choose any number as a replacement for x or y. Since y is isolated it is generally easiest to substitute for x and then calculate y. Note that when x is a multiple of 3, fractional values for y are avoided. To find one solution we choose to replace x with -3.

$$\begin{aligned} y &= \frac{1}{3}x + 2 \\ y &= \frac{1}{3}(-3) + 2 \quad \text{Substituting } -3 \text{ for } x \\ y &= -1 + 2 \\ y &= 1 \end{aligned}$$

Thus, $(-3, 1)$ is one solution of $y = \frac{1}{3}x + 2$.

To find a second solution we can replace x with 0.

$$\begin{aligned} y &= \frac{1}{3}x + 2 \\ y &= \frac{1}{3} \cdot 0 + 2 \quad \text{Substituting 0 for } x \\ y &= 2 \end{aligned}$$

Thus, $(0, 2)$ is a second solution of $y = \frac{1}{3}x + 2$.

To find a third solution we can replace x with 3.

$$\begin{aligned} y &= \frac{1}{3}x + 2 \\ y &= \frac{1}{3} \cdot 3 + 2 \quad \text{Substituting 3 for } x \\ y &= 1 + 2 \\ y &= 3 \end{aligned}$$

Thus, $(3, 3)$ is a third solution of $y = \frac{1}{3}x + 2$.

41. We are free to choose any number as a replacement for x or y. Since y is isolated it is generally easiest to substitute for x and then calculate y. Note that when x is a multiple of 4, fractional values for y are avoided. To find one solution we replace x with -4.

$$\begin{aligned} y &= -\frac{3}{4}x + 1 \\ y &= -\frac{3}{4}(-4) + 1 \quad \text{Substituting } -4 \text{ for } x \\ y &= 3 + 1 \\ y &= 4 \end{aligned}$$

Thus, $(-4, 4)$ is one solution of $y = -\frac{3}{4}x + 1$.

To find a second solution we can replace x with 0.

$$\begin{aligned} y &= -\frac{3}{4}x + 1 \\ y &= -\frac{3}{4} \cdot 0 + 1 \quad \text{Substituting 0 for } x \\ y &= 1 \end{aligned}$$

Thus, $(0, 1)$ is a second solution of $y = -\frac{3}{4}x + 1$.

To find a third solution we can replace x with 4.

$$y = -\frac{3}{4}x + 1$$
$$y = -\frac{3}{4} \cdot 4 + 1 \quad \text{Substituting 4 for } x$$
$$y = -3 + 1$$
$$y = -2$$

Thus, $(4, -2)$ is a third solution of $y = -\frac{3}{4}x + 1$.

43. Graph: $x + y = 4$

We make a table of solutions. Then we plot the points, draw the line, and label it.

When $x = 0$: $0 + y = 4$, $y = 4$

When $y = -2$: $x + (-2) = 4$, $x = 6$

When $x = 4$: $4 + y = 4$, $y = 0$

x	y $x + y = 4$	(x, y)
0	4	$(0, 4)$
6	-2	$(6, -2)$
4	0	$(4, 0)$

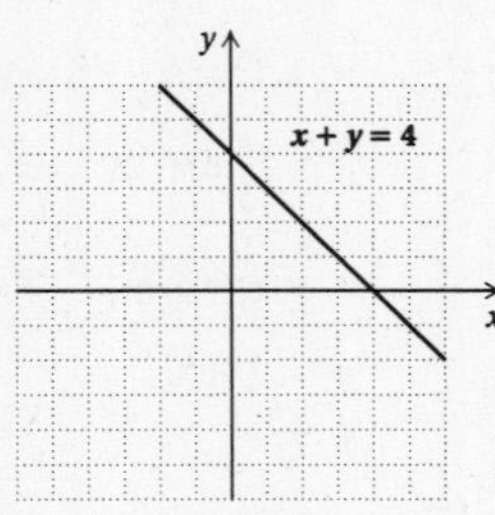

45. Graph: $x - y = 1$

We make a table of solutions. Then we plot the points, draw the line, and label it.

When $x = 5$: $5 - y = 1$, $-y = -4$, $y = 4$

When $x = 0$: $0 - y = 1$, $-y = 1$, $y = -1$

When $y = 0$: $x - 0 = 1$, $x = 1$

x	y $x - y = 1$	(x, y)
5	4	$(5, 4)$
0	-1	$(0, -1)$
1	0	$(1, 0)$

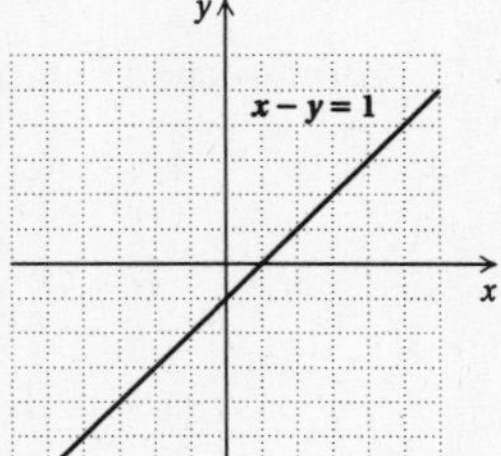

47. Graph: $y = x - 3$

We make a table of solutions. Since y is isolated, it is generally easiest to substitute for x and then calculate y. Then we plot the points, draw the line, and label it.

When $x = -2$, $y = -2 - 3 = -5$.
When $x = 0$, $y = 0 - 3 = -3$.
When $x = 3$, $y = 3 - 3 = 0$.

x	y $y = x - 3$	(x, y)
-2	-5	$(-2, -5)$
0	-3	$(0, -3)$
3	0	$(3, 0)$

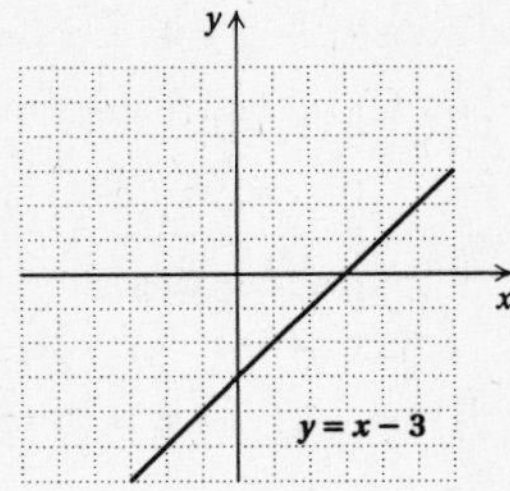

49. Graph: $2x + y = 9$

We make a table of solutions. Then we plot the points, draw the line, and label it.

When $x = 3$: $2 \cdot 3 + y = 9$, $6 + y = 9$, $y = 3$

When $x = 4$: $2 \cdot 4 + y = 9$, $8 + y = 9$, $y = 1$

When $x = 6$: $2 \cdot 6 + y = 9$, $12 + y = 9$, $y = -3$

x	y $2x + y = 9$	(x, y)
3	3	$(3, 3)$
4	1	$(4, 1)$
6	-3	$(6, -3)$

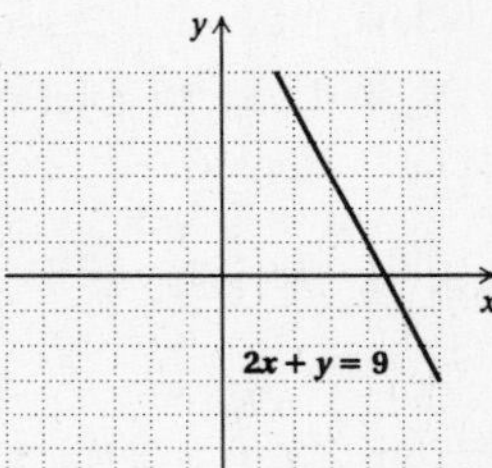

51. Graph: $y = x$

We make a table of solutions. Then we plot the points, draw the line, and label it.

When $x = -3$, $y = -3$.
When $x = 0$, $y = 0$.
When $x = 2$, $y = 2$.

x	y $y = x$	(x, y)
-3	-3	$(-3, -3)$
0	0	$(0, 0)$
2	2	$(2, 2)$

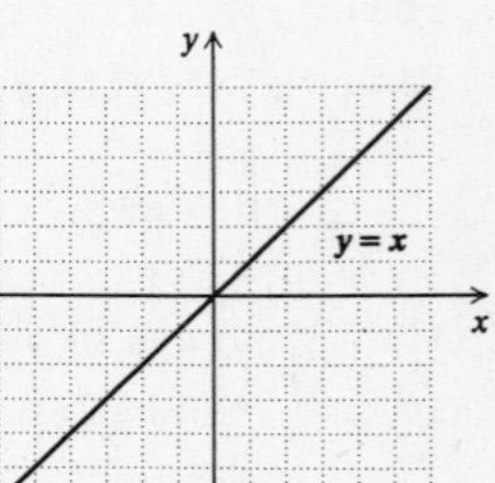

53. Graph: $y = \frac{1}{3}x$

We make a table of solutions. Since y is isolated, it is generally easiest to substitute for x and then calculate y. Note that when x is a multiple of 3, fractional values for y

are avoided. We plot the points, draw the line, and label it.

When $x = -3$, $y = \frac{1}{3}(-3) = -1$.

When $x = 0$, $y = \frac{1}{3} \cdot 0 = 0$.

When $x = 3$, $y = \frac{1}{3} \cdot 3 = 1$.

x	y $y = \frac{1}{3}x$	(x, y)
-3	-1	$(-3, -1)$
0	0	$(0, 0)$
3	1	$(3, 1)$

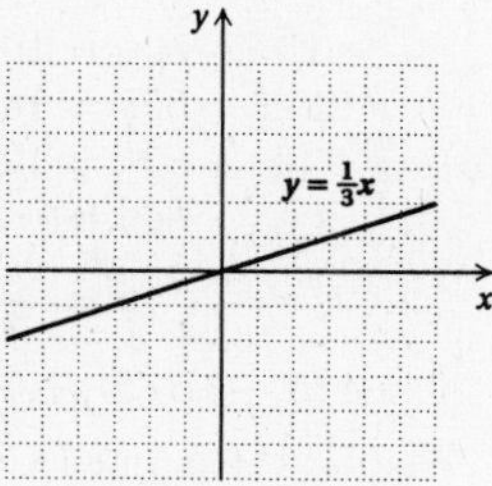

55. Graph: $y = 2x - 7$

We make a table of solutions. Since y is isolated, it is generally easiest to substitute for x and then calculate y. Then we plot the points, draw the line, and label it.

When $x = 0$, $y = 2 \cdot 0 - 7 = 0 - 7 = -7$.
When $x = 2$, $y = 2 \cdot 2 - 7 = 4 - 7 = -3$.
When $x = 5$, $y = 2 \cdot 5 - 7 = 10 - 7 = 3$.

x	y $y = 2x - 7$	(x, y)
0	-7	$(0, -7)$
2	-3	$(2, -3)$
5	3	$(5, 3)$

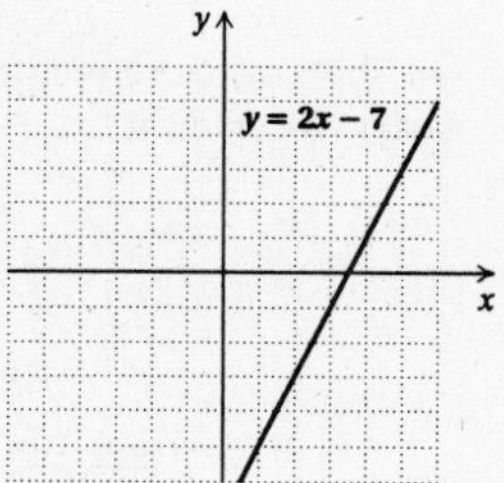

57. Graph: $y = -4x$

We make a table of solutions. Since y is isolated, it is generally easiest to substitute for x and then calculate y. Then we plot the points, draw the line, and label it.

When $x = -1$, $y = -4(-1) = 4$.
When $x = 0$, $y = -4 \cdot 0 = 0$.
When $x = 1$, $y = -4 \cdot 1 = -4$.

x	y $y = -4x$	(x, y)
-1	4	$(-1, 4)$
0	0	$(0, 0)$
1	-4	$(1, -4)$

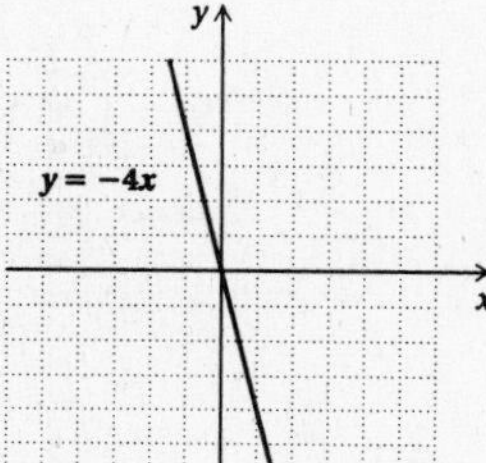

59. Graph: $y = \frac{1}{2}x + 3$

We make a table of solutions. Since y is isolated, it is generally easiest to substitute for x and then calculate y. Note that when x is a multiple of 2, fractional values for y are avoided. We plot the points, draw the line, and label it.

When $x = -4$, $y = \frac{1}{2}(-4) + 3 = -2 + 3 = 1$.

When $x = 0$, $y = \frac{1}{2} \cdot 0 + 3 = 0 + 3 = 3$.

When $x = 2$, $y = \frac{1}{2} \cdot 2 + 3 = 1 + 3 = 4$.

x	y $y = \frac{1}{2}x + 3$	(x, y)
-4	1	$(-4, 1)$
0	3	$(0, 3)$
2	4	$(2, 4)$

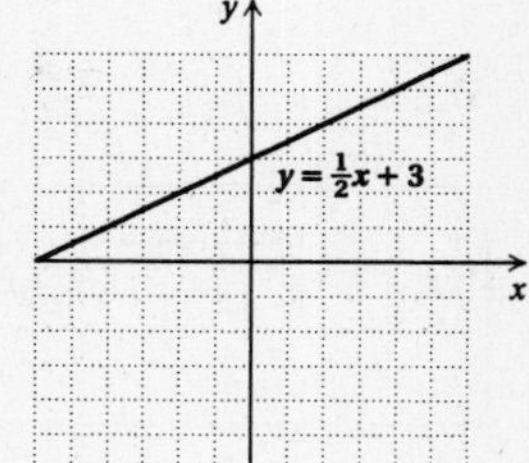

61. Graph: $y = -x + 5$

We make a table of solutions. Since y is isolated, it is generally easiest to substitute for x and then calculate y. Then we plot the points, draw the line, and label it.

When $x = -1$, $y = -(-1) + 5 = 1 + 5 = 6$.
When $x = 2$, $y = -2 + 5 = 3$.
When $x = 5$, $y = -5 + 5 = 0$.

x	y $y = -x + 5$	(x, y)
-1	6	$(-1, 6)$
2	3	$(2, 3)$
5	0	$(5, 0)$

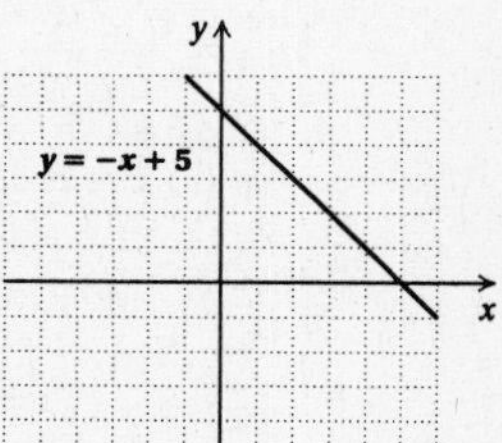

63. Graph: $y = \frac{2}{3}x - 4$

We make a table of solutions. Since y is isolated, it is generally easiest to substitute for x and then calculate y. Note that when x is a multiple of 3, fractional values for y are avoided. We plot the points, draw the line, and label it.

When $x = -3$, $y = \frac{2}{3}(-3) - 4 = -2 - 4 = -6$.

When $x = 0$, $y = \frac{2}{3} \cdot 0 - 4 = 0 - 4 = -4$.

When $x = 3$, $y = \frac{2}{3} \cdot 3 - 4 = 2 - 4 = -2$.

x	y $y = \frac{2}{3}x - 4$	(x, y)
-3	-6	$(-3, -6)$
0	-4	$(0, -4)$
3	-2	$(3, -2)$

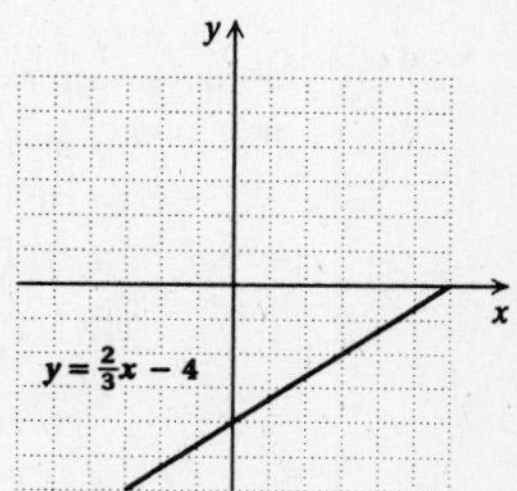

65. Graph: $y = -\frac{1}{2}x + 3$

We make a table of solutions. Since y is isolated, it is generally easiest to substitute for x and then calculate y. Note that when x is a multiple of 2, fractional values for y are avoided. We plot the points, draw the line, and label it.

When $x = -2$, $y = -\frac{1}{2}(-2) + 3 = 1 + 3 = 4$.

When $x = 2$, $y = -\frac{1}{2} \cdot 2 + 3 = -1 + 3 = 2$.

When $x = 4$, $y = -\frac{1}{2} \cdot 4 + 3 = -2 + 3 = 1$.

x	y $y = -\frac{1}{2}x + 3$	(x, y)
-2	4	$(-2, 4)$
2	2	$(2, 2)$
4	1	$(4, 1)$

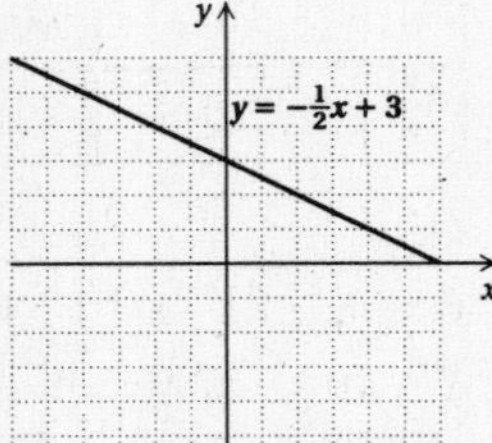

67. ***Familiarize***. We let y = the amount of vinegar needed to make $2\frac{1}{2}$ batches of chili.

Translate. We write a multiplication sentence that fits the situation.

$$y = 2\frac{1}{2} \cdot \left(\frac{3}{4}\right)$$

Solve. We do the computation.

$$y = 2\frac{1}{2} \cdot \left(\frac{3}{4}\right)$$

$$y = \frac{5}{2} \cdot \frac{3}{4} \qquad \text{Writing } 2\frac{1}{2} \text{ as } \frac{5}{2}$$

$$y = \frac{15}{8}, \text{ or } 1\frac{7}{8}$$

Check. We repeat the computation. The result checks.

State. $1\frac{7}{8}$ cups of vinegar are needed to make $2\frac{1}{2}$ batches of chili.

69.
$$\begin{aligned} -8 - 5^2 \cdot 2(3-4) &= -8 - 5^2 \cdot 2(-1) \\ &= -8 - 5 \cdot 5 \cdot 2(-1) \\ &= -8 - 25 \cdot 2(-1) \\ &= -8 - 50(-1) \\ &= -8 + 50 \\ &= 42 \end{aligned}$$

71. 

73. To find one solution we choose to replace x with -2:

$$\begin{aligned} 25x + 80y &= 100 \\ 25(-2) + 80y &= 100 \\ -50 + 80y &= 100 \\ 80y &= 150 \\ y &= \frac{150}{80} = 1.875 \end{aligned}$$

Thus, $(-2, 1.875)$ is one solution of $25x + 80y = 100$.

To find a second solution, we can replace y with 2:

$$\begin{aligned} 25x + 80y &= 100 \\ 25x + 80 \cdot 2 &= 100 \\ 25x + 160 &= 100 \\ 25x &= -60 \\ x &= \frac{-60}{25} = -2.4 \end{aligned}$$

Thus, $(-2.4, 2)$ is a second solution of $25x + 80y = 100$.

To find a third solution, we can replace x with 5:

$$\begin{aligned} 25x + 80y &= 100 \\ 25 \cdot 5 + 80y &= 100 \\ 125 + 80y &= 100 \\ 80y &= -25 \\ y &= \frac{-25}{80} = -0.3125 \end{aligned}$$

Thus, $(5, -0.3125)$ is a third solution of $25x + 80y = 100$.

We plot the points $(-2, 1.875)$, $(-2.4, 2)$, and $(5, -0.3125)$, draw the line, and label it.

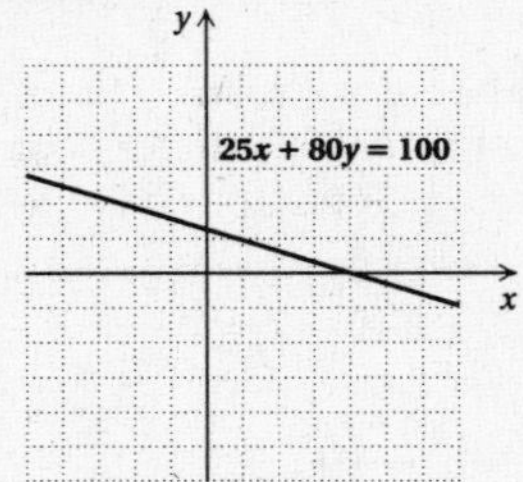

75. On the graph we locate three points whose coordinates are easily determined.

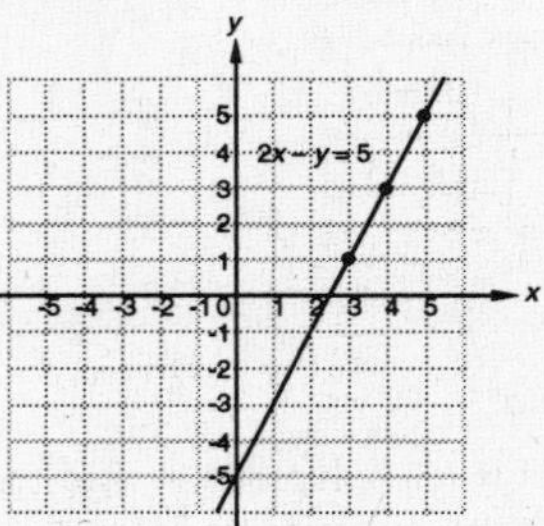

We see that three solutions of $2x - y = 5$ are $(3, 1)$, $(4, 3)$, $(5, 5)$. There are other correct answers.

77. We graph $x + y = 6$ and list the points for which both coordinates are whole numbers.

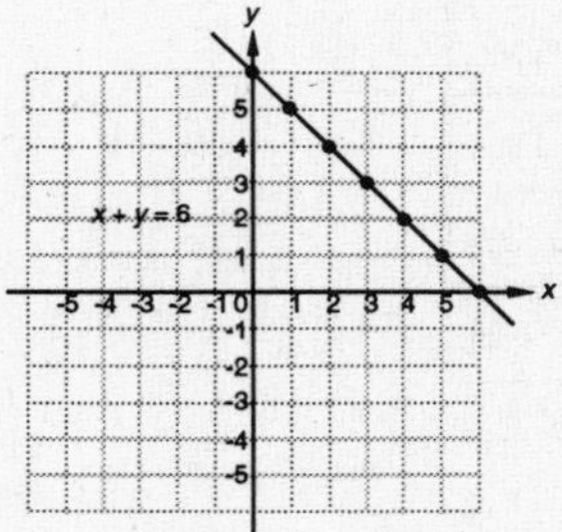

The whole-number solutions are $(0, 6)$, $(1, 5)$, $(2, 4)$, $(3, 3)$, $(4, 2)$, $(5, 1)$, and $(6, 0)$.

Exercise Set 6.5

1. To find the average, add the numbers. Then divide by the number of addends.

$$\frac{16+18+29+14+29+19+15}{7} = \frac{140}{7} = 20$$

The average is 20.

To find the median, first list the numbers in order from smallest to largest. Then locate the middle number.

$$14, 15, 16, 18, 19, 29, 29$$
↑
Middle number

The median is 18.

Find the mode:

The number that occurs most often is 29. The mode is 29.

3. To find the average, add the numbers. Then divide by the number of addends.

$$\frac{5+30+20+20+35+5+25}{7} = \frac{140}{7} = 20$$

The average is 20.

To find the median, first list the numbers in order from smallest to largest. Then locate the middle number.

$$5, 5, 20, 20, 25, 30, 35$$
↑
Middle number

The median is 20.

Find the mode:

There are two numbers that occur most often, 5 and 20. Thus the modes are 5 and 20.

5. Find the average:

$$\frac{1.2+4.3+5.7+7.4+7.4}{5} = \frac{26}{5} = 5.2$$

The average is 5.2.

Find the median:

$$1.2, 4.3, 5.7, 7.4, 7.4$$
↑
Middle number

The median is 5.7.

Find the mode:

The number that occurs most often is 7.4. The mode is 7.4.

7. Find the average:

$$\frac{234+228+234+229+234+278}{6} = \frac{1437}{6} = 239.5$$

The average is 239.5.

Find the median:

$$228, 229, 234, 234, 234, 278$$
↑
Middle number

The median is halfway between 234 and 234. Although it seems clear that this is 234, we can compute it as follows:

$$\frac{234+234}{2} = \frac{468}{2} = 234$$

The median is 234.

Find the mode:

The number that occurs most often is 234. The mode is 234.

9. Find the average:

$$\frac{43^\circ+40^\circ+23^\circ+38^\circ+54^\circ+35^\circ+47^\circ}{7} = \frac{280^\circ}{7} = 40^\circ$$

The average temperature was 40°.

Find the median:

$$23^\circ, 35^\circ, 38^\circ, 40^\circ, 43^\circ, 47^\circ, 54^\circ$$
↑
Middle number

The median is 40°.

Find the mode:

No number repeats, so no mode exists.

11. We divide the total number of miles, 261, by the number of gallons, 9.

$$\frac{261}{9} = 29$$

The average was 29 miles per gallon.

13. To find the GPA we first add the grade point values for each hour taken. This is done by first multiplying the grade point value by the number of hours in the course and then adding as follows:

$$\begin{array}{ll} \text{B} & 3.00\cdot 4 = 12 \\ \text{B} & 3.00\cdot 5 = 15 \\ \text{B} & 3.00\cdot 3 = \ \ 9 \\ \text{C} & 2.00\cdot 4 = \ \ \underline{8} \\ & \qquad\quad\ \ 44 \ \text{(Total)} \end{array}$$

The total number of hours taken is

$4+5+3+4$, or 16.

We divide 44 by 16 and round to the nearest tenth.

$$\frac{44}{16} = 2.75 \approx 2.8$$

The student's grade point average is 2.8.

15. Find the average price per pound:

$$\frac{\$7.99+\$9.49+\$9.99+\$7.99+\$10.49}{5} = \frac{\$45.95}{5} = \$9.19$$

The average price per pound of Atlantic salmon was $9.19.

Find the median price per pound:

List the prices in order:

$7.99, $7.99, $9.49, $9.99, $10.49

↑ Middle number

The median is $9.49.

Find the mode:

The number that occurs most often is $7.99. The mode is $7.99.

17. We can find the total of the five scores needed as follows:

$$80 + 80 + 80 + 80 + 80 = 400.$$

The total of the scores on the first four tests is

$$80 + 74 + 81 + 75 = 310.$$

Thus Rich needs to get at least

$$400 - 310, \text{ or } 90$$

to get a B. We can check this as follows:

$$\frac{80 + 74 + 81 + 75 + 90}{5} = \frac{400}{5} = 80.$$

19. We can find the total number of days needed as follows:

$$266 + 266 + 266 + 266 = 1064.$$

The total number of days for Marta's first three pregnancies is

$$270 + 259 + 272 = 801.$$

Thus, Marta's fourth pregnancy must last

$$1064 - 801 = 263 \text{ days}$$

in order to equal the worldwide average.

We can check this as follows:

$$\frac{270 + 259 + 272 + 263}{4} = \frac{1064}{4} = 266.$$

21. $-144 \div (-9) = \dfrac{-144}{-9} = 16$

(The signs are the same, so the answer is positive.)

23.
$$7.4x = 41.44$$
$$\frac{7.4x}{7.4} = \frac{41.44}{7.4} \quad \text{Dividing by 7.4 on both sides}$$
$$x = 5.6$$

The solution is 5.6.

25.

27. Divide the total by the number of games. Use a calculator.

$$\frac{547}{3} \approx 182.33$$

Drop the amount to the right of the decimal point.

$$\underline{182}\,.\,\boxed{33}$$

This is the average. (182) — Drop this amount. (33)

The bowler's average is 182.

29. We can find the total number of home runs needed over Aaron's 22-yr career as follows:

$$22 \cdot 34\frac{7}{22} = 22 \cdot \frac{755}{22} = \frac{22 \cdot 755}{22} = \frac{22}{22} \cdot \frac{755}{1} = 755.$$

The total number of home runs during the first 21 years of Aaron's career was

$$21 \cdot 35\frac{10}{21} = 21 \cdot \frac{745}{21} = \frac{21 \cdot 745}{21} = \frac{21}{21} \cdot \frac{745}{1} = 745.$$

Then Aaron hit

$$755 - 745 = 10 \text{ home runs}$$

in his final year.

Chapter 7

Ratio and Proportion

Exercise Set 7.1

1. The ratio of 4 to 5 is $\frac{4}{5}$.

3. The ratio of 178 to 572 is $\frac{178}{572}$.

5. The ratio of 0.4 to 12 is $\frac{0.4}{12}$.

7. The ratio of 3.8 to 7.4 is $\frac{3.8}{7.4}$.

9. The ratio of 56.78 to 98.35 is $\frac{56.78}{98.35}$.

11. The ratio of $8\frac{3}{4}$ to $9\frac{5}{6}$ is $\frac{8\frac{3}{4}}{9\frac{5}{6}}$.

13. The ratio of those who play an instrument to the total number of people is $\frac{1}{4}$.

If one person in four plays an instrument, then $4 - 1$, or 3, do not play an instrument. Thus the ratio of those who do not play an instrument to those who do is $\frac{3}{1}$.

15. If four of every five fatal accidents involving a Corvette do not involve another vehicle, then $5 - 4$, or 1, involves a Corvette and at least one other vehicle. Thus, the ratio of fatal accidents involving just a Corvette to those involving a Corvette and at least one other vehicle is $\frac{4}{1}$.

17. The ratio of the longest length to the shortest length is $\frac{13}{5}$.

The perimeter of the triangle is $5 + 12 + 13$, or 30. Thus the ratio of the longest length to the perimeter is $\frac{13}{30}$.

19. The ratio of 4 to 6 is $\frac{4}{6} = \frac{2\cdot 2}{2\cdot 3} = \frac{2}{2}\cdot\frac{2}{3} = \frac{2}{3}$.

21. The ratio of 18 to 24 is $\frac{18}{24} = \frac{3\cdot 6}{4\cdot 6} = \frac{3}{4}\cdot\frac{6}{6} = \frac{3}{4}$.

23. The ratio of 4.8 to 10 is $\frac{4.8}{10} = \frac{4.8}{10}\cdot\frac{10}{10} = \frac{48}{100} = \frac{4\cdot 12}{4\cdot 25} = \frac{4}{4}\cdot\frac{12}{25} = \frac{12}{25}$.

25. The ratio of 2.8 to 3.6 is $\frac{2.8}{3.6} = \frac{2.8}{3.6}\cdot\frac{10}{10} = \frac{28}{36} = \frac{4\cdot 7}{4\cdot 9} = \frac{4}{4}\cdot\frac{7}{9} = \frac{7}{9}$.

27. The ratio is $\frac{20}{30} = \frac{2\cdot 10}{3\cdot 10} = \frac{2}{3}\cdot\frac{10}{10} = \frac{2}{3}$.

29. The ratio is $\frac{56}{100} = \frac{4\cdot 14}{4\cdot 25} = \frac{4}{4}\cdot\frac{14}{25} = \frac{14}{25}$.

31. The ratio is $\frac{128}{256} = \frac{1\cdot 128}{2\cdot 128} = \frac{1}{2}\cdot\frac{128}{128} = \frac{1}{2}$.

33. The ratio is $\frac{0.48}{0.64} = \frac{0.48}{0.64}\cdot\frac{100}{100} = \frac{48}{64} = \frac{3\cdot 16}{4\cdot 16} = \frac{3}{4}\cdot\frac{16}{16} = \frac{3}{4}$.

35. The ratio is $\frac{54}{100} = \frac{2\cdot 27}{2\cdot 50} = \frac{2}{2}\cdot\frac{27}{50} = \frac{27}{50}$.

37. The ratio is $\frac{6.4}{20.2} = \frac{6.4}{20.2}\cdot\frac{10}{10} = \frac{64}{202} = \frac{2\cdot 32}{2\cdot 101} = \frac{2}{2}\cdot\frac{32}{101} = \frac{32}{101}$.

39. $\frac{12}{8} \quad \frac{7}{4}$

The LCD is 8.

$\frac{7}{4}\cdot\frac{2}{2} = \frac{14}{8}$

Since $12 < 14$, it follows that $\frac{12}{8} < \frac{14}{8}$, so $\frac{12}{8} < \frac{7}{4}$.

41. $-\frac{2}{3} \quad -\frac{3}{4}$

The LCD is 12.

$\frac{-2}{3}\cdot\frac{4}{4} = \frac{-8}{12}$

$\frac{-3}{4}\cdot\frac{3}{3} = \frac{-9}{12}$

Since $-8 > -9$, it follows that $\frac{-8}{12} > \frac{-9}{12}$, so $-\frac{2}{3} > -\frac{3}{4}$.

43. ◈

45. 5,950.7 billion $= 5,950.7 \times 1,000,000,000 =$ 5,950,700,000,000

254.7 million $= 254.7 \times 1,000,000 = 254,700,000$

The ratio of GDP to people in the United States is

$$\frac{\$5,950,700,000,000}{254,700,000}.$$

Using a calculator to simplify this ratio, we find that it is approximately $23,364 per person.

47. $\frac{3\frac{3}{4}}{5\frac{7}{8}} = \frac{\frac{15}{4}}{\frac{47}{8}} = \frac{15}{4}\cdot\frac{8}{47} = \frac{15\cdot 8}{4\cdot 47} =$

$\frac{15\cdot 2\cdot 4}{4\cdot 47} = \frac{4}{4}\cdot\frac{15\cdot 2}{47} = \frac{30}{47}$

49. We divide each number in the ratio by 5. Since $5 \div 5 = 1$, $10 \div 5 = 2$, and $15 \div 5 = 3$, we have $1 : 2 : 3$.

Exercise Set 7.2

1. $\frac{120 \text{ km}}{3 \text{ hr}}$, or $40 \frac{\text{km}}{\text{hr}}$

3. $\frac{440 \text{ m}}{40 \text{ sec}}$, or $11 \frac{\text{m}}{\text{sec}}$

5. $\frac{342 \text{ yd}}{2.25 \text{ days}}$, or $152 \frac{\text{yd}}{\text{day}}$

$$\begin{array}{r} 152. \\ 2.25_{\wedge}\overline{)\,342.00_{\wedge}} \\ \underline{225\,00} \\ 117\,00 \\ \underline{112\,50} \\ 4\,50 \\ \underline{4\,50} \\ 0 \end{array}$$

7. $\frac{500 \text{ km}}{20 \text{ hr}} = 25 \frac{\text{km}}{\text{hr}}$

$\frac{20 \text{ hr}}{500 \text{ km}} = 0.04 \frac{\text{hr}}{\text{km}}$

9. $\frac{623 \text{ gal}}{1000 \text{ sq ft}} = 0.623 \frac{\text{gal}}{\text{sq ft}}$

11. $\frac{\$5.75}{10 \text{ min}} = \frac{575¢}{10 \text{ min}} = 57.5 \frac{¢}{\text{min}}$

13. $\frac{11,160,000 \text{ mi}}{1 \text{ min}} = \frac{11,160,000 \text{ mi}}{60 \text{ sec}} =$

$\frac{11,160,000}{60} \frac{\text{mi}}{\text{sec}} = 186,000 \frac{\text{mi}}{\text{sec}}$

15. $\frac{310 \text{ km}}{2.5 \text{ hr}} = 124 \frac{\text{km}}{\text{hr}}$

17. First we find the number of hours in 3 days:

$3 \text{ days} = 3 \text{ days} \cdot \frac{24 \text{ hr}}{1 \text{ day}} = 3 \cdot 24 \cdot \text{hr} \cdot \frac{\text{days}}{\text{days}} = 72 \text{ hr}$

Then $\frac{5040 \text{ mi}}{72 \text{ hr}} = \frac{5040}{72} \frac{\text{mi}}{\text{hr}} = 70 \frac{\text{mi}}{\text{hr}}$, or 70 mph.

19. Unit price $= \frac{\text{Price}}{\text{Number of units}} = \frac{\$165.75}{8.5 \text{ yd}} = 19.5 \frac{\text{dollars}}{\text{yd}}$, or \$19.50/yd

21. We need to find the number of ounces in 2 pounds:

$2 \text{ lb} = 2 \times 1 \text{ lb} = 2 \times 16 \text{ oz} = 32 \text{ oz}$

Unit price $= \frac{\text{Price}}{\text{Number of units}} = \frac{\$8.59}{32 \text{ oz}} = \frac{859¢}{32 \text{ oz}} \approx 26.84 \frac{¢}{\text{oz}}$

23. Unit price $= \frac{\text{Price}}{\text{Number of units}} = \frac{\$1.75}{1\frac{1}{4} \text{ lb}} =$

$\frac{\$1.75}{1.25 \text{ lb}} = \frac{1.75}{1.25} \cdot \frac{\$}{\text{lb}} = 1.4$ dollars per pound, or \$1.40/lb

25. Compare the unit prices.

For Sizzle: $\frac{\$1.79}{18 \text{ oz}} \approx \$0.099/\text{oz}$

For Josie's: $\frac{\$1.65}{16 \text{ oz}} \approx \$0.103/\text{oz}$

Thus, Sizzle has the lower unit price.

27. Compare the unit prices. Recall that 1 qt = 32 oz, so 2 qt = 2 × 1 qt = 2 × 32 oz = 64 oz.

For Bart's: $\frac{\$2.69}{64 \text{ oz}} \approx \$0.042/\text{oz}$

For Sunshine: $\frac{\$1.97}{48 \text{ oz}} \approx \$0.041/\text{oz}$

Thus, Sunshine has the lower unit price.

29. Compare the unit prices. Note that two 10.5-oz cans contain 2(10.5 oz), or 21 oz, of soup.

Big Chunk: $\frac{\$1.59}{21 \text{ oz}} \approx \$0.076/\text{oz}$

Bert's: $\frac{\$0.82}{11 \text{ oz}} \approx \$0.075/\text{oz}$

Thus, Bert's has the lower unit price.

31. Compare the unit prices. Recall that 1 lb = 16 oz, so 1 lb, 8 oz = 16 oz + 8 oz = 24 oz.

For Big Net: $\frac{\$1.29}{7 \text{ oz}} \approx \$0.184/\text{oz}$

For Charlie's: $\frac{\$3.96}{24 \text{ oz}} = \$0.165/\text{oz}$

Thus, Charlie's has the lower unit price.

33. Compare the unit prices.

10-oz size: $\frac{\$2.48}{10 \text{ oz}} = \$0.248/\text{oz}$

15-oz size: $\frac{\$3.69}{15 \text{ oz}} = \$0.246/\text{oz}$

Thus, the 15-oz size has the lower unit price.

35. Six 10-oz bottles contain 6 × 10 oz = 60 oz of sparkling water; four 12-oz bottles contain 4 × 12 oz = 48 oz of sparkling water.

Compare the unit prices.

Six 10-oz bottles: $\frac{\$3.09}{60 \text{ oz}} = \$0.0515/\text{oz}$

Four 12-oz bottles: $\frac{\$2.39}{48 \text{ oz}} \approx \$0.0498/\text{oz}$

Thus, four 12-oz bottles have the lower unit price.

37. ***Familiarize***. We visualize the situation. We let p = the number by which the number of piano players exceeds the number of guitar players, in millions.

18.9 million	p
20.6 million	

Translate. This is a "how-much-more" situation.

$$\underbrace{\text{Number of guitar players}}_{18.9} + \underbrace{\text{Additional number of piano players}}_{p} = \underbrace{\text{Number of piano players}}_{20.6}$$

Solve. To solve the equation we subtract 18.9 on both sides.

$p = 20.6 - 18.9$
$p = 1.7$

$$\begin{array}{r} {\scriptstyle 1\ 9\ 16} \\ \not{2}\,\not{0}.\not{6} \\ -\ 1\ 8.\ 9 \\ \hline 1.\ 7 \end{array}$$

Check. We repeat the calculation.

State. There are 1.7 million more piano players than guitar players.

39. $1\underline{00} \times 678.19$ 678.19.

2 zeros   Move 2 places to the right.

$100 \times 678.19 = 67{,}819$

41. The first coordinate is positive and the second coordinate is negative, so the point is in quadrant IV.

43. 

45. We find the area of a 14-in. pizza:

$r = 14 \text{ in.} \div 2 = 7 \text{ in.}$

$A = \pi r^2 = 3.14(7 \text{ in.})^2 = 3.14(49 \text{ in}^2) = 153.86 \text{ in}^2$

Now we find the area of a 16-in. pizza:

$r = 16 \text{ in.} \div 2 = 8 \text{ in.}$

$A = \pi r^2 = 3.14(8 \text{ in.})^2 = 3.14(64 \text{ in}^2) = 200.96 \text{ in}^2$

We compare the unit prices.

14-in. pizza: $\dfrac{\$10.50}{153.86 \text{ in}^2} \approx \$0.068/\text{in}^2$

16-in. pizza: $\dfrac{\$11.95}{200.96 \text{ in}^2} \approx \$0.059/\text{in}^2$

The 16-in. pizza is a better buy.

47. We compare the unit prices. Recall that 1 lb = 16 oz.

1 lb box: $\dfrac{85¢}{16 \text{ oz}} = 5.3125\dfrac{¢}{\text{oz}}$

14-oz box: $\dfrac{85¢}{14 \text{ oz}} \approx 6.0714\dfrac{¢}{\text{oz}}$

We subtract to find the change in the unit price.

$6.0714 - 5.3125 = 0.7589 \approx 0.76$

The unit price increases $0.76\dfrac{¢}{\text{oz}}$.

49. $25 \text{ mi} = 25 \text{ mi} \cdot \dfrac{5280 \text{ ft}}{1 \text{ mi}} = 25 \cdot 5280 \cdot \text{ft} \cdot \dfrac{\text{mi}}{\text{mi}} = 132{,}000 \text{ ft}$

From Exercise 14, we know that sound travels at a speed of $1100\dfrac{\text{ft}}{\text{sec}}$. Then $132{,}000 \text{ ft} \div 1100\dfrac{\text{ft}}{\text{sec}} =$

$132{,}000 \text{ ft} \cdot \dfrac{1}{1100} \cdot \dfrac{\text{sec}}{\text{ft}} = \dfrac{132{,}000}{1100} \cdot \text{sec} \cdot \dfrac{\text{ft}}{\text{ft}} = 120 \text{ sec.}$

Thus, it will take 120 sec for you to hear the crack of thunder.

From Exercise 13, we know that light travels at a speed of $186{,}000\dfrac{\text{mi}}{\text{sec}}$. Then $25 \text{ mi} \div 186{,}000\dfrac{\text{mi}}{\text{sec}} =$

$25 \text{ mi} \cdot \dfrac{1}{186{,}000} \cdot \dfrac{\text{sec}}{\text{mi}} = \dfrac{25}{186{,}000} \cdot \text{sec} \cdot \dfrac{\text{mi}}{\text{mi}} \approx$ 0.000134 sec. Thus, it will take about 0.000134 sec for you to see the flash of light.

Exercise Set 7.3

1. We can use cross-products:

$5 \cdot 9 = 45$ (5, 7 / 6, 9) $6 \cdot 7 = 42$

Since the cross-products are not the same, $45 \neq 42$, we know that the numbers are not proportional.

3. We can use cross-products:

$1 \cdot 20 = 20$ (1, 10 / 2, 20) $2 \cdot 10 = 20$

Since the cross-products are the same, $20 = 20$, we know that $\dfrac{1}{2} = \dfrac{10}{20}$, so the numbers are proportional.

5. We can use cross-products:

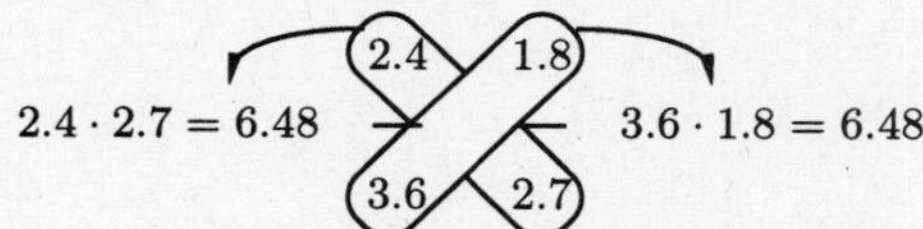

Since the cross-products are the same, $6.48 = 6.48$, we know that $\dfrac{2.4}{3.6} = \dfrac{1.8}{2.7}$, so the numbers are proportional.

7. We can use cross-products:

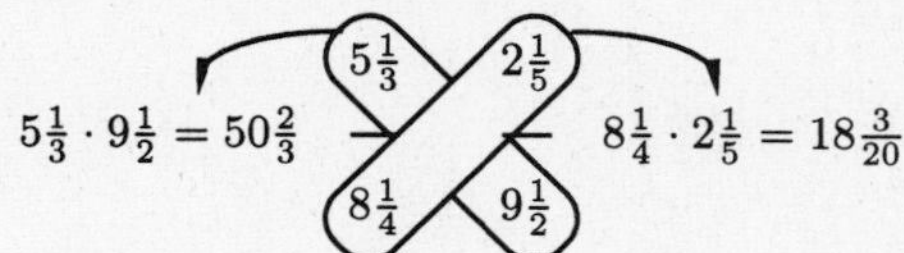

Since the cross-products are not the same, $50\frac{2}{3} \neq 18\frac{3}{20}$, we know that the numbers are not proportional.

9.

$\dfrac{18}{4} = \dfrac{x}{10}$

$18 \cdot 10 = 4 \cdot x$ Equating cross-products

$\dfrac{18 \cdot 10}{4} = x$ Dividing by 4

$\dfrac{180}{4} = x$ Multiplying

$45 = x$ Dividing

11.

$\dfrac{x}{8} = \dfrac{9}{6}$

$6 \cdot x = 8 \cdot 9$ Equating cross-products

$x = \dfrac{8 \cdot 9}{6}$ Dividing by 6

$x = \dfrac{72}{6}$ Multiplying

$x = 12$ Dividing

13. $\frac{t}{12} = \frac{5}{6}$

$6 \cdot t = 12 \cdot 5$

$t = \frac{12 \cdot 5}{6}$

$t = \frac{60}{6}$

$t = 10$

15. $\frac{2}{5} = \frac{8}{n}$

$2 \cdot n = 5 \cdot 8$

$n = \frac{5 \cdot 8}{2}$

$n = \frac{40}{2}$

$n = 20$

17. $\frac{n}{15} = \frac{10}{30}$

$30 \cdot n = 15 \cdot 10$

$n = \frac{15 \cdot 10}{30}$

$n = \frac{150}{30}$

$n = 5$

19. $\frac{16}{12} = \frac{24}{x}$

$16 \cdot x = 12 \cdot 24$

$x = \frac{12 \cdot 24}{16}$

$x = \frac{288}{16}$

$x = 18$

21. $\frac{6}{11} = \frac{12}{x}$

$6 \cdot x = 11 \cdot 12$

$x = \frac{11 \cdot 12}{6}$

$x = \frac{132}{6}$

$x = 22$

23. $\frac{20}{7} = \frac{80}{x}$

$20 \cdot x = 7 \cdot 80$

$x = \frac{7 \cdot 80}{20}$

$x = \frac{560}{20}$

$x = 28$

25. $\frac{12}{9} = \frac{x}{7}$

$12 \cdot 7 = 9 \cdot x$

$\frac{12 \cdot 7}{9} = x$

$\frac{84}{9} = x$

$\frac{28}{3} = x$ Simplifying

$9\frac{1}{3} = x$ Writing a mixed numeral

27. $\frac{x}{13} = \frac{2}{9}$

$9 \cdot x = 13 \cdot 2$

$x = \frac{13 \cdot 2}{9}$

$x = \frac{26}{9}$, or $2\frac{8}{9}$

29. $\frac{t}{0.16} = \frac{0.15}{0.40}$

$0.40 \times t = 0.16 \times 0.15$

$t = \frac{0.16 \times 0.15}{0.40}$

$t = \frac{0.024}{0.40}$

$t = 0.06$

31. $\frac{100}{25} = \frac{20}{n}$

$100 \cdot n = 25 \cdot 20$

$n = \frac{25 \cdot 20}{100}$

$n = \frac{500}{100}$

$n = 5$

33. $\frac{7}{\frac{1}{4}} = \frac{28}{x}$

$7 \cdot x = \frac{1}{4} \cdot 28$

$x = \frac{\frac{1}{4} \cdot 28}{7}$

$x = \frac{7}{7}$

$x = 1$

35. $\frac{\frac{1}{4}}{\frac{1}{2}} = \frac{\frac{1}{2}}{x}$

$\frac{1}{4} \cdot x = \frac{1}{2} \cdot \frac{1}{2}$

$x = \frac{\frac{1}{2} \cdot \frac{1}{2}}{\frac{1}{4}}$

$x = \frac{\frac{1}{4}}{\frac{1}{4}}$

$x = 1$

37.
$$\begin{aligned} \frac{x}{3} &= \frac{0}{9} \\ x\cdot 9 &= 3\cdot 0 \\ x &= \frac{3\cdot 0}{9} \\ x &= 0 \end{aligned}$$

39.
$$\begin{aligned} \frac{\frac{2}{7}}{\frac{3}{4}} &= \frac{\frac{5}{6}}{y} \\ \frac{2}{7}\cdot y &= \frac{3}{4}\cdot\frac{5}{6} \\ y &= \frac{3}{4}\cdot\frac{5}{6}\cdot\frac{7}{2} && \text{Dividing by } \frac{2}{7} \\ y &= \frac{\cancel{3}}{4}\cdot\frac{5}{2\cdot\cancel{3}}\cdot\frac{7}{2} \\ y &= \frac{5\cdot 7}{4\cdot 2\cdot 2} \\ y &= \frac{35}{16}, \text{ or } 2\frac{3}{16} \end{aligned}$$

41.
$$\begin{aligned} \frac{2\frac{1}{2}}{3\frac{1}{3}} &= \frac{x}{4\frac{1}{4}} \\ 2\frac{1}{2}\cdot 4\frac{1}{4} &= 3\frac{1}{3}\cdot x \\ \frac{5}{2}\cdot\frac{17}{4} &= \frac{10}{3}\cdot x \\ \frac{3}{10}\cdot\frac{5}{2}\cdot\frac{17}{4} &= x && \text{Dividing by } \frac{10}{3} \\ \frac{3}{\cancel{5}\cdot 2}\cdot\frac{\cancel{5}}{2}\cdot\frac{17}{4} &= x \\ \frac{3\cdot 17}{2\cdot 2\cdot 4} &= x \\ \frac{51}{16} &= x, \text{ or} \\ 3\frac{3}{16} &= x \end{aligned}$$

43.
$$\begin{aligned} \frac{1.28}{3.76} &= \frac{4.28}{y} \\ 1.28\times y &= 3.76\times 4.28 \\ y &= \frac{3.76\times 4.28}{1.28} \\ y &= \frac{16.0928}{1.28} \\ y &= 12.5725 \end{aligned}$$

45. The first coordinate is positive and the second coordinate is negative, so the point is in quadrant IV.

47.
$$\begin{array}{r} 5\,0 \\ 4\,\overline{)\,2\,0\,0} \\ \underline{2\,0\,0} \\ 0 \end{array}$$

49.
$$\begin{array}{r} 1\,4.\,5 \\ 1\,6\,\overline{)\,2\,3\,2.\,0} \\ \underline{1\,6\,0} \\ 7\,2 \\ \underline{6\,4} \\ 8\,0 \\ \underline{8\,0} \\ 0 \end{array}$$

51.

53.
$$\begin{aligned} \frac{1728}{5643} &= \frac{836.4}{x} \\ 1728\cdot x &= 5643\cdot 836.4 \\ x &= \frac{5643\cdot 836.4}{1728} \\ x &\approx 2731.4 && \text{Using a calculator to multiply and divide} \end{aligned}$$

Exercise Set 7.4

1. Let $d =$ the distance traveled in 42 days.

$$\begin{matrix} \text{Distance} \rightarrow \\ \text{Time} \rightarrow \end{matrix} \frac{234}{14} = \frac{d}{42} \begin{matrix} \leftarrow \text{Distance} \\ \leftarrow \text{Time} \end{matrix}$$

$$\begin{aligned} \text{Solve: } 234\cdot 42 &= 14\cdot d && \text{Equating cross-products} \\ \frac{234\cdot 42}{14} &= d && \text{Dividing by 14} \\ \frac{234\cdot 3\cdot 14}{14} &= d && \text{Factoring} \\ 234\cdot 3 &= d && \text{Simplifying} \\ 702 &= d && \text{Multiplying} \end{aligned}$$

Monica would travel 702 mi in 42 days.

3. Let $x =$ the cost of 9 tee shirts.

$$\begin{matrix} \text{Tee shirts} \rightarrow \\ \text{Dollars} \rightarrow \end{matrix} \frac{2}{18.80} = \frac{9}{x} \begin{matrix} \leftarrow \text{Tee shirts} \\ \leftarrow \text{Dollars} \end{matrix}$$

$$\begin{aligned} \text{Solve: } 2\cdot x &= 18.80\cdot 9 && \text{Equating cross-products} \\ x &= \frac{18.80\cdot 9}{2} && \text{Dividing by 2} \\ x &= \frac{2\cdot 9.40\cdot 9}{2} && \text{Factoring} \\ x &= 9.40\cdot 9 && \text{Simplifying} \\ x &= 84.60 && \text{Multiplying} \end{aligned}$$

Thus, 9 tee shirts would cost \$84.60.

5. We let $g =$ the number of gallons of sealant Bonnie should buy.

$$\begin{matrix} \text{Area} \rightarrow \\ \text{Sealant} \rightarrow \end{matrix} \frac{450}{2} = \frac{1200}{g} \begin{matrix} \leftarrow \text{Area} \\ \leftarrow \text{Sealant} \end{matrix}$$

Solve: $450 \cdot g = 2 \cdot 1200$ Equating cross-products

$g = \frac{2 \cdot 1200}{450}$ Dividing by 450

$g = \frac{2 \cdot 3 \cdot 8 \cdot 50}{3 \cdot 3 \cdot 50}$ Factoring

$g = \frac{16}{3}$

$g = 5\frac{1}{3}$

Bonnie needs 5 entire gallons of sealant and $\frac{1}{3}$ of a sixth gallon, so she should buy 6 gal of sealant.

7. Let w = the weight of 40 books.

$$\begin{array}{l} \text{Books} \rightarrow \\ \text{Weight} \rightarrow \end{array} \frac{24}{37} = \frac{40}{w} \begin{array}{l} \leftarrow \text{Books} \\ \leftarrow \text{Weight} \end{array}$$

Solve: $24 \cdot w = 37 \cdot 40$

$w = \frac{37 \cdot 40}{24}$

$w = \frac{37 \cdot 5 \cdot 8}{3 \cdot 8}$

$w = \frac{185}{3}$, or $61\frac{2}{3}$

The weight of 40 books is $61\frac{2}{3}$ lb.

9. Let s = the amount of sap needed.

$$\begin{array}{l} \text{Sap} \rightarrow \\ \text{Syrup} \rightarrow \end{array} \frac{38}{2} = \frac{s}{9} \begin{array}{l} \leftarrow \text{Sap} \\ \leftarrow \text{Syrup} \end{array}$$

Solve: $38 \cdot 9 = 2 \cdot s$

$\frac{38 \cdot 9}{2} = s$

$\frac{2 \cdot 19 \cdot 9}{2} = s$

$19 \cdot 9 = s$

$171 = s$

171 gal of sap is needed to produce 9 gal of syrup.

11. Let t = the number of trees required to produce 375 pounds of coffee.

$$\begin{array}{l} \text{Trees} \rightarrow \\ \text{Pounds} \rightarrow \end{array} \frac{14}{17} = \frac{t}{375} \begin{array}{l} \leftarrow \text{Trees} \\ \leftarrow \text{Pounds} \end{array}$$

Solve: $14 \cdot 375 = 17 \cdot t$

$\frac{14 \cdot 375}{17} = t$

$\frac{5250}{17} = t$

$308\frac{14}{17} = t$

Because it doesn't make sense to talk about a fractional part of a tree, we round up to the nearest whole tree. Thus, 309 trees are required to produce 375 pounds of coffee.

13. Let s = the number of sections you would expect to see offered. (The student-section ratio would be the same as the student-faculty ratio.)

$$\begin{array}{l} \text{Students} \rightarrow \\ \text{Sections} \rightarrow \end{array} \frac{14}{1} = \frac{56}{s} \begin{array}{l} \leftarrow \text{Students} \\ \leftarrow \text{Sections} \end{array}$$

Solve: $14 \cdot s = 1 \cdot 56$

$s = \frac{1 \cdot 56}{14}$

$s = \frac{1 \cdot 4 \cdot 14}{14}$

$s = 1 \cdot 4$

$s = 4$

You would expect to see 4 sections offered.

15. Let d = the number of defective bulbs in a lot of 22,000.

$$\begin{array}{l} \text{Defective bulbs} \rightarrow \\ \text{Bulbs in lot} \rightarrow \end{array} \frac{18}{200} = \frac{d}{22,000} \begin{array}{l} \leftarrow \text{Defective bulbs} \\ \leftarrow \text{Bulbs in lot} \end{array}$$

Solve: $18 \cdot 22,000 = 200 \cdot d$

$\frac{18 \cdot 22,000}{200} = d$

$\frac{18 \cdot 110 \cdot 200}{200} = d$

$18 \cdot 110 = d$

$1980 = d$

There will be 1980 defective bulbs in a lot of 22,000.

17. Let d = the actual distance between the cities.

$$\begin{array}{l} \text{Map distance} \rightarrow \\ \text{Actual distance} \rightarrow \end{array} \frac{1}{15.6} = \frac{3.5}{d} \begin{array}{l} \leftarrow \text{Map distance} \\ \leftarrow \text{Actual distance} \end{array}$$

Solve: $1 \cdot d = 15.6 \cdot 3.5$

$d = 54.6$

The cities are 54.6 km apart.

19. Let w = the number of inches of water to which $5\frac{1}{2}$ ft of snow will melt.

$$\begin{array}{l} \text{Snow} \rightarrow \\ \text{Water} \rightarrow \end{array} \frac{1\frac{1}{2}}{2} = \frac{5\frac{1}{2}}{w} \begin{array}{l} \leftarrow \text{Snow} \\ \leftarrow \text{Water} \end{array}$$

Solve: $1\frac{1}{2} \cdot w = 2 \cdot 5\frac{1}{2}$

$\frac{3}{2} \cdot w = \frac{2}{1} \cdot \frac{11}{2}$ Writing fractional notation

$w = \frac{2}{1} \cdot \frac{11}{2} \cdot \frac{2}{3}$ Dividing by $\frac{3}{2}$

$w = \frac{2 \cdot 11 \cdot 2}{1 \cdot 2 \cdot 3}$

$w = \frac{11 \cdot 2}{1 \cdot 3}$ Simplifying

$w = \frac{22}{3}$, or $7\frac{1}{3}$

Thus, $5\frac{1}{2}$ ft of snow will melt to $7\frac{1}{3}$ in. of water.

21. To plot $(-3, 2)$, we locate -3 on the first, or horizontal, axis. Then we go up 2 units and make a dot.

To plot $(4, 5)$, we locate 4 on the first, or horizontal, axis. Then we go up 5 units and make a dot.

To plot $(-4, -1)$, we locate -4 on the first, or horizontal, axis. Then we go down 1 unit and make a dot.

To plot $(0, 3)$, we locate 0 on the first, or horizontal, axis. Then we go up 3 units and make a dot.

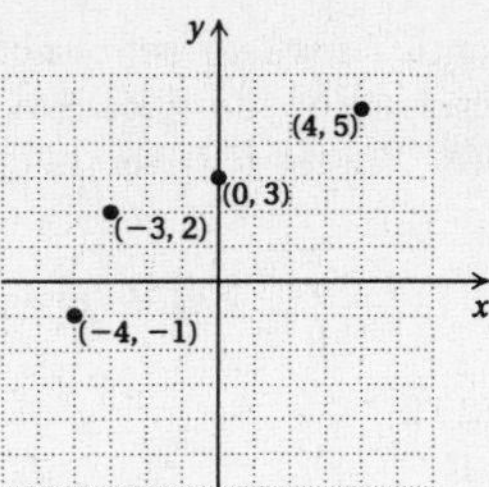

23. First consider $169.36 \div 23.2$:

$$\begin{array}{r} 7.3 \\ 23.2_{\wedge}\overline{)\,169.3_{\wedge}6} \\ 1624\ 0 \\ \hline 69\ 6 \\ 69\ 6 \\ \hline 0 \end{array}$$

Since a positive number divided by a negative number is negative, the answer is -7.3.

25. 

27. Let $f =$ the number of faculty positions required to maintain the current student-to-faculty ratio after the university expands.

$$\begin{array}{l} \text{Students} \rightarrow \\ \text{Faculty} \rightarrow \end{array} \frac{2700}{217} = \frac{2900}{f} \begin{array}{l} \leftarrow \text{Students} \\ \leftarrow \text{Faculty} \end{array}$$

Solve: $2700 \cdot f = 217 \cdot 2900$

$$f = \frac{217 \cdot 2900}{2700}$$

$$f = \frac{217 \cdot 29 \cdot 100}{27 \cdot 100}$$

$$f = \frac{217 \cdot 29}{27}$$

$$f = \frac{6293}{27}, \text{ or } 233\frac{2}{27}$$

Since it is impossible to create a fractional part of a position, we round up to the nearest whole position. Thus, 234 positions will be required after the university expands. We subtract to find how many new positions should be created:

$$234 - 217 = 17$$

17 new faculty positions should be created.

29. First we find the area of the wall.

$A = l \times w = 100 \text{ ft} \times 30 \text{ ft} = 3000 \text{ ft}^2$

Now we let $p =$ the number of gallons of paint Sue should buy.

$$\begin{array}{l} \text{Area} \rightarrow \\ \text{Paint} \rightarrow \end{array} \frac{950}{2} = \frac{3000}{p} \begin{array}{l} \leftarrow \text{Area} \\ \leftarrow \text{Paint} \end{array}$$

Solve: $950 \cdot p = 2 \cdot 3000$

$$p = \frac{2 \cdot 3000}{950}$$

$$p = \frac{2 \cdot 50 \cdot 60}{19 \cdot 50}$$

$$p = \frac{2 \cdot 60}{19}$$

$$p = \frac{120}{19}, \text{ or } 6\frac{6}{19}$$

Assuming that Sue is buying paint in one gallon cans, she will have to buy 7 gal of paint.

Exercise Set 7.5

1. The ratio of h to 5 is the same as the ratio of 45 to 9. We have the proportion

$$\frac{h}{5} = \frac{45}{9}.$$

Solve: $\frac{h}{5} = 5$ Simplifying

$h = 5 \cdot 5$ Multiplying by 5 on both sides

$h = 25$ Simplifying

The missing length h is 25.

3. The ratio of x to 2 is the same as the ratio of 2 to 3. We have the proportion

$$\frac{x}{2} = \frac{2}{3}.$$

Solve: $3 \cdot x = 2 \cdot 2$ Equating cross-products

$x = \frac{2 \cdot 2}{3}$ Dividing by 3 on both sides

$x = \frac{4}{3}$, or $1\frac{1}{3}$

The missing length x is $\frac{4}{3}$, or $1\frac{1}{3}$. We could also have used $\frac{x}{2} = \frac{1}{1\frac{1}{2}}$ to find x.

5. First we find x. The ratio of x to 9 is the same as the ratio of 6 to 8. We have the proportion

$$\frac{x}{9} = \frac{6}{8}.$$

Solve: $\frac{x}{9} = \frac{3}{4}$ Rewriting $\frac{6}{8}$ as $\frac{3}{4}$

$4 \cdot x = 9 \cdot 3$ Equating cross-products

$$x = \frac{9 \cdot 3}{4}$$

$$x = \frac{27}{4}, \text{ or } 6\frac{3}{4}$$

The missing length x is $\frac{27}{4}$, or $6\frac{3}{4}$.

Next we find y. The ratio of y to 12 is the same as the ratio of 6 to 8. We have the proportion

$$\frac{y}{12} = \frac{6}{8}.$$

Solve: $\frac{y}{12} = \frac{3}{4}$ Rewriting $\frac{6}{8}$ as $\frac{3}{4}$

$4 \cdot y = 12 \cdot 3$ Equating cross-products

$y = \frac{12 \cdot 3}{4}$

$y = 9$

The missing length y is 9.

7. First we find x. The ratio of x to 2.5 is the same as the ratio of 2.1 to 0.7. We have the proportion

$$\frac{x}{2.5} = \frac{2.1}{0.7}.$$

Solve: $0.7 \cdot x = 2.5 \cdot 2.1$

$x = \frac{2.5 \cdot 2.1}{0.7}$

$x = 7.5$

The missing length x is 7.5.

Next we find y. The ratio of y to 2.4 is the same as the ratio of 2.1 to 0.7. We have the proportion

$$\frac{y}{2.4} = \frac{2.1}{0.7}.$$

Solve: $0.7 \cdot y = 2.4 \cdot 2.1$

$y = \frac{2.4 \cdot 2.1}{0.7}$

$y = 7.2$

The missing length y is 7.2.

(Note that in each solution above our first step could have been to rewrite $\frac{2.1}{0.7}$ as 3).

9. If we use the sun's rays to represent the third side of a triangle in a drawing of the situation, we see that we have similar triangles. We let s = the length of a shadow cast by a person 2 m tall.

Sun's rays 8 m
5 m
Sun's rays 2 m
s

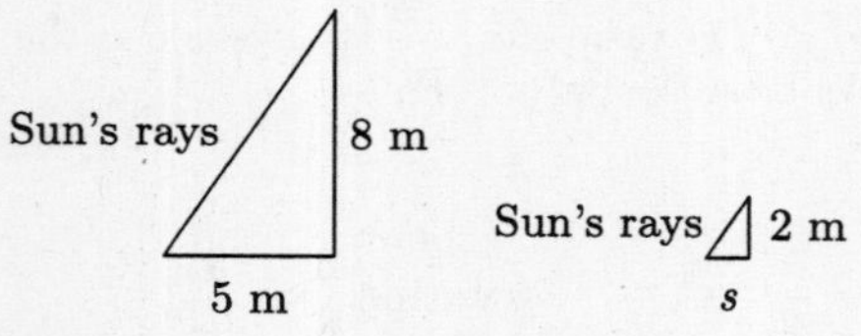

The ratio of s to 5 is the same as the ratio of 2 to 8. We have the proportion

$$\frac{s}{5} = \frac{2}{8}.$$

Solve: $\frac{s}{5} = \frac{1}{4}$ Rewriting $\frac{2}{8}$ as $\frac{1}{4}$

$4 \cdot s = 5 \cdot 1$

$s = \frac{5 \cdot 1}{4}$

$s = \frac{5}{4}$, or 1.25

The length of a shadow cast by a person 2 m tall is 1.25 m.

11. If we use the sun's rays to represent the third side of a triangle in a drawing of the situation, we see that we have similar triangles. We let h = the height of the tree.

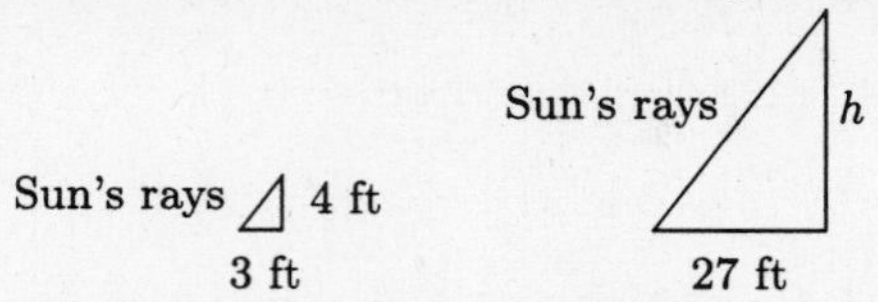

Sun's rays h
Sun's rays 4 ft
3 ft
27 ft

The ratio of h to 4 is the same as the ratio of 27 to 3. We have the proportion

$$\frac{h}{4} = \frac{27}{3}.$$

Solve: $\frac{h}{4} = 9$ Simplifying

$h = 4 \cdot 9$

$h = 36$

The tree is 36 ft tall.

13. $\frac{\text{Width} \rightarrow 6}{\text{Length} \rightarrow 9} = \frac{x \leftarrow \text{Width}}{6 \leftarrow \text{Length}}$

Solve: $\frac{2}{3} = \frac{x}{6}$ Rewriting $\frac{6}{9}$ as $\frac{2}{3}$

$2 \cdot 6 = 3 \cdot x$ Equating cross-products

$\frac{2 \cdot 6}{3} = x$

$\frac{2 \cdot 2 \cdot 3}{3} = x$

$2 \cdot 2 = x$

$4 = x$

The missing length x is 4.

15. $\frac{\text{Width} \rightarrow 4}{\text{Length} \rightarrow 7} = \frac{6 \leftarrow \text{Width}}{x \leftarrow \text{Length}}$

Solve: $4 \cdot x = 7 \cdot 6$ Equating cross-products

$x = \frac{7 \cdot 6}{4}$

$x = \frac{7 \cdot 2 \cdot 3}{2 \cdot 2}$

$x = \frac{7 \cdot 3}{2}$

$x = \frac{21}{2}$, or $10\frac{1}{2}$

The missing length x is $10\frac{1}{2}$.

17. First we find x. The ratio of x to 8 is the same as the ratio of 2 to 3. We have the proportion

$$\frac{x}{8} = \frac{2}{3}.$$

Solve: $3 \cdot x = 8 \cdot 2$

$x = \frac{8 \cdot 2}{3}$

$x = \frac{16}{3}$, or $5\frac{1}{3}$

The missing length x is $\frac{16}{3}$, or $5\frac{1}{3}$.

Next we find y. The ratio of y to 7 is the same as the ratio of 2 to 3. We have the proportion

$$\frac{y}{7} = \frac{2}{3}.$$

Solve: $3 \cdot y = 7 \cdot 2$

$$y = \frac{7 \cdot 2}{3}$$

$$y = \frac{14}{3} = 4\frac{2}{3}$$

The missing length y is $\frac{14}{3}$, or $4\frac{2}{3}$.

Finally we find z. The ratio of z to 8 is the same as the ratio of 2 to 3. We have the proportion

$$\frac{z}{9} = \frac{2}{3}.$$

This is the same proportion we solved above when we found x. Then the missing length z is $5\frac{1}{3}$.

19. First we find x. The ratio of x to 5 is the same as the ratio of 3 to 4. We have the proportion

$$\frac{x}{5} = \frac{3}{4}.$$

Solve: $4 \cdot x = 5 \cdot 3$

$$x = \frac{5 \cdot 3}{4}$$

$$x = \frac{15}{4}, \text{ or } 3\frac{3}{4}$$

The missing length x is $\frac{15}{4}$, or $3\frac{3}{4}$.

Next we find y. The ratio of y to 6 is the same as the ratio of 3 to 4. We have the proportion

$$\frac{y}{6} = \frac{3}{4}.$$

Solve: $4 \cdot y = 6 \cdot 3$

$$y = \frac{6 \cdot 3}{4}$$

$$y = \frac{2 \cdot 3 \cdot 3}{2 \cdot 2}$$

$$y = \frac{3 \cdot 3}{2}$$

$$y = \frac{9}{2}, \text{ or } 4\frac{1}{2}$$

The missing length y is $\frac{9}{2}$, or $4\frac{1}{2}$.

21. $\frac{\text{Height} \rightarrow h}{\text{Width} \rightarrow 32} = \frac{5 \leftarrow \text{Height}}{8 \leftarrow \text{Width}}$

Solve: $8 \cdot h = 32 \cdot 5$

$$h = \frac{32 \cdot 5}{8}$$

$$h = \frac{4 \cdot 8 \cdot 5}{8}$$

$$h = 4 \cdot 5$$

$$h = 20$$

The missing length is 20 ft.

23. ***Familiarize***. This is a multistep problem.

First we find the total cost of the purchases. We let c = this amount.

Translate and Solve.

Price of book	plus	Price of CD	plus	Price of sweatshirt	is	Total cost
↓	↓	↓	↓	↓	↓	↓
\$49.95	+	\$14.88	+	\$29.95	=	c

To solve the equation we carry out the addition.

$$\begin{array}{r} \scriptstyle 2\ 2\ 1 \\ 49.95 \\ 14.88 \\ +\ 29.95 \\ \hline 94.78 \end{array}$$

Thus, c = \$94.78.

Now we find how much more money the student needs to make these purchases. We let m = this amount.

Money student has	plus	How much more money	is	Total cost of purchases
↓	↓	↓	↓	↓
\$34.97	+	m	=	\$94.78

To solve the equation we subtract 34.97 on both sides.

$m = 94.78 - 34.97$
$m = 59.81$

$$\begin{array}{r} \scriptstyle 13 \\ \scriptstyle 8\ \not{3}\ 17 \\ \not{9}\not{4}.\not{7}8 \\ -\ 34.97 \\ \hline 59.81 \end{array}$$

Check. We repeat the calculations.

State. The student needs \$59.81 more to make the purchases.

25. Multiplying the absolute values, we have

$$\begin{array}{r} \scriptstyle 7\ 7\ 1 \\ \scriptstyle 3\ 3 \\ 80.892 \\ \times\ \ 8.4 \\ \hline 323568 \\ 6471360 \\ \hline 679.4928 \end{array}$$

Since the product of a negative number and a positive number is negative, the answer is -679.4928.

27. $1\underline{00} \times 274.568$ 274.56.8

2 zeros   Move 2 places to the right.

$100 \times 274.568 = 27,456.8$

29. ◈

31. 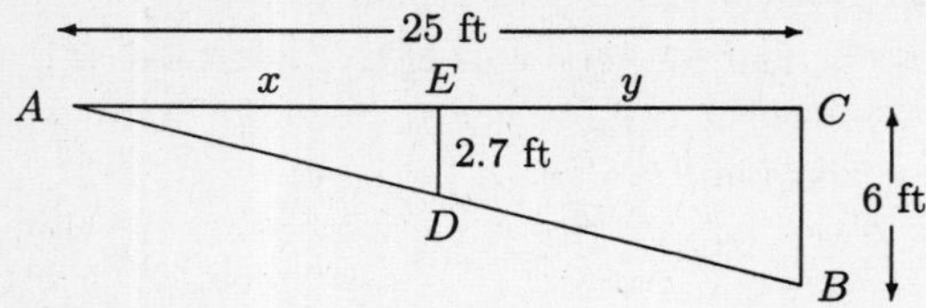

We note that triangle ADE is similar to triangle ABC and use this information to find the length x.

$$\frac{x}{25} = \frac{2.7}{6}$$
$$6 \cdot x = 25 \cdot 2.7$$
$$x = \frac{25 \cdot 2.7}{6}$$
$$x = 11.25$$

Thus the goalie should be 11.25 ft from point A. We subtract to find how far from the goal the goalie should be located.

$$25 - 11.25 = 13.75$$

The goalie should stand 13.75 ft from the goal.

33. Since the ratio of d to 25 ft is the same as the ratio of 40 ft to 10 ft, we have the proportion

$$\frac{d}{25} = \frac{40}{10}.$$

Solve: $10 \cdot d = 25 \cdot 40$
$$d = \frac{25 \cdot 40}{10}$$
$$d = 100$$

The distance across the river is 100 ft.

35. We let $h =$ the height of the model hoop. Then we translate to a proportion.

$$\begin{array}{l} \text{Width} \to \\ \text{Height} \to \end{array} \frac{12}{h} = \frac{116}{10} \begin{array}{l} \leftarrow \text{Width} \\ \leftarrow \text{Height} \end{array}$$

Solve: $12 \cdot 10 = h \cdot 116$ Equating cross-products
$$\frac{12 \cdot 10}{116} = h$$
$1.034 \approx h$ Calculating and rounding

The model hoop should be 1.034 cm high.

Chapter 8

Percent Notation

Exercise Set 8.1

1. Since the circle is divided into 100 sections, we can think of it as a pie cut into 100 equally sized pieces. We shade a wedge equal in size to 4 of these pieces to represent 4%. Then we shade wedges equal in size to 30, 34, 30, and 2 of these pieces to represent 30%, 34%, 30%, and 2%, respectively.

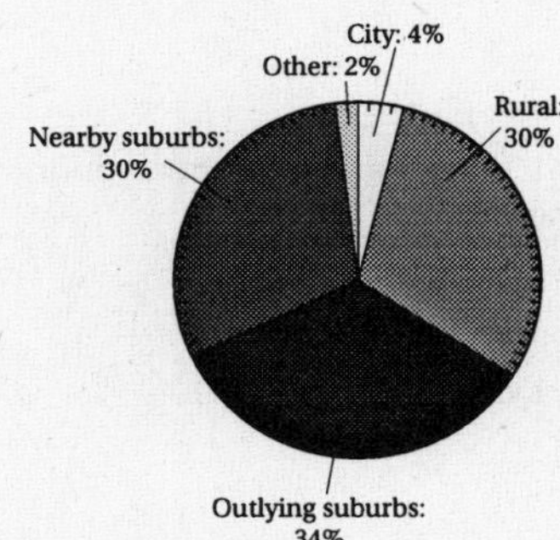

3. $90\% = \dfrac{90}{100}$ A ratio of 90 to 100

$90\% = 90 \times \dfrac{1}{100}$ Replacing % with $\times \dfrac{1}{100}$

$90\% = 90 \times 0.01$ Replacing % with $\times 0.01$

5. $12.5\% = \dfrac{12.5}{100}$ A ratio of 12.5 to 100

$12.5\% = 12.5 \times \dfrac{1}{100}$ Replacing % with $\times \dfrac{1}{100}$

$12.5\% = 12.5 \times 0.01$ Replacing % with $\times 0.01$

7. 19%

a) Replace the percent symbol with $\times 0.01$.

19×0.01

b) Multiply to move the decimal point two places to the left.

0.19.

Thus, 19% = 0.19.

9. 45.6%

a) Replace the percent symbol with $\times 0.01$.

45.6×0.01

b) Multiply to move the decimal point two places to the left.

0.45.6

Thus, 45.6% = 0.456.

11. 59.01%

a) Replace the percent symbol with $\times 0.01$.

59.01×0.01

b) Multiply to move the decimal point two places to the left.

0.59.01

Thus, 59.01% = 0.5901.

13. 10%

a) Replace the percent symbol with $\times 0.01$.

10×0.01

b) Multiply to move the decimal point two places to the left.

0.10.

Thus, 10% = 0.1.

15. 1%

a) Replace the percent symbol with $\times 0.01$.

1×0.01

b) Multiply to move the decimal point two places to the left.

0.01.

Thus, 1% = 0.01.

17. 400%

a) Replace the percent symbol with $\times 0.01$.

400×0.01

b) Multiply to move the decimal point two places to the left.

4.00.

Thus, 400%=4.

19. 0.1%

a) Replace the percent symbol with $\times 0.01$.

0.1×0.01

b) Multiply to move the decimal point two places to the left.

0.00.1

Thus, 0.1% = 0.001.

21. 0.09%

a) Replace the percent symbol with $\times 0.01$.

0.09×0.01

b) Multiply to move the decimal point two places to the left.

0.00.09

Thus, 0.09% = 0.0009.

23. 0.18%

a) Replace the percent symbol with $\times 0.01$.

0.18×0.01

b) Multiply to move the decimal point two places to the left.

0.00.18

Thus, 0.18% = 0.0018.

25. 23.19%

a) Replace the percent symbol with $\times 0.01$.

23.19×0.01

b) Multiply to move the decimal point two places to the left.

0.23.19

Thus, 23.19% = 0.2319.

27. 90%

a) Replace the percent symbol with $\times 0.01$.

90×0.01

b) Multiply to move the decimal point two places to the left.

0.90.

Thus, 90% = 0.9.

29. 10.8%

a) Replace the percent symbol with $\times 0.01$.

10.8×0.01

b) Multiply to move the decimal point two places to the left.

0.10.8

Thus, 10.8% = 0.108.

31. 45.8%

a) Replace the percent symbol with $\times 0.01$.

45.8×0.01

b) Multiply to move the decimal point two places to the left.

0.45.8

Thus, 45.8% = 0.458.

33. 0.74

a) Multiply by 100 to move the decimal point two places to the right.

0.74.

b) Write a percent symbol: 74%

Thus, 0.74 = 74%.

35. 0.03

a) Multiply by 100 to move the decimal point two places to the right.

0.03.

b) Write a percent symbol: 3%

Thus, 0.03 = 3%.

37. 1.00

a) Multiply by 100 to move the decimal point two places to the right.

1.00.

b) Write a percent symbol: 100%

Thus, 1.00 = 100%.

39. 0.334

a) Multiply by 100 to move the decimal point two places to the right.

0.33.4

b) Write a percent symbol: 33.4%

Thus, 0.334 = 33.4%.

41. 0.4

a) Multiply by 100 to move the decimal point two places to the right.

0.40.

b) Write a percent symbol: 40%

Thus, 0.4 = 40%.

43. 0.006

a) Multiply by 100 to move the decimal point two places to the right.

0.00.6

b) Write a percent symbol: 0.6%

Thus, 0.006 = 0.6%.

45. 0.2718

a) Multiply by 100 to move the decimal point two places to the right.

0.27.18

b) Write a percent symbol: 27.18%

Thus, 0.2718 = 27.18%.

47. 0.0239

a) Multiply by 100 to move the decimal point two places to the right.

```
0.02.39
 └─┘↑
```

b) Write a percent symbol: 2.39%

Thus, 0.0239 = 2.39%.

49. 0.025

a) Multiply by 100 to move the decimal point two places to the right.

```
0.02.5
 └─┘↑
```

b) Write a percent symbol: 2.5%

Thus, 0.025 = 2.5%.

51. 0.24

a) Multiply by 100 to move the decimal point two places to the right.

```
0.24.
 └─┘↑
```

b) Write a percent symbol: 24%

Thus, 0.24 = 24%.

53. We use the definition of percent as a ratio.

$$\frac{41}{100} = 41\%$$

55. We use the definition of percent as a ratio.

$$\frac{5}{100} = 5\%$$

57. We multiply by 1 to get 100 in the denominator.

$$\frac{2}{10} = \frac{2}{10} \cdot \frac{10}{10} = \frac{20}{100} = 20\%$$

59. We multiply by 1 to get 100 in the denominator.

$$\frac{9}{10} = \frac{9}{10} \cdot \frac{10}{10} = \frac{90}{100} = 90\%$$

61. $\frac{3}{4} = \frac{3}{4} \cdot \frac{25}{25} = \frac{75}{100} = 75\%$

63. Find decimal notation by division.

```
    0.6 2 5
 8 ⟌5.0 0 0
    4 8
      2 0
      1 6
        4 0
        4 0
          0
```

$$\frac{5}{8} = 0.625$$

Convert to percent notation.

```
0.62.5
 └─┘↑
```

$$\frac{5}{8} = 62.5\%, \text{ or } 62\tfrac{1}{2}\%$$

65. $\frac{2}{5} = \frac{2}{5} \cdot \frac{20}{20} = \frac{40}{100} = 40\%$

67. Find decimal notation by division.

```
    0.6 6 6
 3 ⟌2.0 0 0
    1 8
      2 0
      1 8
        2 0
        1 8
          2
```

We get a repeating decimal: $\frac{2}{3} = 0.66\overline{6}$

Convert to percent notation.

```
0.66.6̄
 └─┘↑
```

$$\frac{2}{3} = 66.\overline{6}\%, \text{ or } 66\tfrac{2}{3}\%$$

69.

```
    0.1 6 6
 6 ⟌1.0 0 0
      6
      4 0
      3 6
        4 0
        3 6
          4
```

We get a repeating decimal: $\frac{1}{6} = 0.16\overline{6}$

Convert to percent notation.

```
0.16.6̄
 └─┘↑
```

$$\frac{1}{6} = 16.\overline{6}\%, \text{ or } 16\tfrac{2}{3}\%$$

71. $\frac{4}{25} = \frac{4}{25} \cdot \frac{4}{4} = \frac{16}{100} = 16\%$

73. $\frac{31}{50} = \frac{31}{50} \cdot \frac{2}{2} = \frac{62}{100} = 62\%$

75. $\frac{3}{20} = \frac{3}{20} \cdot \frac{5}{5} = \frac{15}{100} = 15\%$

77. $\frac{9}{25} = \frac{9}{25} \cdot \frac{4}{4} = \frac{36}{100} = 36\%$

79. $85\% = \frac{85}{100}$ Definition of percent

$$= \frac{5 \cdot 17}{5 \cdot 20} = \frac{5}{5} \cdot \frac{17}{20} = \frac{17}{20} \quad \text{Simplifying}$$

81. $62.5\% = \frac{62.5}{100}$ Definition of percent

$= \frac{62.5}{100} \cdot \frac{10}{10}$ Multiplying by 1 to eliminate the decimal point in the numerator

$= \frac{625}{1000}$

$= \frac{5 \cdot 125}{8 \cdot 125}$

$= \frac{5}{8} \cdot \frac{125}{125}$ Simplifying

$= \frac{5}{8}$

83. $33\frac{1}{3}\% = \frac{100}{3}\%$ Converting from mixed numeral to fractional notation

$= \frac{100}{3} \times \frac{1}{100}$ Definition of percent

$= \frac{100 \cdot 1}{3 \cdot 100}$ Multiplying

$= \frac{1}{3} \cdot \frac{100}{100}$ Simplifying

$= \frac{1}{3}$

85. $16.\overline{6}\% = 16\frac{2}{3}\%$ $\left(16.\overline{6} = 16\frac{2}{3}\right)$

$= \frac{50}{3}\%$ Converting from mixed numeral to fractional notation

$= \frac{50}{3} \times \frac{1}{100}$ Definition of percent

$= \frac{50 \cdot 1}{3 \cdot 50 \cdot 2}$ Multiplying

$= \frac{1}{2 \cdot 3} \cdot \frac{50}{50}$ Simplifying

$= \frac{1}{6}$

87. $7.25\% = \frac{7.25}{100} = \frac{7.25}{100} \cdot \frac{100}{100}$

$= \frac{725}{10,000} = \frac{29 \cdot 25}{400 \cdot 25} = \frac{29}{400} \cdot \frac{25}{25}$

$= \frac{29}{400}$

89. $0.8\% = \frac{0.8}{100} = \frac{0.8}{100} \cdot \frac{10}{10}$

$= \frac{8}{1000} = \frac{1 \cdot 8}{125 \cdot 8} = \frac{1}{125} \cdot \frac{8}{8}$

$= \frac{1}{125}$

91. The bar graph shows that 55% of people greatly enjoy Italian food.

$55\% = \frac{55}{100} = \frac{5 \cdot 11}{5 \cdot 20} = \frac{5}{5} \cdot \frac{11}{20} = \frac{11}{20}$

93. The bar graph shows that 38% of people greatly enjoy Chinese food.

$38\% = \frac{38}{100} = \frac{2 \cdot 19}{2 \cdot 50} = \frac{2}{2} \cdot \frac{19}{50} = \frac{19}{50}$

95. The bar graph shows that 11% of people greatly enjoy French food.

$11\% = \frac{11}{100}$

97. $24\% = \frac{24}{100} = \frac{4 \cdot 6}{4 \cdot 25} = \frac{4}{4} \cdot \frac{6}{25} = \frac{6}{25}$

99. $27.5\% = \frac{27.5}{100} = \frac{27.5}{100} \cdot \frac{10}{10} = \frac{275}{1000} = \frac{25 \cdot 11}{25 \cdot 40} = \frac{25}{25} \cdot \frac{11}{40} = \frac{11}{40}$

101. $\frac{1}{8} = 1 \div 8$

$$\begin{array}{r} 0.125 \\ 8\overline{)1.000} \\ \underline{8} \\ 20 \\ \underline{16} \\ 40 \\ \underline{40} \\ 0 \end{array}$$

$\mathbf{\frac{1}{8} = 0.125 = 12\frac{1}{2}\%}$**, or 12.5%**

$\frac{1}{6} = 1 \div 6$

$$\begin{array}{r} 0.166 \\ 6\overline{)1.000} \\ \underline{6} \\ 40 \\ \underline{36} \\ 40 \\ \underline{36} \\ 4 \end{array}$$

We get a repeating decimal: $0.1\overline{6}$

$0.16.\overline{6}$ $0.1\overline{6} = 16.\overline{6}\%$

$\mathbf{\frac{1}{6} = 0.1\overline{6} = 16.\overline{6}\%}$**, or** $\mathbf{16\frac{2}{3}\%}$

$20\% = \frac{20}{100} = \frac{1}{5} \cdot \frac{20}{20} = \frac{1}{5}$

$0.20.$ $20\% = 0.2$

$\mathbf{\frac{1}{5} = 0.2 = 20\%}$

$0.25.$ $0.25 = 25\%$

$25\% = \frac{25}{100} = \frac{1}{4} \cdot \frac{25}{25} = \frac{1}{4}$

$\mathbf{\frac{1}{4} = 0.25 = 25\%}$

$33\frac{1}{3}\% = \frac{100}{3}\% = \frac{100}{3} \times \frac{1}{100} = \frac{100}{300} = \frac{1}{3} \cdot \frac{100}{100} = \frac{1}{3}$

$0.33.\overline{3}$ $\quad 33.\overline{3}\% = 0.33\overline{3}$, or $0.\overline{3}$

$\mathbf{\frac{1}{3} = 0.\overline{3} = 33\frac{1}{3}\%}$**, or** $\mathbf{33.\overline{3}\%}$

$37.5\% = \frac{37.5}{100} = \frac{37.5}{100} \cdot \frac{10}{10} = \frac{375}{1000} = \frac{3}{8} \cdot \frac{125}{125} = \frac{3}{8}$

$0.37.5$ $\quad 37.5\% = 0.375$

$\mathbf{\frac{3}{8} = 0.375 = 37\frac{1}{2}\%}$**, or** $\mathbf{37.5\%}$

$40\% = \frac{40}{100} = \frac{2}{5} \cdot \frac{20}{20} = \frac{2}{5}$

$0.40.$ $\quad 40\% = 0.4$

$\mathbf{\frac{2}{5} = 0.4 = 40\%}$

103. $0.50.$ $\quad 0.5 = 50\%$

$50\% = \frac{50}{100} = \frac{1}{2} \cdot \frac{50}{50} = \frac{1}{2}$

$\mathbf{\frac{1}{2} = 0.5 = 50\%}$

$\frac{1}{3} = 1 \div 3$

$$\begin{array}{r} 0.3 \\ 3\overline{)1.0} \\ \underline{9} \\ 1 \end{array}$$

We get a repeating decimal: $0.\overline{3}$

$0.33.\overline{3}$ $\quad 0.\overline{3} = 33.\overline{3}\%$

$\mathbf{\frac{1}{3} = 0.\overline{3} = 33.\overline{3}\%}$**, or** $\mathbf{33\frac{1}{3}\%}$

$25\% = \frac{25}{100} = \frac{25}{25} \cdot \frac{1}{4} = \frac{1}{4}$

$0.25.$ $\quad 25\% = 0.25$

$\mathbf{\frac{1}{4} = 0.25 = 25\%}$

$16\frac{2}{3}\% = \frac{50}{3}\% = \frac{50}{3} \times \frac{1}{100} = \frac{50 \cdot 1}{3 \cdot 2 \cdot 50} = \frac{50}{50} \cdot \frac{1}{6} = \frac{1}{6}$

$\frac{1}{6} = 1 \div 6$

$$\begin{array}{r} 0.1\,6 \\ 6\overline{)1.0\,0} \\ \underline{6} \\ 4\,0 \\ \underline{3\,6} \\ 4 \end{array}$$

We get a repeating decimal: $0.1\overline{6}$

$\mathbf{\frac{1}{6} = 0.1\overline{6} = 16\frac{2}{3}\%}$**, or** $\mathbf{16.\overline{6}\%}$

$0.12.5$ $\quad 0.125 = 12.5\%$

$12.5\% = \frac{12.5}{100} = \frac{12.5}{100} \cdot \frac{10}{10} = \frac{125}{1000} = \frac{125}{125} \cdot \frac{1}{8} = \frac{1}{8}$

$\mathbf{\frac{1}{8} = 0.125 = 12.5\%}$**, or** $\mathbf{12\frac{1}{2}\%}$

$\frac{3}{4} = \frac{3}{4} \cdot \frac{25}{25} = \frac{75}{100} = 75\%$

$0.75.$ $\quad 75\% = 0.75$

$\mathbf{\frac{3}{4} = 0.75 = 75\%}$

$0.8\overline{3} = 0.83.\overline{3}$ $\quad 0.8\overline{3} = 83.\overline{3}\%$

$83.\overline{3}\% = 83\frac{1}{3}\% = \frac{250}{3}\% = \frac{250}{3} \times \frac{1}{100} = \frac{5 \cdot 50}{3 \cdot 2 \cdot 50} = \frac{5}{6} \cdot \frac{50}{50} = \frac{5}{6}$

$\mathbf{\frac{5}{6} = 0.8\overline{3} = 83.\overline{3}\%}$**, or** $\mathbf{83\frac{1}{3}\%}$

$\frac{3}{8} = 3 \div 8$

$$\begin{array}{r} 0.3\,7\,5 \\ 8\overline{)3.0\,0\,0} \\ \underline{2\,4} \\ 6\,0 \\ \underline{5\,6} \\ 4\,0 \\ \underline{4\,0} \\ 0 \end{array}$$

$\frac{3}{8} = 0.375$

$0.37.5$ $\quad 0.375 = 37.5\%$

$\mathbf{\frac{3}{8} = 0.375 = 37.5\%}$**, or** $\mathbf{37\frac{1}{2}\%}$

105.
$$\begin{array}{r} 3\,3 \\ 3\overline{)1\,0\,0} \\ \underline{9\,0} \\ 1\,0 \\ \underline{9} \\ 1 \end{array}$$

$\frac{100}{3} = 33\frac{1}{3}$

107. $0.05 \times b = 20$

$$\frac{0.05 \times b}{0.05} = \frac{20}{0.05}$$

$$b = 400$$

109. $\frac{24}{37} = \frac{15}{x}$

$24 \cdot x = 37 \cdot 15$ Equating cross-products

$$x = \frac{37 \cdot 15}{24}$$

$$x = 23.125$$

111. 

113. Use a calculator.

$$\frac{41}{369} = 0.11.\overline{1} = 11.\overline{1}\%$$

115. $\frac{14}{9}\% = \frac{14}{9} \times \frac{1}{100} = \frac{2 \cdot 7 \cdot 1}{9 \cdot 2 \cdot 50} = \frac{2}{2} \cdot \frac{7}{450} = \frac{7}{450}$

To find decimal notation for $\frac{7}{450}$ we divide.

```
         0.0 1 5 5
  4 5 0 ) 7.0 0 0 0 0
          4 5 0
          2 5 0 0
          2 2 5 0
            2 5 0 0
            2 2 5 0
              2 5 0
```

We get a repeating decimal: $\frac{14}{9}\% = 0.01\overline{5}$

Exercise Set 8.2

1. What is 37% of 74?

What → amount; 37% → number of hundredths; 74 → base

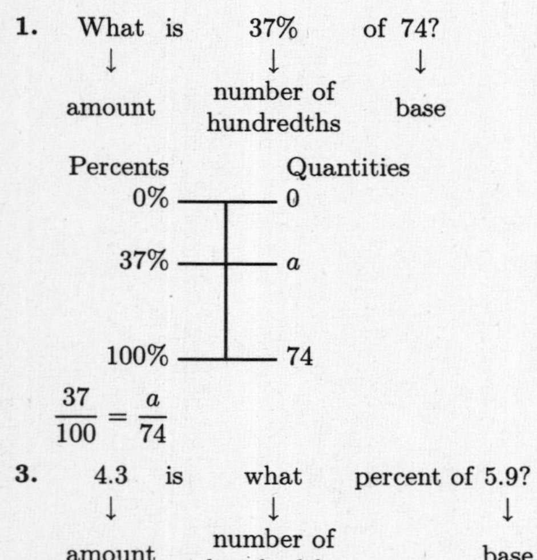

$$\frac{37}{100} = \frac{a}{74}$$

3. 4.3 is what percent of 5.9?

4.3 → amount; what → number of hundredths; 5.9 → base

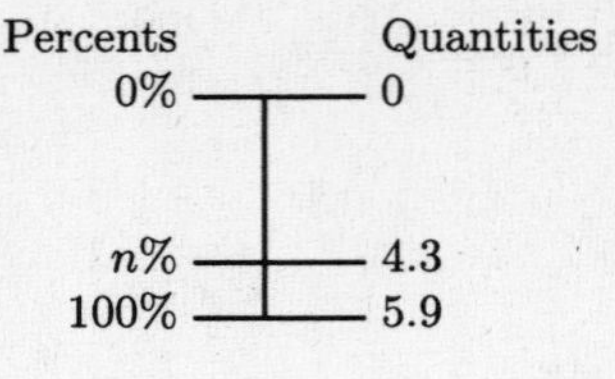

$$\frac{n}{100} = \frac{4.3}{5.9}$$

5. 14 is 25% of what?

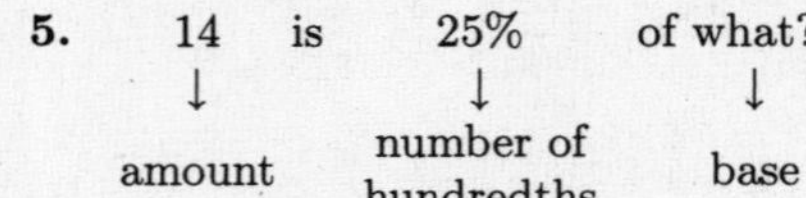

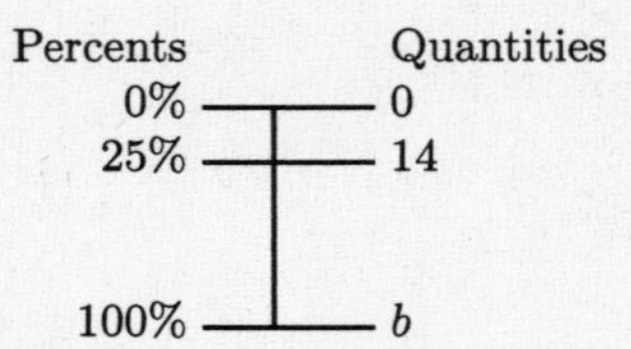

$$\frac{25}{100} = \frac{14}{b}$$

7. What is 176% of 125?

What → amount; 176% → number of hundredths; 125 → base

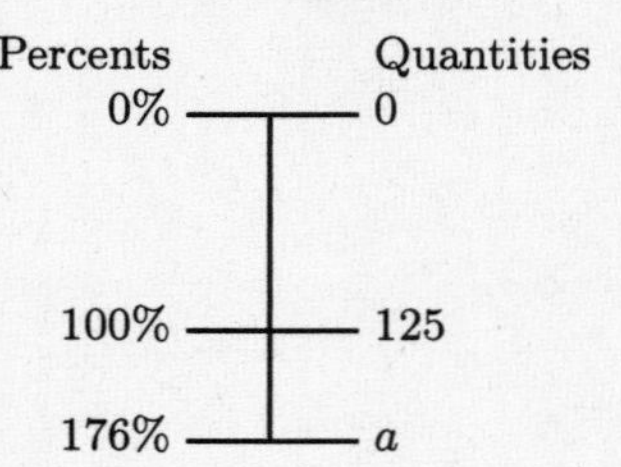

$$\frac{176}{100} = \frac{a}{125}$$

9. 70% of 660 is what?

70% → number of hundredths; 660 → base; what → amount

Percents — Quantities
0% — 0
70% — a
100% — 660

$$\frac{70}{100} = \frac{a}{660}$$

11. What is 4% of 1000?

What → amount; 4% → number of hundredths; 1000 → base

Percents — Quantities
0% — 0
4% — a
100% — 1000

Translate: $\frac{4}{100} = \frac{a}{1000}$

Solve:
$4 \cdot 1000 = 100 \cdot a$ Equating cross-products
$\frac{4 \cdot 1000}{100} = a$ Dividing by 100
$\frac{4000}{100} = a$
$40 = a$ Simplifying

40 is 4% of 1000. The answer is 40.

13. 4.8% of 60 is what?
4.8% ↓ number of hundredths; 60 ↓ base; what ↓ amount

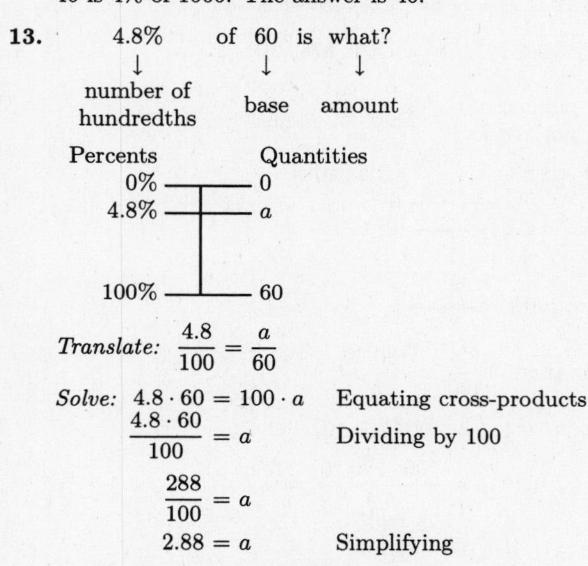

Translate: $\frac{4.8}{100} = \frac{a}{60}$

Solve:
$4.8 \cdot 60 = 100 \cdot a$ Equating cross-products
$\frac{4.8 \cdot 60}{100} = a$ Dividing by 100
$\frac{288}{100} = a$
$2.88 = a$ Simplifying

4.8% of 60 is 2.88. The answer is 2.88.

15. \$24 is what percent of \$96?
\$24 ↓ amount; what ↓ number of hundredths; \$96 ↓ base

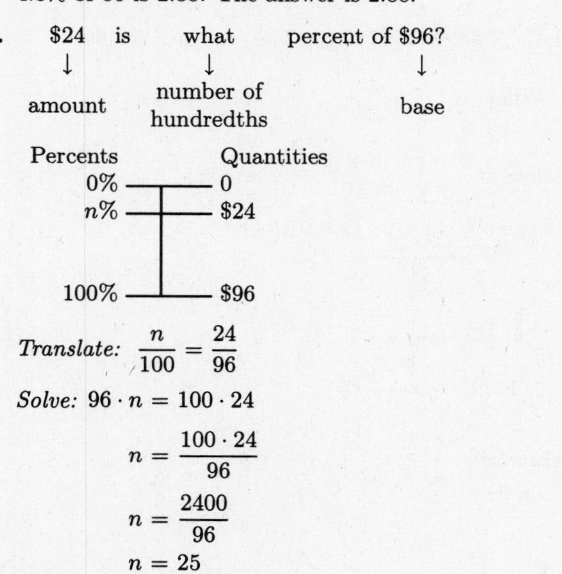

Translate: $\frac{n}{100} = \frac{24}{96}$

Solve: $96 \cdot n = 100 \cdot 24$
$n = \frac{100 \cdot 24}{96}$
$n = \frac{2400}{96}$
$n = 25$

\$24 is 25% of \$96. The answer is 25%.

17. 102 is what percent of 100?
102 ↓ amount; what ↓ number of hundredths; 100 ↓ base

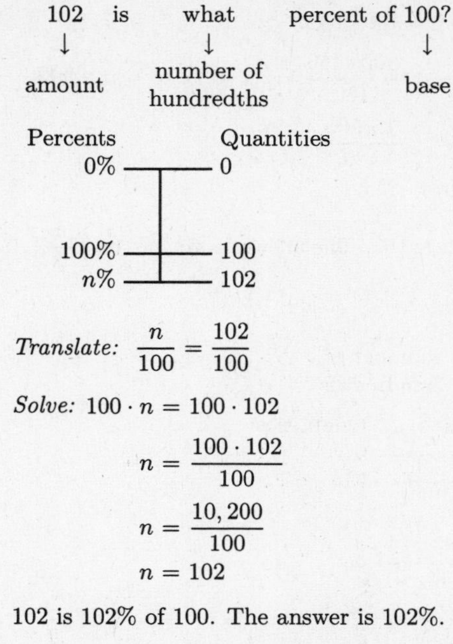

Translate: $\frac{n}{100} = \frac{102}{100}$

Solve: $100 \cdot n = 100 \cdot 102$
$n = \frac{100 \cdot 102}{100}$
$n = \frac{10,200}{100}$
$n = 102$

102 is 102% of 100. The answer is 102%.

19. What percent of \$480 is \$120?
What ↓ number of hundredths; \$480 ↓ base; \$120 ↓ amount

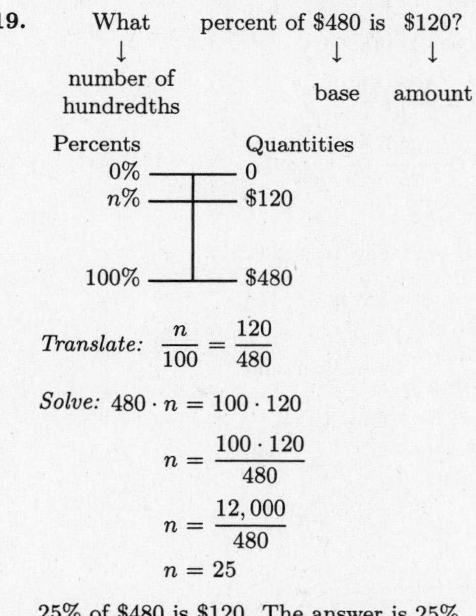

Translate: $\frac{n}{100} = \frac{120}{480}$

Solve: $480 \cdot n = 100 \cdot 120$
$n = \frac{100 \cdot 120}{480}$
$n = \frac{12,000}{480}$
$n = 25$

25% of \$480 is \$120. The answer is 25%.

21. What percent of 180 is 150?
What ↓ number of hundredths; 180 ↓ base; 150 ↓ amount

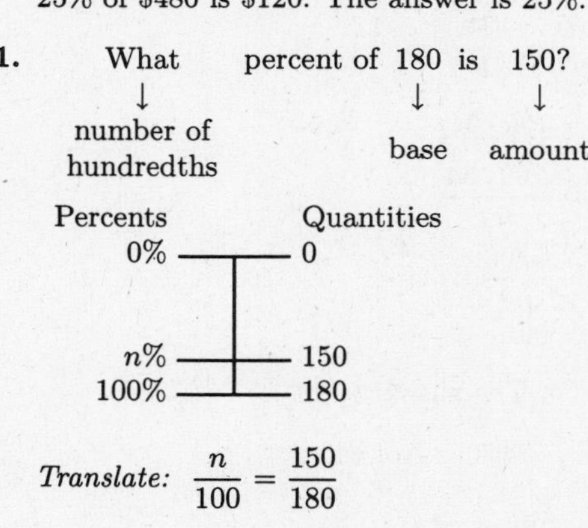

Translate: $\frac{n}{100} = \frac{150}{180}$

Solve: $180 \cdot n = 100 \cdot 150$

$$n = \frac{100 \cdot 150}{180}$$

$$n = \frac{15,000}{180}$$

$$n = 83.\overline{3}$$

$83.\overline{3}\%$ of 180 is 150. The answer is $83.\overline{3}\%$, or $83\frac{1}{3}\%$.

23.

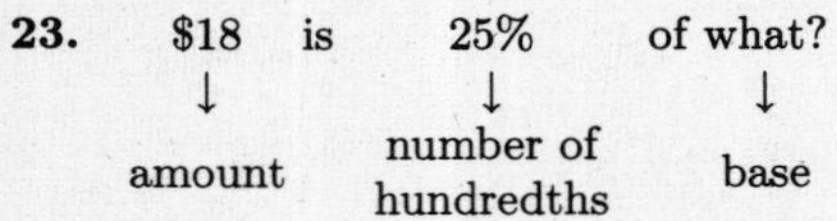

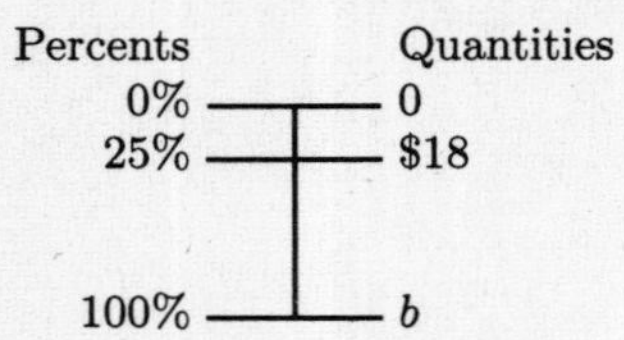

Translate: $\frac{25}{100} = \frac{18}{b}$

Solve: $25 \cdot b = 100 \cdot 18$

$$b = \frac{100 \cdot 18}{25}$$

$$b = \frac{1800}{25}$$

$$b = 72$$

$18 is 25% of $72. The answer is $72.

25.

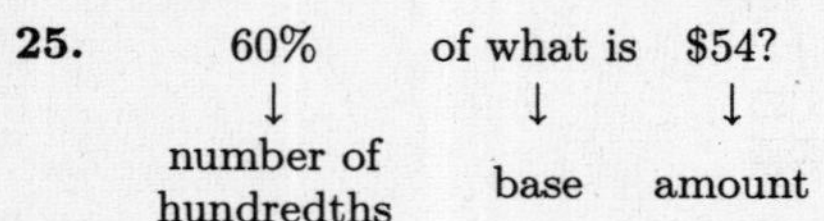

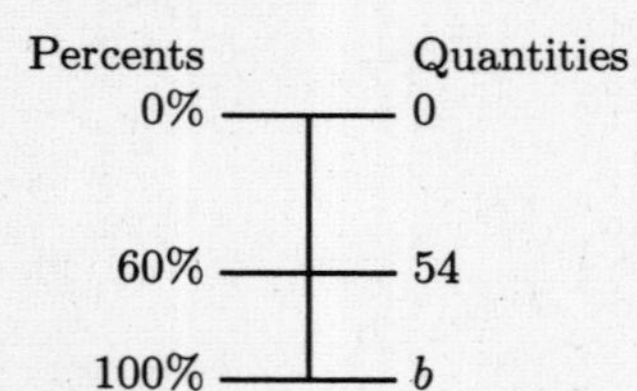

Translate: $\frac{60}{100} = \frac{54}{b}$

Solve: $60 \cdot b = 100 \cdot 54$

$$b = \frac{100 \cdot 54}{60}$$

$$b = \frac{5400}{60}$$

$$b = 90$$

60% of 90 is 54. The answer is 90.

27. 65.12 is 74% of what?

↓ amount; ↓ number of hundredths; ↓ base

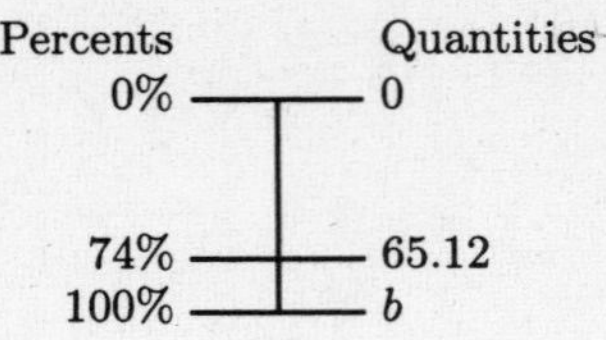

Translate: $\frac{74}{100} = \frac{65.12}{b}$

Solve: $74 \cdot b = 100 \cdot 65.12$

$$b = \frac{100 \cdot 65.12}{74}$$

$$b = \frac{6512}{74}$$

$$b = 88$$

65.12 is 74% of 88. The answer is 88.

29. 9.4% of what is $780.20?

↓ number of hundredths; ↓ base; ↓ amount

Percents	Quantities
0%	0
9.4%	780.20
100%	b

Translate: $\frac{9.4}{100} = \frac{780.20}{b}$

Solve: $9.4 \cdot b = 100 \cdot 780.20$

$$b = \frac{100 \cdot 780.20}{9.4}$$

$$b = \frac{78,020}{9.4}$$

$$b = 8300$$

9.4% of $8300 is $780.20. The answer is $8300.

31.

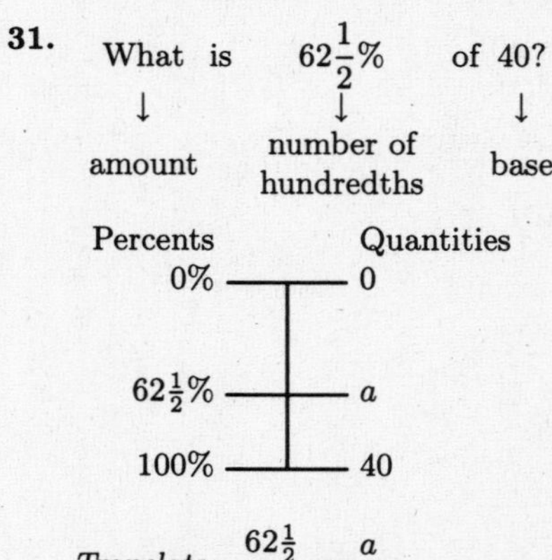

Translate: $\frac{62\frac{1}{2}}{100} = \frac{a}{40}$

Solve: $62\frac{1}{2} \cdot 40 = 100 \cdot a$

$$\frac{125}{2} \cdot \frac{40}{1} = 100 \cdot a$$

$$2500 = 100 \cdot a$$

$$\frac{2500}{100} = a$$

$$25 = a$$

25 is $62\frac{1}{2}\%$ of 40. The answer is 25.

33. Graph: $y = -\frac{1}{2}x$

We make a table of solutions. Note that when x is a multiple of 2, fractional values for y are avoided. Next we plot the points, draw the line, and label it.

When $x = -4$, $y = -\frac{1}{2}(-4) = \frac{4}{2} = 2$.

When $x = 0$, $y = -\frac{1}{2} \cdot 0 = 0$.

When $x = 2$, $y = -\frac{1}{2} \cdot 2 = -\frac{2}{2} = -1$.

x	y $y = -\frac{1}{2}x$	(x, y)
-4	2	$(-4, 2)$
0	0	$(0, 0)$
2	-1	$(2, -1)$

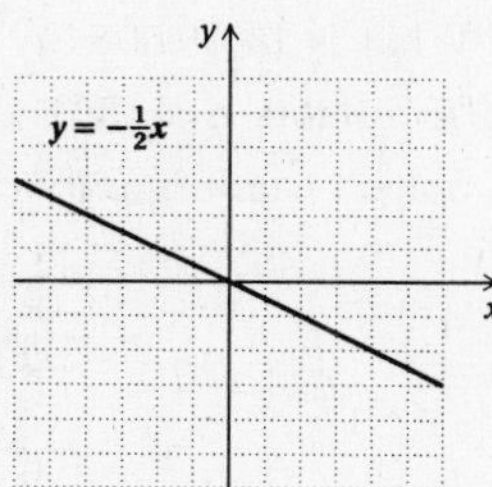

35. Graph: $y = 2x - 4$

We make a table of solutions. Then we plot the points, draw the line, and label it.

When $x = -1$, $y = 2(-1) - 4 = -2 - 4 = -6$.
When $x = 1$, $y = 2 \cdot 1 - 4 = 2 - 4 = -2$.
When $x = 4$, $y = 2 \cdot 4 - 4 = 8 - 4 = 4$.

x	y $y = 2x - 4$	(x, y)
-1	-6	$(-1, -6)$
1	-2	$(1, -2)$
4	4	$(4, 4)$

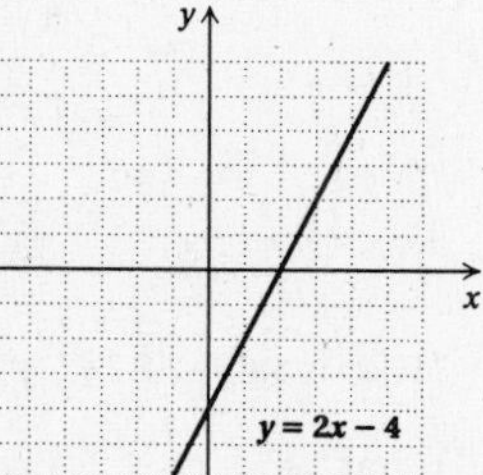

37.

39. Estimate: Round 8.85% to 9%, and $12,640 to $12,600.

What is 9% of $12,600?
↓ ↓ ↓
amount — number of hundredths — base

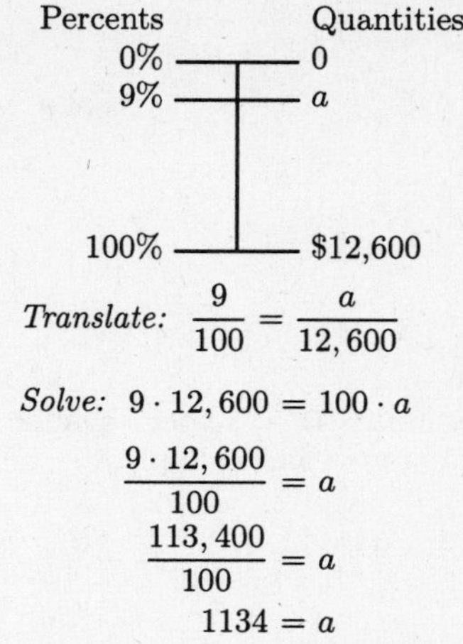

Translate: $\frac{9}{100} = \frac{a}{12,600}$

Solve: $9 \cdot 12,600 = 100 \cdot a$

$$\frac{9 \cdot 12,600}{100} = a$$

$$\frac{113,400}{100} = a$$

$$1134 = a$$

$1134 is about 8.85% of $12,640. (Answers may vary.)

Calculate:

What is 8.85% of $12,640?
↓ ↓ ↓
amount — number of hundredths — base

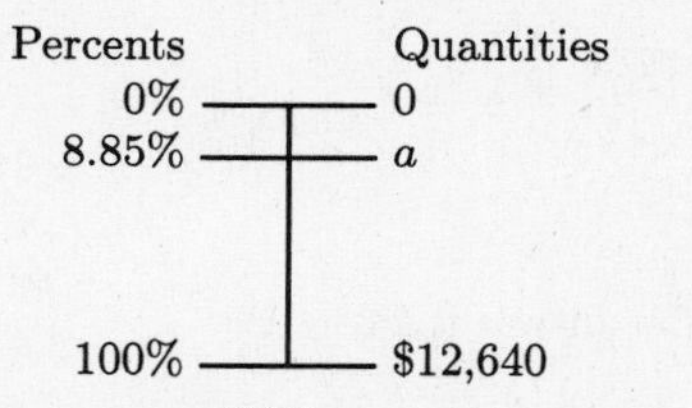

Translate: $\frac{8.85}{100} = \frac{a}{12,640}$

Solve: $8.85 \cdot 12,640 = 100 \cdot a$

$$\frac{8.85 \cdot 12,640}{100} = a$$

$$\frac{111,864}{100} = a \quad \text{Use a calculator to multiply and divide.}$$

$$1118.64 = a$$

$1118.64 is 8.85% of $12,640.

Exercise Set 8.3

1. What is 29% of 53?
↓ ↓ ↓ ↓ ↓
$a = 29\% \times 53$

3. 89 is what percent of 99?
↓ ↓ ↓ ↓ ↓
$89 = n \times 99$

5. 13 is 25% of what?
↓ ↓ ↓ ↓ ↓
$13 = 25\% \times b$

7. What percent of 43 is 51?
↓ ↓ ↓ ↓ ↓
$n \times 43 = 51$

9. 27% of 9 is what?

$$\begin{array}{l} \text{27\% of 9 is what?} \\ \downarrow \quad \downarrow \downarrow \downarrow \quad \downarrow \\ 27\% \times 9 = \quad a \end{array}$$

11. 82% of what is 19?

$$\begin{array}{l} \text{82\% of what is 19?} \\ \downarrow \quad \downarrow \quad \downarrow \quad \downarrow \downarrow \\ 82\% \times \quad b \quad = 19 \end{array}$$

13. What is 85% of 76?

Translate: $a = 85\% \cdot 76$

Solve: The letter is by itself. To solve the equation we convert 85% to decimal notation and multiply.

$$\begin{array}{rr} & 7\,6 \\ & \times\, 0.\,8\,5 \\ \hline & 3\,8\,0 \\ & 6\,0\,8\,0 \\ \hline a = & 6\,4.\,6\,0 \end{array} \qquad (85\% = 0.85)$$

64.6 is 85% of 76. The answer is 64.6.

15. 150% of 16 is what?

Translate: $150\% \times 16 = a$

Solve: Convert 150% to decimal notation and multiply.

$$\begin{array}{rr} & 1\,6 \\ & \times\, 1.\,5 \\ \hline & 8\,0 \\ & 1\,6\,0 \\ \hline a = & 2\,4.\,0 \end{array} \qquad (150\% = 1.5)$$

150% of 16 is 24. The answer is 24.

17. 20 is what percent of 10?

Translate: $20 = n \times 10$

Solve: To solve the equation we divide on both sides by 10 and convert the answer to percent notation.

$$n \cdot 10 = 20$$
$$\frac{n \cdot 10}{10} = \frac{20}{10}$$
$$n = 2 = 200\%$$

20 is 200% of 10. The answer is 200%.

19. What percent of \$300 is \$180?

Translate: $n \times 300 = 180$

Solve: $n \cdot 300 = 180$

$$\frac{n \cdot 300}{300} = \frac{180}{300}$$
$$n = 0.6 = 60\%$$

60% of \$300 is \$180. The answer is 60%.

21. 40% of what is 16?

Translate: $40\% \times b = 16$

Solve: To solve the equation we divide on both sides by 40%:

$$b = 16 \div 40\%$$
$$b = 16 \div 0.4 \qquad (40\% = 0.4)$$
$$b = 40$$

$$\begin{array}{r} 4\,0\,. \\ 0.\,4_\wedge \overline{)\,1\,6.\,0_\wedge} \\ 1\,6\,0 \\ \hline 0 \\ 0 \\ \hline 0 \end{array}$$

40% of 40 is 16. The answer is 40.

23. 45 is 50% of what?

Translate: $45 = 50\% \times b$

Solve: To solve the equation we divide on both sides by 50%:

$$45 \div 50\% = b$$
$$45 \div 0.5 = b \qquad (50\% = 0.5)$$
$$90 = b$$

$$\begin{array}{r} 9\,0\,. \\ 0.\,5_\wedge \overline{)\,4\,5.\,0_\wedge} \\ 4\,5\,0 \\ \hline 0 \end{array}$$

45 is 50% of 90. The answer is 90.

25. What is 130% of 84?

Translate: $a = 130\% \cdot 84$

Solve: Convert 130% to decimal notation and multiply.

$$\begin{array}{r} 8\,4 \\ \times\, 1.\,3 \\ \hline 2\,5\,2 \\ 8\,4\,0 \\ \hline 1\,0\,9.\,2 \end{array} \qquad (130\% = 1.3)$$

109.2 is 130% of 84. The answer is 109.2.

27. 71.04 is 96% of what?

Translate: $71.04 = 96\% \times b$

Solve: We divide on both sides by 96%.

$$71.04 \div 96\% = b$$
$$71.04 \div 0.96 = b \qquad (96\% = 0.96)$$
$$74 = b$$

$$\begin{array}{r} 7\,4\,. \\ 0.\,9\,6_\wedge \overline{)\,7\,1.\,0\,4_\wedge} \\ 6\,7\,2\,0 \\ \hline 3\,8\,4 \\ 3\,8\,4 \\ \hline 0 \end{array}$$

71.04 is 96% of 74. The answer is 74.

29. What percent of 60 is 72?

Translate: $n \times 60 = 72$

Solve: We divide on both sides by 60 and convert the answer to percent notation.

$$n \times 60 = 72$$
$$\frac{n \times 60}{60} = \frac{72}{60}$$
$$n = 1.2 = 120\%$$

120% of 60 is 72. The answer is 120%.

31. $35\frac{1}{4}\%$ of 1200 is what?

Translate: $35\frac{1}{4}\% \times 1200 = a$

Solve: Convert $35\frac{1}{4}\%$ to decimal notation and multiply.

$$\begin{array}{r} 1200 \\ \times\ 0.3525 \\ \hline 6000 \\ 24000 \\ 600000 \\ 3600000 \\ \hline 423.0000 \end{array} \quad (35\tfrac{1}{4}\% = 35.25\% = 0.3525)$$

$35\frac{1}{4}\%$ of 1200 is 423. The answer is 423.

33. 32 is what percent of 48?

Translate: $32 = n \times 48$

Solve: We divide on both sides by 48 and convert the answer to percent notation.

$$n \times 48 = 32$$
$$\frac{n \times 48}{48} = \frac{32}{48}$$
$$n = 0.66\overline{6} = 66.\overline{6}\%$$

32 is $66.\overline{6}\%$ of 48. The answer is $66.\overline{6}\%$, or $66\frac{2}{3}\%$.

35. $33\frac{1}{3}\%$ of what is 21.4?

Translate: $33\frac{1}{3}\% \times b = 21.4$

Solve: We express $33\frac{1}{3}\%$ as $\frac{1}{3}$ and then multiply by 3 on both sides of the equation.

$$\frac{1}{3} \times b = 21.4$$
$$b = 3(21.4)$$
$$b = 64.2$$

$33\frac{1}{3}\%$ of 64.2 is 21.4 The answer is 64.2.

37. $\frac{17}{1000}$ 0.017.

3 zeros Move 3 places.

$\frac{17}{1000} = 0.017$

39. $\frac{253}{100}$ 2.53.

2 zeros Move 2 places.

$\frac{253}{100} = 2.53$

41. 

43. Estimate: Round 7.75% to 8% and \$10,880 to \$11,000. Then translate:

What is 8% of \$11,000?

$$\downarrow \quad \downarrow \quad \downarrow \quad \downarrow \quad \downarrow$$
$$a \quad = 8\% \times \quad 11,000$$

We convert 8% to decimal notation and multiply.

$$\begin{array}{r} 11,000 \\ \times\ 0.08 \\ \hline 880.00 \end{array} \quad (8\% = 0.08)$$

\$880 is about 7.75% of \$10,880. (Answers may vary.)

Calculate: First we translate.

What is 7.75% of \$10,880?

$$\downarrow \quad \downarrow \quad \downarrow \quad \downarrow \quad \downarrow$$
$$a \quad = 7.75\% \times \quad 10,880$$

Use a calculator to multiply:

$$0.0775 \times 10,880 = 843.2$$

\$843.20 is 7.75% of \$10,880.

45. We reword the problem as two questions, translate each to an equation, and solve the equations.

What is 40% of 270?

$$\downarrow \quad \downarrow \quad \downarrow \quad \downarrow \quad \downarrow$$
$$a = 40\% \times 270$$
$$a = 0.4 \times 270$$
$$a = 108$$

What is 50% of 270?

$$\downarrow \quad \downarrow \quad \downarrow \quad \downarrow \quad \downarrow$$
$$a = 50\% \times 270$$
$$a = 0.5 \times 270$$
$$a = 135$$

108 tons to 135 tons of the trash is recyclable.

Exercise Set 8.4

1. ***Familiarize.*** The pie chart in the text helps us familiarize ourselves with the problem. First we find the amount spent each week for advertising. We let a = this amount.

Translate. We rephrase the question and translate.

What is 5% of \$8000?

$$\downarrow \quad \downarrow \quad \downarrow \quad \downarrow \quad \downarrow$$
$$a \quad = 5\% \times \quad 8000$$

Solve. We convert 5% to decimal notation and multiply.

$$a = 5\% \times 8000 = 0.05 \times 8000 = 400$$

Now we find the amount spent for other expenses. We let x = this amount.

$$\underbrace{\text{Operating budget}} \ \text{minus} \ \underbrace{\text{Advertising spending}} \ \text{is} \ \underbrace{\text{Other expenses}}$$
$$8000 \quad - \quad 400 \quad = \quad x$$

To solve the equation we carry out the subtraction.

$$x = 8000 - 400 = 7600$$

Check. To check the amount spent on advertising, we solve the same problem using a proportion.

$$\frac{5}{100} = \frac{a}{8000}$$
$$5 \cdot 8000 = 100 \cdot a \quad \text{Equating cross-products}$$
$$\frac{5 \cdot 8000}{100} = a \quad \text{Dividing by 100}$$
$$\frac{40,000}{100} = a$$
$$400 = a$$

Since we get the same answer both ways, we have a check for the advertising amount.

Note that $7600 + 400 = 8000$, so the amount of other expenses checks also.

State. Ariel Electronics should spend \$400 for advertising and \$7600 for other expenses each week.

3. ***Familiarize.*** Since $99.9\% \approx 100\%$, we would expect about $100\% \times 240$ L, or 240 L to be pure. Similarly, we would expect the amount of impurities to be close to 0 L. First we find the number of liters that would be pure. We let p = this amount.

Translate. We rephrase the question and translate.

What is 99.9% of 240?
↓ ↓ ↓ ↓ ↓
$p = 99.9\% \times 240$

Solve. We convert 99.9% to decimal notation and multiply.

$$p = 99.9\% \times 240 = 0.999 \times 240 = 239.76$$

Now we find the amount of impurities. We let y = this amount.

Total amount minus Pure amount is Impure amount
↓ ↓ ↓ ↓ ↓
$240 - 239.76 = y$

To solve the equation we carry out the subtraction.

$$y = 240 - 239.76 = 0.24$$

Check. To check the amount that is pure, we could solve the same problem using a proportion. Instead we note that the answer, 239.76 L, is close to 240 L as estimated in the Familiarize step. We also note that the amount of impurities, 0.24 L, is close to 0 as estimated.

State. 239.76 L would be pure, and 0.24 L would be impurities.

5. ***Familiarize.*** The question asks for percents. We know that 10% of 40 is 4. Since $13 \approx 3 \times 4$, we would expect the percent of at bats that are hits to be close to 30%. Then we would also expect the percent of at bats that are not hits to be 70%. First we find h, the percent that are hits.

Translate. We rephrase the question and translate.

13 is what percent of 40?
↓ ↓ ↓ ↓ ↓
$13 = h \times 40$

Solve. We divide on both sides by 40 and convert the result to percent notation.

$$13 = h \times 40$$
$$\frac{13}{40} = h$$
$$0.325 = h$$
$$32.5\% = h \qquad \text{Finding percent notation}$$

Now we find the percent of at bats that are not hits. We let n = this percent.

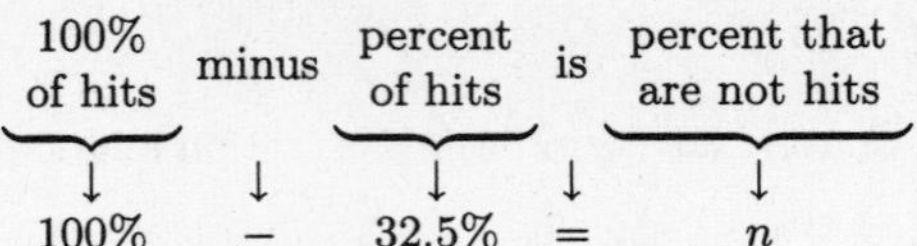

$100\% - 32.5\% = n$

To solve the equation we carry out the subtraction.

$$n = 100\% - 32.5\% = 67.5\%$$

Check. To check the percent of at bats that are hits, we could solve the problem using a proportion. Instead we note that the answer, 32.5%, is close to 30% as estimated in the Familiarize step. We also note that the percent of at bats that are not hits, 67.5%, is close to 70% as estimated.

State. 32.5% of at bats are hits, and 67.5% are not hits.

7. ***Familiarize.*** First we find the amount of the solution that is acid. We let a = this amount.

Translate. We rephrase the question and translate.

What is 3% of 680?
↓ ↓ ↓ ↓ ↓
$a = 3\% \text{ of } 680$

Solve. We convert 3% to decimal notation and multiply.

$$a = 3\% \times 680 = 0.03 \times 680 = 20.4$$

Now we find the amount that is water. We let w = this amount.

Total amount minus Amount of acid is Amount of water
↓ ↓ ↓ ↓ ↓
$680 - 20.4 = w$

To solve the equation we carry out the subtraction.

$$w = 680 - 20.4 = 659.6$$

Check. To check the amount of the solution that is acid, we solve the same problem using a proportion.

$$\frac{3}{100} = \frac{a}{680}$$
$$3 \cdot 680 = 100 \cdot a$$
$$\frac{3 \cdot 680}{100} = a$$
$$\frac{2040}{100} = a$$
$$20.4 = a$$

Since we get the same answer both ways, we have a check for the amount of acid. Note that $20.4 + 659.6 = 680$, so the amount of water checks also.

State. The solution contains 20.4 mL of acid and 659.6 mL of water.

9. ***Familiarize.*** The question asks for a percent. We know that 10% of 800 is 80 and 56 is less than 80, so we would expect the answer to be less than 10%. We let p = the percent of people who will catch a cold.

Translate. We rephrase the question and translate.

56 is what percent of 800?
↓ ↓ ↓ ↓ ↓
$56 = p \times 800$

Solve. We divide on both sides by 800 and convert to percent notation.

$$56 = p \times 800$$
$$\frac{56}{800} = p$$
$$0.07 = p$$
$$7\% = p$$

Check. To check, we could solve the same problem using a proportion. Instead, we note that the answer, 7%, is less than 10% as estimated in the Familiarize step.

State. 7% of the people who kiss someone who has a cold will catch a cold.

11. a) ***Familiarize***. We let b = the percent of the cost it takes to operate refinery B.

Translate.

$\underbrace{\text{100\% of cost}}$ minus $\underbrace{\text{A's percent of cost}}$ is $\underbrace{\text{B's percent of cost}}$

$$100\% - 37.5\% = b$$

Solve. We carry out the subtraction.

$$b = 100\% - 37.5\% = 62.5\%$$

Check. Since $37.5\% + 62.5\% = 100\%$, the answer checks.

State. It takes 62.5% of the cost to run refinery B.

b) We let a = the cost of operating refinery A and c = the cost of operating refinery B.

Translate. We rephrase each question and translate.

What is 37.5% of \$40,000?

$$a = 37.5\% \times 40,000$$

What is 62.5% of \$40,000?

$$c = 62.5\% \times 40,000$$

Solve. We convert each percent to decimal notation and carry out the multiplications.

$$a = 37.5\% \times 40,000 = 0.375 \times 40,000 = 15,000$$

$$c = 62.5\% \times 40,000 = 0.625 \times 40,000 = 25,000$$

Check. We could solve the problem again using proportions. Instead, we note that $15,000 + 25,000 = 40,000$. Our answer checks.

State. It costs \$15,000 a day to operate refinery A and \$25,000 to operate refinery B.

13. ***Familiarize***. We note that the increase in population was $882 - 840 = 42$. A drawing can help us visualize the situation. We let n = the percent of increase.

840	
840	42
100%	
100%	?%

Translate. We rephrase the question and translate.

42 is $\underbrace{\text{what percent}}$ of 840?

$$42 = n \times 840$$

Solve. We divide on both sides by 840.

$$\frac{42}{840} = n$$
$$0.05 = n$$
$$5\% = n \quad \text{Finding percent notation}$$

Check. To check, we solve the problem using a proportion.

$$\frac{42}{840} = \frac{n}{100}$$
$$42 \cdot 100 = 840 \cdot n$$
$$\frac{42 \cdot 100}{840} = n$$
$$5 = n$$

Both methods give an answer of 5%, so we have a check.

State. The population increased 5%

15. ***Familiarize***. We note that the reduction is $\$70 - \$56 = \$14$. We make a drawing.

\$70	
\$56	\$14
100%	
	?%

We let n = the percent of decrease.

Translate. We rephrase the question and translate.

\$14 is $\underbrace{\text{what percent}}$ of \$70?

$$14 = n \times 70$$

Solve. To solve the equation, we divide on both sides by 70.

$$\frac{14}{70} = n$$
$$0.2 = n$$
$$20\% = n \quad \text{Finding percent notation}$$

Check. To check, we solve the problem using a proportion.

$$\frac{14}{70} = \frac{n}{100}$$
$$14 \cdot 100 = 70 \cdot n$$
$$\frac{14 \cdot 100}{70} = n$$
$$20 = n$$

Both methods give an answer of 20%, so we have a check.

State. The percent of decrease was 20%.

17. ***Familiarize.*** We note that the amount of the raise can be found and then added to the old salary. A drawing helps us visualize the situation.

$18,600	$?
100%	5%

We let x = the new salary.

Translate. We rephrase the question and translate.

What is the old salary plus 5% of the old salary?

$$x = 18,600 + 5\% \times 18,600$$

Solve. We convert 5% to a decimal and simplify.

$$\begin{aligned} x &= 18,600 + 0.05 \times 18,600 \\ &= 18,600 + 930 \qquad \text{The raise is \$930.} \\ &= 19,530 \end{aligned}$$

Check. To check, we note that the new salary is 100% of the old salary plus 5% of the old salary, or 105% of the old salary. Since $1.05 \times 18,600 = 19,530$, our answer checks.

Check. The new salary is $19,530.

19. ***Familiarize.*** We note that the amount of the reduction can be found and then subtracted from the old salary. A drawing helps us visualize the situation.

$55,000	
	$?
100%	
89%	11%

We let s = the new salary.

Translate. We rephrase the question and translate.

What is the old salary minus 11% of the old salary?

$$s = 55,000 - 11\% \times 55,000$$

Solve. We convert 11% to a decimal and simplify.

$$\begin{aligned} s &= 55,000 - 0.11 \times 55,000 \\ s &= 55,000 - 6050 \\ s &= 48,950 \end{aligned}$$

Check. To check, we note that the new salary is 100% of the old salary minus 11% of the old salary, or 89% of the old salary. Since $0.89 \times 55,000 = 48,950$, our answer checks.

State. The reduced salary was $48,950.

21. ***Familiarize.*** We note that the amount of the reduction can be found and then subtracted from the old bill. A drawing helps us visualize the situation.

$106.00	
	$?
100%	
50%	50%

We let b = the amount of the new bill.

Translate. We rephrase the question and translate.

What is the old bill minus 50% of the old bill?

$$b = 106 - 50\% \times 106$$

Solve. We convert 50% to a decimal and simplify.

$$\begin{aligned} b &= 106 - 0.5 \times 106 \\ b &= 106 - 53 \\ b &= 53 \end{aligned}$$

Check. To check, we note that the new bill is 100% of the old bill minus 50% of the old bill, or 50% of the old bill. Since $0.5 \times 106 = 53$, our answer checks.

State. The new bill would be $53.

23. ***Familiarize.*** We use the formula $A = l \times w$ to find the area of a cross-section of the finished piece of lumber:

$$A = 1\frac{1}{2} \times 3\frac{1}{2} = \frac{3}{2} \times \frac{7}{2} = \frac{21}{4} = 5.25 \text{ in}^2$$

We also find the area of a cross-section of the rough board:

$$A = 2 \times 4 = 8 \text{ in}^2$$

Now we subtract to find the area that is removed in planing and drying:

$$8 \text{ in}^2 - 5.25 \text{ in}^2 = 2.75 \text{ in}^2$$

We let w = the percent of the wood that is removed.

Translate. We rephrase the question and translate.

2.75 is what percent of 8?

$$2.75 = w \times 8$$

Solve. We divide on both sides by 8.

$$\begin{aligned} \frac{2.75}{8} &= w \\ 0.34375 &= w \\ 34.375\% &= w \qquad \text{Finding percent notation} \end{aligned}$$

Check. To check, we solve the problem using a proportion.

$$\begin{aligned} \frac{2.75}{8} &= \frac{w}{100} \\ 2.75 \cdot 100 &= 8 \cdot w \\ \frac{2.75 \cdot 100}{8} &= w \\ 34.375 &= w \end{aligned}$$

Both methods give an answer of 34.375%, so we have a check.

State. 34.375%, or $34\frac{3}{8}\%$, of the wood is removed in planing and drying.

25. First we find the maximum heart rate for a 25 year old person.

Familiarize. Note that $220 - 25 = 195$. We let $x =$ the maximum heart rate for a 25 year old person.

Translate. We rephrase the question and translate.

What is 85% of 195?
$\downarrow \quad \downarrow \; \downarrow \; \downarrow \; \downarrow$
$x \quad = 85\% \times 195$

Solve. We convert 85% to a decimal and simplify.

$$x = 0.85 \times 195 = 165.75 \approx 166$$

Check. We solve the problem using a proportion.

$$\frac{x}{195} = \frac{85}{100}$$
$$100 \cdot x = 195 \cdot 85$$
$$x = \frac{195 \cdot 85}{100}$$
$$x = 165.75 \approx 166$$

Both methods give an answer of $165.75 \approx 166$, so we have a check.

State. The maximum heart rate for a 25 year old person is 166 beats per minute.

Next we find the maximum heart rate for a 36 year old person.

Familiarize. Note that $220 - 36 = 184$. We let $x =$ the maximum heart rate for a 36 year old person.

Translate. We rephrase the question and translate.

What is 85% of 184?
$\downarrow \quad \downarrow \; \downarrow \; \downarrow \; \downarrow$
$x \quad = 85\% \times 184$

Solve. We convert 85% to a decimal and simplify.

$$x = 0.85 \times 184 = 156.4 \approx 156$$

Check. We solve the problem using a proportion.

$$\frac{x}{184} = \frac{85}{100}$$
$$100 \cdot x = 184 \cdot 85$$
$$x = \frac{184 \cdot 85}{100}$$
$$x = 156.4 \approx 156$$

Both methods give an answer of $156.4 \approx 156$, so we have a check.

State. The maximum heart rate for a 36 year old person is 156 beats per minute.

Next we find the maximum heart rate for a 48 year old person.

Familiarize. Note that $220 - 48 = 172$. We let $x =$ the maximum heart rate for a 48 year old person.

Translate. We rephrase the question and translate.

What is 85% of 172?
$\downarrow \quad \downarrow \; \downarrow \; \downarrow \; \downarrow$
$x \quad = 85\% \times 172$

Solve. We convert 85% to a decimal and simplify.

$$x = 0.85 \times 172 = 146.2 \approx 146$$

Check. We solve the problem using a proportion.

$$\frac{x}{172} = \frac{85}{100}$$
$$100 \cdot x = 172 \cdot 85$$
$$x = \frac{172 \cdot 85}{100}$$
$$x = 146.2 \approx 146$$

Both methods give an answer of $146.2 \approx 146$, so we have a check.

State. The maximum heart rate for a 48 year old person is 146 beats per minute.

We find the maximum heart rate for a 60 year old person.

Familiarize. Note that $220 - 60 = 160$. We let $x =$ the maximum heart rate for a 60 year old person.

Translate. We rephrase the question and translate.

What is 85% of 160?
$\downarrow \quad \downarrow \; \downarrow \; \downarrow \; \downarrow$
$x \quad = 85\% \times 160$

Solve. We convert 85% to a decimal and simplify.

$$x = 0.85 \times 160 = 136$$

Check. We solve the problem using a proportion.

$$\frac{x}{160} = \frac{85}{100}$$
$$100 \cdot x = 160 \cdot 85$$
$$x = \frac{160 \cdot 85}{100}$$
$$x = 136$$

Both methods give an answer of 136, so we have a check.

State. The maximum heart rate for a 60 year old person is 136 beats per minute.

Finally we find the maximum heart rate for a 76 year old person.

Familiarize. Note that $220 - 76 = 144$. We let $x =$ the maximum heart rate for a 76 year old person.

Translate. We rephrase the question and translate.

What is 85% of 144?
$\downarrow \quad \downarrow \; \downarrow \; \downarrow \; \downarrow$
$x \quad = 85\% \times 144$

Solve. We convert 85% to a decimal and simplify.

$$x = 0.85 \times 144 = 122.4 \approx 122$$

Check. We solve the problem using a proportion.

$$\frac{x}{144} = \frac{85}{100}$$
$$100 \cdot x = 144 \cdot 85$$
$$x = \frac{144 \cdot 85}{100}$$
$$x = 122.4 \approx 122$$

Both methods give an answer of $122.4 \approx 122$, so we have a check.

State. The maximum heart rate for a 76 year old person is 122 beats per minute.

27. ***Familiarize***. First we use the formula $A = l \times w$ to find the area of the strike zone:

$$A = 40 \times 15 = 600 \text{ in}^2$$

When a 2-in. border is added to the outside of the strike zone, the dimensions of the larger zone are 19 in. by 44 in. The area of this zone is

$$A = 44 \times 19 = 836 \text{ in}^2$$

We subtract to find the increase in area:

$$836 \text{ in}^2 - 600 \text{ in}^2 = 236 \text{ in}^2$$

We let a = the percent of increase in the area.

Translate. We rephrase the question and translate.

$$\underbrace{236}\ \text{is}\ \underbrace{\text{what percent}}\ \text{of}\ 600?$$
$$\downarrow \quad \downarrow \qquad \downarrow \qquad \downarrow \quad \downarrow$$
$$236 = \qquad a \qquad \times\ 600$$

Solve. We divide by 600 on both sides.

$$\frac{236}{600} = a$$
$$0.39\overline{3} = a$$
$$39.\overline{3}\% = a \quad \text{Finding percent notation}$$

Check. We solve the problem using a proportion.

$$\frac{236}{600} = \frac{a}{100}$$
$$236 \cdot 100 = 600 \cdot a$$
$$\frac{236 \cdot 100}{600} = a$$
$$39.\overline{3} = a$$

Both methods give an answer of $39.\overline{3}\%$, so we have a check.

State. The area of the strike zone is increased by $39.\overline{3}\%$.

29. First we divide 572 by 28.

$$\begin{array}{r} 20 \\ 28\overline{)572} \\ \underline{560} \\ 12 \end{array} \qquad \frac{572}{28} = 20\frac{12}{28} = 20\frac{3}{7}$$

Because a negative number divided by a positive number is negative, the answer is $-20\frac{3}{7}$.

31. First we divide 48 by 7.

$$\begin{array}{r} 6 \\ 7\overline{)48} \\ \underline{42} \\ 6 \end{array} \qquad \frac{48}{7} = 6\frac{6}{7}$$

Because the quotient of two negative numbers is positive, the answer is $6\frac{6}{7}$.

33. 

35. Let S = the original salary. After a 3% raise, the salary becomes $103\% \cdot S$, or $1.03S$. After a 6% raise, the new salary is $1.06\% \cdot 1.03S$, or $1.06(1.03S)$. Finally, after a 9% raise, the salary is $109\% \cdot 1.06(1.03S)$, or $1.09(1.06)(1.03S)$. Multiplying, we get $1.09(1.06)(1.03S) = 1.190062S$. This is equivalent to $119.0062\% \cdot S$, so the original salary has increased by 19.0062%.

37. ***Familiarize***. We will express 4 ft, 8 in. as 56 in. (4 ft + 8 in. = $4 \cdot 12$ in. + 8 in. = 48 in. + 8 in. = 56 in.) We let h = Cynthia's final adult height.

Translate. We rephrase the question and translate.

$$\underbrace{56 \text{ in.}}\ \text{is}\ 84.4\%\ \text{of what?}$$
$$\downarrow \quad \downarrow \quad \downarrow \quad \downarrow \quad \downarrow$$
$$56 \ = 84.4\% \times \ h$$

Solve. First we convert 84.4% to a decimal.

$$56 = 0.844 \times h$$
$$\frac{56}{0.844} = h \qquad \text{Dividing by } h \text{ on both sides}$$
$$66.4 \approx h$$

Check. We solve the same problem using a proportion.

$$\frac{84.4}{100} = \frac{56}{h}$$
$$84.4 \cdot h = 100 \cdot 56$$
$$h = \frac{100 \cdot 56}{84.4}$$
$$h \approx 66.4$$

Since we get the same answer both ways, we have a check.

State. Cynthia's final adult height will be about 66.4 in., or 5 ft, 6.4 in.

39. ***Familiarize***. We note that in each case the amount of the tip can be found and then added to the price of the meal. We let x, y, and z represent the amounts charged for the \$15, \$34, and \$49 meals, respectively.

Translate. We rephrase the question and translate to three equations.

$$\text{What is}\ \underbrace{\text{the price of the meal}}\ \text{plus 15\% of}\ \underbrace{\text{the price of the meal?}}$$
$$x = 15 + 15\% \times 15$$

$$\text{What is}\ \underbrace{\text{the price of the meal}}\ \text{plus 15\% of}\ \underbrace{\text{the price of the meal?}}$$
$$y = 34 + 15\% \times 34$$

$$\text{What is}\ \underbrace{\text{the price of the meal}}\ \text{plus 15\% of}\ \underbrace{\text{the price of the meal?}}$$
$$z = 49 + 15\% \times 49$$

Solve. To solve each equation we convert 15% to a decimal and simplify.

$$x = 15 + 0.15 \times 15$$
$$x = 15 + 2.25$$
$$x = 17.25$$

$$y = 34 + 0.15 \times 34$$
$$y = 34 + 5.1$$
$$y = 39.1$$

$$z = 49 + 0.15 \times 49$$
$$z = 49 + 7.35$$
$$z = 56.35$$

Check. To check we note that each amount charged is 100% of the price of the meal plus 15% of the price of the meal, or 115% of the price of the meal. Since $1.15 \times 15 = 17.25$, $1.15 \times 34 = 39.1$, and $1.15 \times 49 = 56.35$, our answers check.

State. The amounts charged for the \$15 meal, the \$34 meal, and the \$49 meal would be \$17.25, \$39.10, and \$56.35, respectively.

41. ***Familiarize***. We let $c =$ the cost of the dinner without the tip and the coupon. Note that the bill was computed by first adding the tip to the full price of the dinner and then subtracting \$10.

Translate.

$$\underbrace{\text{Cost of dinner}}_{c} \ \underbrace{\text{plus}}_{+} \ \underbrace{15\%}_{15\%} \ \underbrace{\text{of}}_{\times} \ \underbrace{\text{cost of dinner}}_{c} \ \underbrace{\text{minus}}_{-} \ \underbrace{\$10}_{10} \ \underbrace{\text{is}}_{=} \ \underbrace{\$44.05.}_{44.05}$$

Solve. We convert 15% to a decimal and solve the equation.

$$c + 0.15c - 10 = 44.05$$
$$1.15c - 10 = 44.05$$
$$1.15c = 44.05 + 10$$
$$1.15c = 54.05$$
$$c = \frac{54.05}{1.15}$$
$$c = 47$$

Check. Note that 15% of 47 is 7.05. Then $\$47 + \$7.05 = \$54.05$ is the full price of the dinner with the tip and $\$54.05 - \$10 = \$44.05$, is the bill presented after the tip was added and the coupon amount was subtracted. Our answer checks.

State. The dinner would have cost \$47 without the tip and the coupon.

Exercise Set 8.5

1. a) The sales tax on an item costing \$586 is

$$\underbrace{\text{Sales tax rate}}_{5\%} \ \times \ \underbrace{\text{Purchase price}}_{\$586,}$$

or 0.05×586, or 29.3. Thus the tax is \$29.30.

b) The total price is given by the purchase price plus the sales tax:

$$\$586 + \$29.30, \text{ or } \$615.30.$$

To check, note that the total price is the purchase price plus 5% of the purchase price. Thus the total price is 105% of the purchase price. Since $1.05 \times 586 = 615.3$, we have a check. The total price is \$615.30.

3. a) We first find the cost of the telephones. It is

$$5 \times \$53 = \$265.$$

b) The sales tax on items costing \$265 is

$$\underbrace{\text{Sales tax rate}}_{6.25\%} \ \times \ \underbrace{\text{Purchase price}}_{\$265,}$$

or 0.0625×265, or 16.5625. Thus the tax is \$16.56.

c) The total price is given by the purchase price plus the sales tax:

$$\$265 + \$16.56, \text{ or } \$281.56.$$

To check, note that the total price is the purchase price plus 6.25% of the purchase price. Thus the total price is 106.25% of the purchase price. Since $1.0625 \times \$265 = \$281.56$ (rounded to the nearest cent), we have a check. The total price is \$281.56.

5. *Rephrase:* Sales tax is what percent of purchase price?

Translate:
$$\underbrace{\text{Sales tax}}_{48} \ \underbrace{\text{is}}_{=} \ \underbrace{\text{what percent}}_{r} \ \underbrace{\text{of}}_{\times} \ \underbrace{\text{purchase price?}}_{960}$$

To solve the equation, we divide on both sides by 960.

$$48 \div 960 = r$$
$$0.05 = r$$
$$5\% = r$$

The sales tax rate is 5%.

7. *Rephrase:* Sales tax is what percent of purchase price?

Translate:
$$\underbrace{\text{Sales tax}}_{35.80} \ \underbrace{\text{is}}_{=} \ \underbrace{\text{what percent}}_{r} \ \underbrace{\text{of}}_{\times} \ \underbrace{\text{purchase price?}}_{895}$$

To solve the equation, we divide on both sides by 895.

$$35.80 \div 895 = r$$
$$0.04 = r$$
$$4\% = r$$

The sales tax rate is 4%.

9. *Rephrase:* Sales tax is 5% of what?

Translate:
$$\underbrace{\text{Sales tax}}_{100} \ \underbrace{\text{is}}_{=} \ \underbrace{5\%}_{5\%} \ \underbrace{\text{of}}_{\times} \ \underbrace{\text{what?}}_{b,} \quad \text{or}$$
$$100 = 0.05 \times b$$

To solve the equation, we divide on both sides by 0.05.

$$100 \div 0.05 = b$$
$$2000 = b$$

$$\begin{array}{r} 2\,0\,0\,0\,. \\ 0.\,0\,5_\wedge \overline{)\,1\,0\,0.\,0\,0_\wedge} \\ \underline{1\,0\,0\,0\,0} \\ 0 \end{array}$$

The purchase price is \$2000.

11. *Rephrase:* Sales tax is 3.5% of what?

Translate:
$$\underbrace{\text{Sales tax}}_{28} \ \underbrace{\text{is}}_{=} \ \underbrace{3.5\%}_{3.5\%} \ \underbrace{\text{of}}_{\times} \ \underbrace{\text{what?}}_{b,} \quad \text{or}$$
$$28 = 0.035 \times b$$

To solve the equation, we divide on both sides by 0.035.

$$28 \div 0.035 = b$$
$$800 = b$$

$$\begin{array}{r} 800. \\ 0.035_{\wedge}\overline{)28.000_{\wedge}} \\ \underline{28000} \\ 0 \end{array}$$

The purchase price is \$800.

13. a) We first find the cost of the shower units. It is

$$2 \times \$332.50 = \$665.$$

b) The total tax rate is the city tax rate plus the state tax rate, or $1\% + 6\% = 7\%$. The sales tax paid on items costing \$665 is

$$\underbrace{\text{Sales tax rate}}_{\downarrow \ 7\%} \ \times \ \underbrace{\text{Purchase price}}_{\downarrow \ \$665,}$$

or 0.07×665, or 46.55. Thus the tax is \$46.55.

c) The total price is given by the purchase price plus the sales tax:

$$\$665 + \$46.55 = \$711.55.$$

To check, note that the total price is the purchase price plus 7% of the purchase price. Thus the total price is 107% of the purchase price. Since $1.07 \times 665 = 711.55$, we have a check. The total amount paid for the 2 shower units is \$711.55.

15. *Rephrase:* $\underbrace{\text{Sales tax}}$ is $\underbrace{\text{what percent}}$ of $\underbrace{\text{purchase price?}}$

Translate: $1030.40 = r \times 18,400$

To solve the equation, we divide on both sides by 18,400.

$$1030.40 \div 18,400 = r$$
$$0.056 = r$$
$$5.6\% = r$$

The sales tax rate is 5.6%.

17. Commission = Commission rate × Sales

$$C = 6\% \times 45,000$$

This tells us what to do. We multiply.

$$\begin{array}{r} 45,000 \\ \times \quad 0.06 \\ \hline 2700.00 \end{array} \qquad (6\% = 0.06)$$

The commission is \$2700.

19. Commission = Commission rate × Sales

$$120 = r \times 2400$$

To solve this equation we divide on both sides by 2400:

$$120 \div 2400 = r$$

We can divide, but this time we simplify by removing a factor of 1:

$$r = \frac{120}{2400} = \frac{1}{20} \cdot \frac{120}{120} = \frac{1}{20} = 0.05 = 5\%$$

The commission rate is 5%.

21. Commission = Commission rate × Sales

$$392 = 40\% \times S$$

To solve this equation we divide on both sides by 40%:

$$392 \div 40\% = S$$
$$392 \div 0.4 = S$$
$$980 = S$$

$$\begin{array}{r} 980. \\ 0.4_{\wedge}\overline{)392.0_{\wedge}} \\ \underline{3600} \\ 320 \\ \underline{320} \\ 0 \\ \underline{0} \\ 0 \end{array}$$

\$980 worth of artwork was sold.

23. Commission = Commission rate × Sales

$$C = 6\% \times 98,000$$

This tells us what to do. We multiply.

$$\begin{array}{r} 98,000 \\ \times \quad 0.06 \\ \hline 5880.00 \end{array} \qquad (6\% = 0.06)$$

The commission is \$5880.

25. Commission = Commission rate × Sales

$$280.80 = r \times 2340$$

To solve this equation we divide on both sides by 2340.

$$280.80 \div 2340 = r$$
$$0.12 = r$$
$$12\% = r$$

$$\begin{array}{r} 0.12 \\ 2340\overline{)280.80} \\ \underline{2340} \\ 4680 \\ \underline{4680} \\ 0 \end{array}$$

The commission is 12%.

27. First we find the commission on the first \$2000 of sales.

Commission = Commission rate × Sales

$$C = 5\% \times 2000$$

This tells us what to do. We multiply.

$$\begin{array}{r} 2000 \\ \times \ 0.05 \\ \hline 100.00 \end{array}$$

The commission on the first \$2000 of sales is \$100.

Next we subtract to find the amount of sales over \$2000.

$$\$6000 - \$2000 = \$4000$$

Miguel had \$4000 in sales over \$2000.

Then we find the commission on the sales over \$2000.

Commission = Commission rate × Sales

$$C = 8\% \times 4000$$

This tells us what to do. We multiply.

$$\begin{array}{r} 4000 \\ \times \ 0.08 \\ \hline 320.00 \end{array}$$

The commission on the sales over \$2000 is \$320.

Finally we add to find the total commission.

$\$100 + \$320 = \$420$

The total commission is \$420.

29. Discount = Rate of discount × Marked price

$D \quad = \quad 10\% \quad \times \quad \$300$

Convert 10% to decimal notation and multiply.

$$\begin{array}{r} 3\,0\,0 \\ \times \quad 0.\,1 \\ \hline 3\,0.\,0 \end{array} \qquad (10\% = 0.10 = 0.1)$$

The discount is \$30.

Sale price = Marked price − Discount

$S \quad = \quad 300 \quad - \quad 30$

We subtract:
$$\begin{array}{r} 3\,0\,0 \\ -\quad 3\,0 \\ \hline 2\,7\,0 \end{array}$$

To check, note that the sale price is 90% of the marked price: $0.9 \times 300 = 270$.

The sale price is \$270.

31. Discount = Rate of discount × Marked price

$D \quad = \quad 15\% \quad \times \quad \$17$

Convert 15% to decimal notation and multiply.

$$\begin{array}{r} 1\,7 \\ \times\, 0.\,1\,5 \\ \hline 8\,5 \\ 1\,7\,0 \\ \hline 2.\,5\,5 \end{array} \qquad (15\% = 0.15)$$

The discount is \$2.55.

Sale price = Marked price − Discount

$S \quad = \quad 17 \quad - \quad 2.55$

We subtract:
$$\begin{array}{r} 1\,7.\,0\,0 \\ -\quad 2.\,5\,5 \\ \hline 1\,4.\,4\,5 \end{array}$$

To check, note that the sale price is 85% of the marked price: $0.85 \times 17 = 14.45$.

The sale price is \$14.45.

33. Discount = Rate of discount × Marked price

$12.50 \quad = \quad 10\% \quad \times \quad M$

To solve the equation we divide on both sides by 10%.

$$\begin{aligned} 12.50 \div 10\% &= M \\ 12.50 \div 0.1 &= M \\ 125 &= M \end{aligned}$$

The marked price is \$125.

Sale price = Marked price − Discount

$S \quad = \quad 125.00 \quad - \quad 12.50$

We subtract:
$$\begin{array}{r} 1\,2\,5.\,0\,0 \\ -\quad 1\,2.\,5\,0 \\ \hline 1\,1\,2.\,5\,0 \end{array}$$

To check, note that the sale price is 90% of the marked price: $0.9 \times 125 = 112.50$.

The sale price is \$112.50.

35. Discount = Rate of discount × Marked price

$240 \quad = \quad r \quad \times \quad 600$

To solve the equation we divide on both sides by 600.

$240 \div 600 = r$

We can simplify by removing a factor of 1:

$$r = \frac{240}{600} = \frac{2}{5} \cdot \frac{120}{120} = \frac{2}{5} = 0.4 = 40\%$$

The rate of discount is 40%.

Sale price = Marked price − Discount

$S \quad = \quad 600 \quad - \quad 240$

We subtract:
$$\begin{array}{r} 6\,0\,0 \\ -\,2\,4\,0 \\ \hline 3\,6\,0 \end{array}$$

To check, note that a 40% discount rate means that 60% of the marked price is paid. Since $\dfrac{360}{600} = 0.6$, or 60%, we have a check.

The sale price is \$360.

37. Discount = Marked price − Sale price

$D \quad = \quad 1275 \quad - \quad 888$

We subtract:
$$\begin{array}{r} 1\,2\,7\,5 \\ -\quad 8\,8\,8 \\ \hline 3\,8\,7 \end{array}$$

The discount is \$387.

Discount = Rate of discount × Marked price

$387 \quad = \quad R \quad \times \quad 1275$

To solve the equation we divide on both sides by 1275.

$$\begin{aligned} 387 \div 1275 &= R \\ 0.30353 &\approx R \\ 30.353\% &\approx R, \text{ or} \\ 30\frac{6}{17}\% &= R \end{aligned}$$

To check, note that a discount rate of $30\frac{6}{17}\%$ means that $69\frac{11}{17}\%$, or about 69.647%, of the marked price is paid: $0.69647 \times 1275 = 887.99925 \approx 888$.

The rate of discount is $30\frac{6}{17}\%$, or about 30.4%.

39.
$$\begin{aligned} \frac{x}{12} &= \frac{24}{16} \\ 16 \cdot x &= 12 \cdot 24 \qquad \text{Equating cross-products} \\ x &= \frac{12 \cdot 24}{16} \qquad \text{Dividing by 16 on both sides} \\ x &= \frac{288}{16} \\ x &= 18 \end{aligned}$$

The solution is 18.

41. Graph: $y = \frac{4}{3}x$

We make a table of solutions. Note that when x is a multiple of 3, fractional values for y are avoided. Next we plot the points, draw the line and label it.

When $x = -3$, $y = \frac{4}{3}(-3) = -\frac{12}{3} = -4$.

When $x = 0$, $y = \frac{4}{3} \cdot 0 = 0$.

When $x = 3$, $y = \frac{4}{3} \cdot 3 = \frac{12}{3} = 4$.

x	y $y = \frac{4}{3}x$	(x, y)
-3	-4	$(-3, -4)$
0	0	$(0, 0)$
3	4	$(3, 4)$

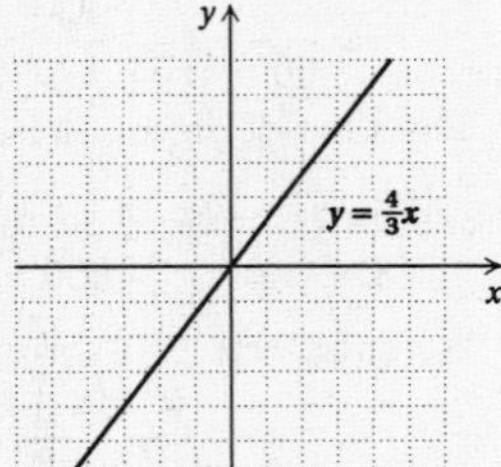

43.

45. Commission = Commission rate × Sales

C = 7.5% × 98,500

We multiply.

```
     9 8, 5 0 0
  ×    0. 0 7 5     (7.5% = 0.075)
   4 9 2 5 0 0
 6 8 9 5 0 0 0
 7 3 8 7. 5 0 0
```

The commission is $7387.50.

We subtract to find how much the seller gets for the house after paying the commission.

$98,500 − $7387.50 = $91,112.50

47. First we find the commission on the first $5000 in sales.

Commission = Commission rate × Sales

C = 10% × 5000

Using a calculator we find that 0.1 × 5000 = 500, so the commission on the first $5000 in sales was $500. We subtract to find the additional commission:

$2405 − $500 = $1905

Now we find the amount of sales required to earn $1905 at a commission rate of 15%.

Commission = Commission rate × Sales

1905 = 15% × S

Using a calculator to divide 1905 by 15%, or 0.15, we get 12,700.

Finally we add to find the total sales:

$5000 + $12,700 = $17,700

Exercise Set 8.6

1. We determine 13% of $200:

$13\% \times 200 = 0.13 \times 200$
$= 26$

```
    2 0 0
 × 0. 1 3
    6 0 0
  2 0 0 0
  2 6. 0 0
```

The interest is $26.

3. First we find the interest for 1 year.

$12.4\% \times 2000 = 0.124 \times 2000$
$= 248$

```
      2 0 0 0
  × 0. 1 2 4
      8 0 0 0
    4 0 0 0 0
  2 0 0 0 0 0
  2 4 8. 0 0 0
```

Next we multiply that amount by $\frac{1}{2}$.

$$\frac{1}{2} \times 248 = \frac{248}{2} = 124.$$

The interest is $124.

(We could have instead found $\frac{1}{2}$ of 12.4% and then multiplied by 2000.)

5. First we find the interest for 1 year.

$14\% \times 4300 = 0.14 \times 4300$
$= 602$

```
    4 3 0 0
 ×    0. 1 4
  1 7 2 0 0
  4 3 0 0 0
  6 0 2. 0 0
```

Next we multiply that amount by $\frac{1}{4}$.

$$\frac{1}{4} \times 602 = \frac{602}{4} = 150.5$$

```
     1 5 0. 5
  4 ) 6 0 2. 0
      4 0 0
      2 0 2
      2 0 0
          2 0
          2 0
            0
```

The interest is $150.50.

(We could have instead found $\frac{1}{4}$ of 14% and then multiplied by 4300.)

7. a) We express 60 days as a fractional part of a year and find the interest.

$$\text{Interest} = (\text{Interest for 1 year}) \times \frac{60}{365}$$
$$= (9\% \times 10,000) \times \frac{60}{365}$$
$$= 0.09 \times 10,000 \times \frac{12}{73} \qquad \left(\frac{60}{365} = \frac{12 \cdot 5}{73 \cdot 5}\right)$$
$$= 900 \times \frac{12}{73}$$
$$= \frac{10,800}{73}$$
$$\approx 147.95 \quad \text{Rounding to the nearest hundredth}$$

The interest due for 60 days is $147.95.

b) The total amount that must be paid after 60 days is the principal plus the interest.

$10,000 + 147.95 = 10,147.95$

The total amount due is $10,147.95.

9. a) We express 90 days as a fractional part of a year and find the interest.

$$\text{Interest} = (\text{Interest for 1 year}) \times \frac{90}{365}$$
$$= (8\% \times 6500) \times \frac{90}{365}$$
$$= 0.08 \times 6500 \times \frac{18}{73} \quad \left(\frac{90}{365} = \frac{18 \cdot 5}{73 \cdot 5}\right)$$
$$= 520 \times \frac{18}{73}$$
$$= \frac{9360}{73}$$
$$\approx 128.22 \quad \text{Rounding to the nearest hundredth}$$

The interest due for 90 days is $128.22.

b) The total amount that must be paid after 90 days is the principal plus the interest.

$$6500 + 128.22 = 6628.22$$

The total amount due is $6628.22.

11. a) We express 30 days as a fractional part of a year and find the interest.

$$\text{Interest} = (\text{Interest for 1 year}) \times \frac{30}{365}$$
$$= (10\% \times 5600) \times \frac{30}{365}$$
$$= 0.1 \times 5600 \times \frac{6}{73} \quad \left(\frac{30}{365} = \frac{6 \cdot 5}{73 \cdot 5}\right)$$
$$= 560 \times \frac{6}{73}$$
$$= \frac{3360}{73}$$
$$\approx 46.03 \quad \text{Rounding to the nearest hundredth}$$

The interest due for 30 days is $46.03.

b) The total amount that must be paid after 30 days is the principal plus the interest.

$$5600 + 46.03 = 5646.03$$

The total amount due is $5646.03.

13. a) After 1 year, the account will contain 110% of $400.

$1.1 \times \$400 = \$440$

$$\begin{array}{r} 400 \\ \times\ 1.1 \\ \hline 400 \\ 4000 \\ \hline 440.0 \end{array}$$

b) At the end of the second year, the account will contain 110% of $440.

$1.1 \times \$440 = \$484$

$$\begin{array}{r} 440 \\ \times\ 1.1 \\ \hline 440 \\ 4400 \\ \hline 484.0 \end{array}$$

The amount in the account after 2 years is $484.

15. a) After 1 year, the account will contain 108.8% of $200.

$1.088 \times \$200 = \$217.60$

$$\begin{array}{r} 1.088 \\ \times\ \ 200 \\ \hline 217.600 \end{array}$$

b) At the end of the second year, the account will contain 108.8% of $217.60.

$1.088 \times \$217.60 = \$236.7488$

$$\begin{array}{r} 217.6 \\ \times 1.088 \\ \hline 17408 \\ 174080 \\ 2176000 \\ \hline 236.7488 \end{array}$$

$\approx \$236.75$ Rounding to the nearest cent

The amount in the account after 2 years is $236.75.

17. a) Since interest is compounded semiannually (twice a year), the rate is 7% ÷ 2, or 3.5%, each compounding period.

b) At two compounding periods per year, there will be $2 \cdot 1$, or 2, periods in 1 year.

c) We substitute.

$$A = P(1+r)^n$$
$$= \$4000(1+0.035)^2$$
$$= \$4000(1.035)^2$$
$$= \$4000(1.071225)$$
$$= \$4284.90$$

After 1 year, the account will contain $4284.90.

19. a) Since interest is compounded semiannually (twice a year), the rate is 9% ÷ 2, or 4.5%, each compounding period.

b) At two compounding periods per year, there will be $2 \cdot 3$, or 6, periods in 3 years.

c) We substitute.

$$A = P(1+r)^n$$
$$= \$2000(1+0.045)^6$$
$$= \$2000(1.045)^6$$
$$\approx \$2000(1.302260125)$$
$$\approx \$2604.52025$$
$$\approx \$2604.52 \quad \text{Rounding to the nearest cent.}$$

After 3 years, the account will contain $2604.52.

21. a) Since there are 12 months in a year, 6% compounded monthly means that 6%÷12, or 0.5% interest, is calculated each month.

b) Since each compounding period is one month, after five months 5 compounding periods have elapsed.

c) We substitute.

$$A = P(1+r)^n$$
$$= \$4000(1+0.005)^5$$
$$= \$4000(1.005)^5$$
$$\approx \$4000(1.025251253)$$
$$\approx \$4101.005013$$
$$\approx \$4101.01 \quad \text{Rounding to the nearest cent.}$$

After 5 months, the account will contain $4101.01.

23. a) Since there are 4 quarters in a year, 10% compounded quarterly means that 10% ÷ 4, or 2.5% interest, is calculated each quarter.

b) Since there are 4 quarters in a year, there are 4 compounding periods in one year.

c) We substitute.

$$\begin{aligned} A &= P(1+r)^n \\ &= \$1200(1+0.025)^4 \\ &= \$1200(1.025)^4 \\ &\approx \$1200(1.103812891) \\ &\approx \$1324.575469 \\ &\approx \$1324.58 \quad \text{Rounding to the nearest cent.} \end{aligned}$$

After 1 year, the account will contain $1324.58.

25.
$$\begin{aligned} \frac{9}{10} &= \frac{x}{5} \\ 9\cdot 5 &= 10\cdot x \\ \frac{9\cdot 5}{10} &= x \\ \frac{45}{10} &= x \\ \frac{9}{2} &= x, \text{ or} \\ 4\frac{1}{2} &= x \end{aligned}$$

27.
$$\begin{aligned} \frac{8}{3} &= \frac{10}{x} \\ 8\cdot x &= 3\cdot 10 \\ x &= \frac{3\cdot 10}{8} \\ x &= \frac{30}{8} \\ x &= \frac{15}{4} \text{ or, } 3\frac{3}{4} \end{aligned}$$

29. First we convert $\frac{64}{17}$ to a mixed numeral.

$$\begin{array}{r} 3 \\ 17\overline{)64} \\ \underline{51} \\ 13 \end{array} \qquad \frac{64}{17} = 3\,\frac{13}{17}$$

Since $\frac{64}{17} = 3\frac{13}{17}$, we have $-\frac{64}{17} = -3\frac{13}{17}$.

31.

33. For a principle P invested at 10% compounded daily, to find the amount in the account at the end of 1 year we would multiply P by $(1 + 0.1/365)^{365}$. Since $(1 + 0.1/365)^{365} \approx 1.105$, the effective yield is approximately 10.5%.

35. At the end of 1 year, the $20,000 spent on the car has been reduced in value by 30% of $20,000, or 0.3 × $20,000, or $6000.

If the $20,000 is invested at 9%, compounded daily, the amount in the account at the end of 1 year is

$$\begin{aligned} A &= \$20,000\left(1+\frac{0.09}{365}\right)^{365} \\ &\approx \$20,000(1.000246575)^{365} \\ &\approx \$20,000(1.094162144) \\ &\approx \$21,883.24. \end{aligned}$$

Then the $20,000 has increased in value by $21,883.24 − $20,000, or $1883.24. All together, the Coniglios have saved the $6000 they would have lost on the value of the car plus the $1883.24 increase in the value of the $20,000 invested at 9%, compounded daily. That is, they have saved $6000 + $1883.24, or $7883.24.

Chapter 9

Geometry and Measures

Exercise Set 9.1

1. 1 foot = 12 in.

This is the relation stated on page 505 of the text.

3. $1\text{ in.} = 1\text{ in.} \times \frac{1\text{ ft}}{12\text{ in.}}$ Multiplying by 1 using $\frac{1\text{ ft}}{12\text{ in.}}$ to eliminate in.

$= \frac{1\text{ in.}}{12\text{ in.}} \times 1\text{ ft}$

$= \frac{1}{12} \times \frac{\text{in.}}{\text{in.}} \times 1\text{ ft}$

$= \frac{1}{12} \times 1\text{ ft}$ The $\frac{\text{in.}}{\text{in.}}$ acts like 1, so we can omit it.

$= \frac{1}{12}\text{ ft}$

5. 1 mi = 5280 ft

This is the relation stated on page 505 of the text.

7. $4\text{ yd} = 4 \times 1\text{ yd}$

$= 4 \times 36\text{ in.}$ Substituting 36 in. for 1 yd

$= 144\text{ in.}$ Multiplying

9. $84\text{ in.} = \frac{84\text{ in.}}{1} \times \frac{1\text{ ft}}{12\text{ in.}}$ Multiplying by 1 using $\frac{1\text{ ft}}{12\text{ in.}}$

$= \frac{84}{12} \times 1\text{ ft}$

$= 7 \times 1\text{ ft}$

$= 7\text{ ft}$

11. $18\text{ in.} = \frac{18\text{ in.}}{1} \times \frac{1\text{ ft}}{12\text{ in.}}$ Multiplying by 1 using $\frac{1\text{ ft}}{12\text{ in.}}$

$= \frac{18}{12} \times 1\text{ ft}$

$= \frac{3}{2} \times 1\text{ ft}$

$= \frac{3}{2}\text{ ft, or } 1\frac{1}{2}\text{ ft}$

13. $5\text{ mi} = 5 \times 1\text{ mi}$

$= 5 \times 5280\text{ ft}$ Substituting 5280 ft for 1 mi

$= 26,400\text{ ft}$ Multiplying

15. $48\text{ in.} = \frac{48\text{ in.}}{1} \times \frac{1\text{ ft}}{12\text{ in.}}$ Multiplying by 1 using $\frac{1\text{ ft}}{12\text{ in.}}$

$= \frac{48}{12} \times 1\text{ ft}$

$= 4 \times 1\text{ ft}$

$= 4\text{ ft}$

17. $19\text{ ft} = 19\text{ ft} \times \frac{1\text{ yd}}{3\text{ ft}}$ Multiplying by 1 using $\frac{1\text{ yd}}{3\text{ ft}}$

$= \frac{19}{3} \times 1\text{ yd}$

$= \frac{19}{3}\text{ yd, or } 6\frac{1}{3}\text{ yd}$

19. $10\text{ mi} = 10 \times 1\text{ mi}$

$= 10 \times 5280\text{ ft}$ Substituting 5280 ft for 1 mi

$= 52,800\text{ ft}$ Multiplying

21. $7\frac{1}{2}\text{ ft} = 7\frac{1}{2}\text{ ft} \times \frac{1\text{ yd}}{3\text{ ft}}$

$= \frac{15}{2}\text{ ft} \times \frac{1\text{ yd}}{3\text{ ft}}$

$= \frac{15}{6} \times 1\text{ yd}$

$= \frac{5}{2} \times 1\text{ yd}$

$= \frac{5}{2}\text{ yd, or } 2\frac{1}{2}\text{ yd}$

23. $360\text{ in.} = 360\text{ in.} \times \frac{1\text{ ft}}{12\text{ in.}} \times \frac{1\text{ yd}}{3\text{ ft}}$

$= \frac{360}{36} \times 1\text{ yd}$

$= 10 \times 1\text{ yd}$

$= 10\text{ yd}$

25. $330\text{ ft} = 330\text{ ft} \times \frac{1\text{ yd}}{3\text{ ft}}$

$= \frac{330}{3} \times 1\text{ yd}$

$= 110 \times 1\text{ yd}$

$= 110\text{ yd}$

27. $3520\text{ yd} = 3520\text{ yd} \times \frac{3\text{ ft}}{1\text{ yd}} \times \frac{1\text{ mi}}{5280\text{ ft}}$

$= \frac{10,560}{5280} \times 1\text{ mi}$

$= 2 \times 1\text{ mi}$

$= 2\text{ mi}$

29. $100\text{ yd} = 100 \times 1\text{ yd}$

$= 100 \times 3\text{ ft}$

$= 300\text{ ft}$

31. $63,360\text{ in.} = 63,360\text{ in.} \times \frac{1\text{ ft}}{12\text{ in.}} \times \frac{1\text{ mi}}{5280\text{ ft}}$

$= \frac{63,360}{63,360} \times 1\text{ mi}$

$= 1 \times 1\text{ mi}$

$= 1\text{ mi}$

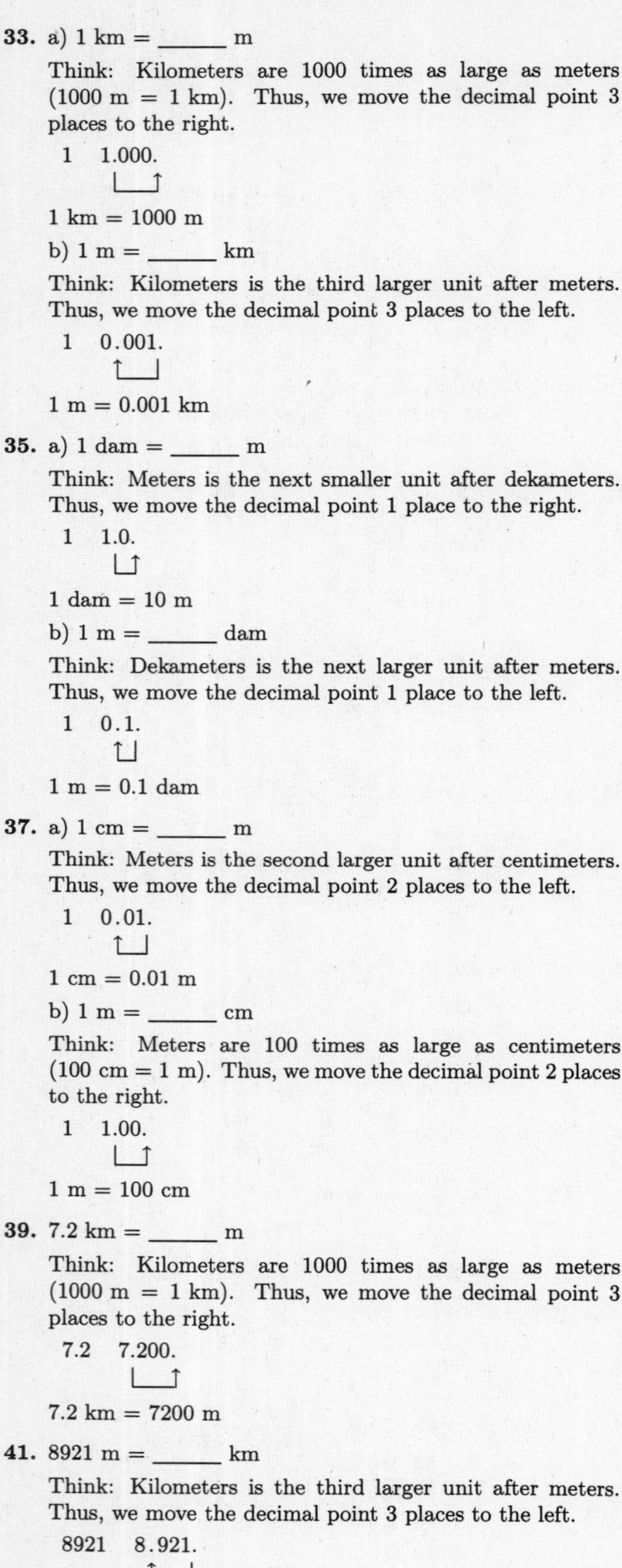

33. a) 1 km = ______ m

Think: Kilometers are 1000 times as large as meters (1000 m = 1 km). Thus, we move the decimal point 3 places to the right.

1 1.000.

1 km = 1000 m

b) 1 m = ______ km

Think: Kilometers is the third larger unit after meters. Thus, we move the decimal point 3 places to the left.

1 0.001.

1 m = 0.001 km

35. a) 1 dam = ______ m

Think: Meters is the next smaller unit after dekameters. Thus, we move the decimal point 1 place to the right.

1 1.0.

1 dam = 10 m

b) 1 m = ______ dam

Think: Dekameters is the next larger unit after meters. Thus, we move the decimal point 1 place to the left.

1 0.1.

1 m = 0.1 dam

37. a) 1 cm = ______ m

Think: Meters is the second larger unit after centimeters. Thus, we move the decimal point 2 places to the left.

1 0.01.

1 cm = 0.01 m

b) 1 m = ______ cm

Think: Meters are 100 times as large as centimeters (100 cm = 1 m). Thus, we move the decimal point 2 places to the right.

1 1.00.

1 m = 100 cm

39. 7.2 km = ______ m

Think: Kilometers are 1000 times as large as meters (1000 m = 1 km). Thus, we move the decimal point 3 places to the right.

7.2 7.200.

7.2 km = 7200 m

41. 8921 m = ______ km

Think: Kilometers is the third larger unit after meters. Thus, we move the decimal point 3 places to the left.

8921 8.921.

8921 m = 8.921 km

43. 732.6 m = ______ km

Think: Kilometers is the third larger unit after meters. Thus, we move the decimal point 3 places to the left.

732.6 0.732.6

732.6 m = 0.7326 km

45. 4528 m = ______ cm

Think: Meters are 100 times as large as centimeters (100 cm = 1 m). Thus, we move the decimal point 2 places to the right.

4528 4528.00.

4528 m = 452,800 cm

47. 477 cm = ______ m

Think: Meters is the second larger unit after centimeters. Thus, we move the decimal point 2 places to the left.

477 4.77.

477 cm = 4.77 m

49. 6.88 m = ______ cm

Think: Meters are 100 times as larger as centimeters (100 cm = 1 m). Thus, we move the decimal point 2 places to the right.

6.88 6.88.

6.88 m = 688 cm

51. 1 mm = ______ cm

Think: Centimeters is the next larger unit after millimeters. Thus, we move the decimal point 1 place to the left.

1 0.1.

1 mm = 0.1 cm

53. 1 km = ______ cm

Think: Kilometers are 100,000 times as large as centimeters (100,000 cm = 1 km). Thus, we move the decimal point 5 places to the right.

1 1.00000.

1 km = 100,000 cm

55. 9.34 cm = ______ mm

Think: Millimeters is the next smaller unit after centimeters. Thus, we move the decimal point 1 place to the right.

9.34 9.3.4

9.34 cm = 93.4 mm

57. 8.2 mm = ______ cm

Think: Centimeters is the next larger unit after millimeters. Thus, we move the decimal point 1 place to the left.

8.2 0.8.2

8.2 mm = 0.82 cm

59. 4500 mm = ______ cm

Think: Centimeters is the next larger unit after millimeters. Thus, we move the decimal point 1 place to the left.

4500 450.0.

4500 mm = 450 cm

61. 0.024 mm = ______ m

Think: Meters is the third larger unit after millimeters. Thus, we move the decimal point 3 places to the left.

0.024 0.000.024

0.024 mm = 0.000024 m

63. 6.88 m = ______ dam

Think: Dekameters is the next larger unit after meters. Thus, we move the decimal point 1 place to the left.

6.88 0.6.88

6.88 m = 0.688 dam

65. 543 dam = ______ km

Think: Kilometers is the second larger unit after dekameters. Thus, we move the decimal point 2 places to the left.

543 5.43.

543 dam = 5.43 km

67. $10 \text{ km} \approx 10 \cancel{\text{km}} \times \dfrac{0.621 \text{ mi}}{1 \cancel{\text{km}}} \approx 6.21 \times 1 \text{ mi} \approx 6.21 \text{ mi}$

69. $14 \text{ cm} \approx 14 \cancel{\text{cm}} \times \dfrac{1 \text{ in.}}{2.54 \cancel{\text{cm}}} \approx \dfrac{14}{2.54} \times 1 \text{ in.} \approx 5.51 \text{ in.}$

71. $36 \text{ yd} \approx 36 \cancel{\text{yd}} \times \dfrac{1 \text{ m}}{1.1 \cancel{\text{yd}}} \approx \dfrac{36}{1.1} \times 1 \text{ m} \approx 32.7 \text{ m}$

73. $330 \text{ ft} \approx 330 \cancel{\text{ft}} \times \dfrac{1 \text{ m}}{3.3 \cancel{\text{ft}}} \approx \dfrac{330}{3.3} \times 1 \text{ m} \approx 100 \text{ m}$

75. $2 \text{ m} \approx 2 \cancel{\text{m}} \times \dfrac{3.3 \text{ ft}}{1 \cancel{\text{m}}} \approx 6.6 \times 1 \text{ ft} \approx 6.6 \text{ ft}$

77. $55 \text{ mph} = 55 \dfrac{\text{mi}}{\text{hr}} = 55 \times \dfrac{1 \text{ mi}}{\text{hr}} \approx 55 \times \dfrac{1.609 \text{ km}}{\text{hr}} \approx$

88.5 km/h

79.
$$\begin{aligned} -7x - 9x &= 24 \\ -16x &= 24 && \text{Collecting like terms} \\ \frac{-16x}{-16} &= \frac{24}{-16} && \text{Dividing by } -16 \text{ on both sides} \\ x &= \frac{3 \cdot 8}{-2 \cdot 8} = \frac{3}{-2} \cdot \frac{8}{8} \\ x &= \frac{3}{-2}, \text{ or } -\frac{3}{2} \end{aligned}$$

81. Let c represent the cost of 7 calculators. We translate to a proportion.

$$\begin{matrix} \text{Number} \rightarrow & 3 & & 7 & \leftarrow \text{Number} \\ & \overline{} & = & \overline{} & \\ \text{Cost} \rightarrow & 43.50 & & c & \leftarrow \text{Cost} \end{matrix}$$

$$\begin{aligned} \text{Solve: } 3 \cdot c &= 43.50 \cdot 7 && \text{Equating cross-products} \\ c &= \frac{43.50 \cdot 7}{3} && \text{Dividing by 3 on both sides} \\ c &= \frac{304.50}{3} && \text{Multiplying} \\ c &= 101.50 && \text{Dividing} \end{aligned}$$

Seven calculators would cost \$101.50.

83. a) Multiply by 100 to move the decimal point two places to the right.

0.47.

b) Write a percent symbol: 47%

Thus, 0.47 = 47%.

85.

87. $2 \text{ mi} \approx 2 \cancel{\text{mi}} \times \dfrac{1.609 \cancel{\text{km}}}{1 \cancel{\text{mi}}} \times \dfrac{100,000 \text{ cm}}{1 \cancel{\text{km}}} \approx$

$321,800 \times 1 \text{ cm} \approx 321,800 \text{ cm}$

89. We find the number of meters of tape that are used in 30 min of playing time.

$$\begin{aligned} &1\frac{7}{8} \, \frac{\text{in.}}{\text{sec}} \\ &\approx 1.875 \frac{\cancel{\text{in.}}}{\cancel{\text{sec}}} \times \frac{1 \text{ m}}{39.37 \cancel{\text{in.}}} \times \frac{60 \cancel{\text{sec}}}{1 \cancel{\text{min}}} \times 30 \cancel{\text{min}} \\ &\approx \frac{3375}{39.37} \text{ m} \\ &\approx 85.725 \text{ m} \end{aligned}$$

About 85.725 m of tape is used for a 60-min cassette.

91.
$$\begin{aligned} &\frac{100 \text{ m}}{9.86 \text{ sec}} \\ &\approx \frac{100 \cancel{\text{m}}}{9.86 \cancel{\text{sec}}} \cdot \frac{60 \cancel{\text{sec}}}{1 \cancel{\text{min}}} \cdot \frac{60 \cancel{\text{min}}}{1 \text{ hr}} \cdot \frac{3.3 \cancel{\text{ft}}}{1 \cancel{\text{m}}} \cdot \frac{1 \text{ mi}}{5280 \cancel{\text{ft}}} \\ &\approx \frac{1,188,000}{52,060.8} \, \frac{\text{mi}}{\text{hr}} \\ &\approx 22.8 \frac{\text{mi}}{\text{hr}}, \text{ or } 22.8 \text{ mph} \end{aligned}$$

(If we first convert 100 m to 0.1 km and then use the conversion factor $\dfrac{1 \text{ mi}}{1.609 \text{ km}}$, the result is about 22.7 mph.)

Exercise Set 9.2

1.
$$\begin{aligned} A &= b \cdot h && \text{Area of a parallelogram} \\ &= 8 \text{ cm} \cdot 4 \text{ cm} && \text{Substituting 8 cm for } b \text{ and 4 cm for } h \\ &= 32 \text{ cm}^2 \end{aligned}$$

3. $A = \frac{1}{2} \cdot h \cdot (a+b)$ Area of a trapezoid
$= \frac{1}{2} \cdot 8 \text{ ft} \cdot (6+20) \text{ ft}$ Substituting 8 ft for h, 6 ft for a, and 20 ft for b
$= \frac{8 \cdot 26}{2} \text{ ft}^2$
$= \frac{\not{2} \cdot 4 \cdot 26}{1 \cdot \not{2}} \text{ ft}^2$
$= 104 \text{ ft}^2$

5. $A = b \cdot h$ Area of a parallelogram
$= 8 \text{ m} \cdot 8 \text{ m}$ Substituting 8 m for b and 8 m for h
$= 64 \text{ m}^2$

7. $A = b \cdot h$ Area of a parallelogram
$= 2.3 \text{ cm} \cdot 3.5 \text{ cm}$ Substituting 2.3 cm for b and 3.5 cm for h
$= 8.05 \text{ cm}^2$

9. $A = \frac{1}{2} \cdot h \cdot (a+b)$ Area of a trapezoid
$= \frac{1}{2} \cdot 9 \text{ mi} \cdot (13+19) \text{ mi}$ Substituting 9 mi for h, 13 mi for a, and 19 mi for b
$= \frac{9 \cdot 32}{2} \text{ mi}^2$
$= \frac{9 \cdot \not{2} \cdot 16}{1 \cdot \not{2}} \text{ mi}^2$
$= 144 \text{ mi}^2$

11. $A = b \cdot h$ Area of a parallelogram
$= 12\frac{1}{4} \text{ ft} \cdot 4\frac{1}{2} \text{ ft}$ Substituting $12\frac{1}{4}$ ft for b and $4\frac{1}{2}$ ft for h
$= \frac{49}{4} \cdot \frac{9}{2} \cdot \text{ ft}^2$
$= \frac{441}{8} \text{ ft}^2$
$= 55\frac{1}{8} \text{ ft}^2$

13. $A = \frac{1}{2} \cdot h \cdot (a+b)$ Area of a trapezoid
$= \frac{1}{2} \cdot 7 \text{ m} \cdot (9+5) \text{ m}$ Substituting 7 m for h, 9 m for a, and 5 m for b
$= \frac{7 \cdot 14}{2} \text{ m}^2$
$= \frac{7 \cdot \not{2} \cdot 7}{1 \cdot \not{2}} \text{ m}^2$
$= 49 \text{ m}^2$

15. $A = b \cdot h$ Area of a parallelogram
$= 12 \text{ cm} \cdot 9 \text{ cm}$ Substituting 12 cm for b and 9 cm for h
$= 108 \text{ cm}^2$

17. $A = \frac{1}{2} \cdot h \cdot (a+b)$ Area of a trapezoid
$= \frac{1}{2} \cdot 8 \text{ yd} \cdot (9.1+7.9) \text{ yd}$ Substituting 8 yd for h, 9.1 yd for a, and 7.9 yd for b
$= \frac{8 \cdot 17}{2} \text{ yd}^2$
$= \frac{\not{2} \cdot 4 \cdot 17}{1 \cdot \not{2}} \text{ yd}^2$
$= 68 \text{ yd}^2$

19. $d = 2 \cdot r$
$= 2 \cdot 7 \text{ cm} = 14 \text{ cm}$

21. $d = 2 \cdot r$
$= 2 \cdot \frac{3}{4} \text{ in.} = \frac{6}{4} \text{ in.} = \frac{3}{2} \text{ in., or } 1\frac{1}{2} \text{ in.}$

23. $r = \frac{d}{2}$
$= \frac{32 \text{ ft}}{2} = 16 \text{ ft}$

25. $r = \frac{d}{2}$
$= \frac{1.4 \text{ cm}}{2} = 0.7 \text{ cm}$

27. $C = 2 \cdot \pi \cdot r$
$\approx 2 \cdot \frac{22}{7} \cdot 7 \text{ cm} \approx \frac{2 \cdot 22 \cdot 7}{7} \text{ cm} \approx 44 \text{ cm}$

29. $C = 2 \cdot \pi \cdot r$
$\approx 2 \cdot \frac{22}{7} \cdot \frac{3}{4} \text{ in.} \approx \frac{2 \cdot 22 \cdot 3}{7 \cdot 4} \text{ in.} \approx \frac{132}{28} \text{ in.} \approx \frac{33}{7} \text{ in.,}$
or $4\frac{5}{7}$ in.

31. $C = \pi \cdot d$
$\approx 3.14 \cdot 32 \text{ ft} \approx 100.48 \text{ ft}$

33. $C = \pi \cdot d$
$\approx 3.14 \cdot 1.4 \text{ cm} \approx 4.396 \text{ cm}$

35. $A = \pi \cdot r \cdot r$
$\approx \frac{22}{7} \cdot 7 \text{ cm} \cdot 7 \text{ cm} \approx \frac{22}{7} \cdot 49 \text{ cm}^2 \approx 154 \text{ cm}^2$

37. $A = \pi \cdot r \cdot r$
$\approx \frac{22}{7} \cdot \frac{3}{4} \text{ in.} \cdot \frac{3}{4} \text{ in.} \approx \frac{22 \cdot 3 \cdot 3}{7 \cdot 4 \cdot 4} \text{ in}^2 \approx \frac{99}{56} \text{ in}^2,$
or $1\frac{43}{56} \text{ in}^2$

39. $A = \pi \cdot r \cdot r$
$\approx 3.14 \cdot 16 \text{ ft} \cdot 16 \text{ ft}$ $(r = \frac{d}{2}; r = \frac{32 \text{ ft}}{2} = 16 \text{ ft})$
$\approx 3.14 \cdot 256 \text{ ft}^2$
$\approx 803.84 \text{ ft}^2$

41. $A = \pi \cdot r \cdot r$
$\approx 3.14 \cdot 0.7 \text{ cm} \cdot 0.7 \text{ cm}$
$(r = \frac{d}{2}; r = \frac{1.4 \text{ cm}}{2} = 0.7 \text{ cm})$
$\approx 3.14 \cdot 0.49 \text{ cm}^2 \approx 1.5386 \text{ cm}^2$

43. $r = \frac{d}{2}$

$= \frac{6 \text{ cm}}{2} = 3 \text{ cm}$

The radius is 3 cm.

$C = \pi \cdot d$

$\approx 3.14 \cdot 6 \text{ cm} \approx 18.84 \text{ cm}$

The circumference is about 18.84 cm.

$A = \pi \cdot r \cdot r$

$\approx 3.14 \cdot 3 \text{ cm} \cdot 3 \text{ cm} \approx 28.26 \text{ cm}^2$

The area is about 28.26 cm^2.

45. $A = \pi \cdot r \cdot r$

$\approx 3.14 \cdot 220 \text{ mi} \cdot 220 \text{ mi} \approx 151,976 \text{ mi}^2$

The broadcast area is about 151,976 mi^2.

47. $C = \pi \cdot d$

$\approx 3.14 \cdot 2.5 \text{ cm} \approx 7.85 \text{ cm}$

The circumference of a quarter is about 7.85 cm.

In order to find the area, we first find the radius.

$r = \frac{d}{2}$

$= \frac{2.5 \text{ cm}}{2} = 1.25 \text{ cm}$

$A = \pi \cdot r \cdot r$

$\approx 3.14 \cdot 1.25 \text{ cm} \cdot 1.25 \text{ cm} \approx 4.90625 \text{ cm}^2$

The area of a quarter is about 4.90625 cm^2.

49. The tree's circumference is 47.1 in.

$C = \pi \cdot d$

$47.1 \approx 3.14 \cdot d$

$\frac{47.1}{3.14} \approx d$

$15 \approx d$

The tree's diameter is about 15 in.

51. Find the area of the larger circle (pool plus walk). Its diameter is 1 yd + 8 yd + 1 yd, or 10 yd. Thus its radius is $\frac{10}{2}$ yd, or 5 yd.

$$A = \pi \cdot r \cdot r$$
$$\approx 3.14 \cdot 5 \text{ yd} \cdot 5 \text{ yd} \approx 78.5 \text{ yd}^2$$

Find the area of the pool. Its diameter is 8 yd. Thus its radius is $\frac{8}{2}$ yd, or 4 yd.

$$A = \pi \cdot r \cdot r$$
$$\approx 3.14 \cdot 4 \text{ yd} \cdot 4 \text{ yd} \approx 50.24 \text{ yd}^2$$

We subtract to find the area of the walk:

$$A = 78.5 \text{ yd}^2 - 50.24 \text{ yd}^2$$
$$= 28.26 \text{ yd}^2$$

The area of the walk is 28.26 yd^2.

53. The perimeter consists of the circumferences of three semicircles, each with diameter 8 ft, and one side of a square of length 8 ft. We first find the circumference of one semicircle. This is one-half the circumference of a circle with diameter 8 ft:

$$\frac{1}{2} \cdot \pi \cdot d \approx \frac{1}{2} \cdot 3.14 \cdot 8 \text{ ft} = 12.56 \text{ ft}$$

Then we multiply by 3:

$$3 \cdot (12.56 \text{ ft}) = 37.68 \text{ ft}$$

Finally we add the circumferences of the semicircles and the length of the side of the square:

$$37.68 \text{ ft} + 8 \text{ ft} = 45.68 \text{ ft}$$

The perimeter is 45.68 ft.

55. The perimeter consists of three-fourths of the perimeter of a square with side of length 10 yd and the circumference of a semicircle with diameter 10 yd. First we find three-fourths of the perimeter of the square:

$$\frac{3}{4} \cdot 4 \cdot s = \frac{3}{4} \cdot 4 \cdot 10 \text{ yd} = 30 \text{ yd}$$

Then we find one-half of the circumference of a circle with diameter 10 yd:

$$\frac{1}{2} \cdot \pi \cdot d \approx \frac{1}{2} \cdot 3.14 \cdot 10 \text{ yd} = 15.7 \text{ yd}$$

Then we add:

$$30 \text{ yd} + 15.7 \text{ yd} = 45.7 \text{ yd}$$

The perimeter is 45.7 yd.

57. The shaded region consists of a circle of radius 8 m, with two circles each of diameter 8 m, removed. First we find the area of the large circle:

$$A = \pi \cdot r \cdot r \approx 3.14 \cdot 8 \text{ m} \cdot 8 \text{ m} = 200.96 \text{ m}^2$$

Then we find the area of one of the small circles:

The radius is $\frac{8 \text{ m}}{2} = 4$ m.

$$A = \pi \cdot r \cdot r \approx 3.14 \cdot 4 \text{ m} \cdot 4 \text{ m} = 50.24 \text{ m}^2$$

We multiply this area by 2 to find the area of the two small circles:

$$2 \cdot 50.24 \text{ m}^2 = 100.48 \text{ m}^2$$

Finally we subtract to find the area of the shaded region:

$$200.96 \text{ m}^2 - 100.48 \text{ m}^2 = 100.48 \text{ m}^2$$

The area of the shaded region is 100.48 m^2.

59. The shaded region consists of one-half of a circle with diameter 2.8 cm and a triangle with base 2.8 cm and height 2.8 cm. First we find the area of the semicircle. The radius is $\frac{2.8 \text{ cm}}{2} = 1.4$ cm.

$$A = \frac{1}{2} \cdot \pi \cdot r \cdot r \approx \frac{1}{2} \cdot 3.14 \cdot 1.4 \text{ cm} \cdot 1.4 \text{ cm} = 3.0772 \text{ cm}^2$$

Then we find the area of the triangle:

$$A = \frac{1}{2} \cdot b \cdot h = \frac{1}{2} \cdot 2.8 \text{ cm} \cdot 2.8 \text{ cm} = 3.92 \text{ cm}^2$$

Finally we add to find the area of the shaded region:

$$3.0772 \text{ cm}^2 + 3.92 \text{ cm}^2 = 6.9972 \text{ cm}^2$$

The area of the shaded region is 6.9972 cm^2.

61.
$$\begin{aligned} 9.25\% &= \frac{9.25}{100} \\ &= \frac{9.25}{100} \cdot \frac{100}{100} \\ &= \frac{925}{10,000} \\ &= \frac{25 \cdot 37}{25 \cdot 400} \\ &= \frac{25}{25} \cdot \frac{37}{400} \\ &= \frac{37}{400} \end{aligned}$$

63. a) First find decimal notation by division.

$$\begin{array}{r} 1.375 \\ 8\overline{)11.000} \\ \underline{8} \\ 30 \\ \underline{24} \\ 60 \\ \underline{56} \\ 40 \\ \underline{40} \\ 0 \end{array}$$

$$\frac{11}{8} = 1.375$$

b) Convert the decimal notation to percent notation. Move the decimal point two places to the right and write a % symbol.

1.37.5

$$\frac{11}{8} = 137.5\%$$

65. $\frac{5}{4} = \frac{5}{4} \cdot \frac{25}{25} = \frac{125}{100} = 125\%$

67.

69. If we remove the top and bottom and "unroll" the can, we have two circles, each with a diameter of 2.5 in., and a rectangle whose length is the circumference of the can and whose width is the height of the can. First we find the area of the circles. Their radius is $\frac{d}{2} = \frac{2.5 \text{ in.}}{2} = 1.25$ in. Then the area of each circle is

$$\begin{aligned} A &= \pi \cdot r \cdot r \\ &\approx 3.14 \cdot 1.25 \text{ in.} \cdot 1.25 \text{ in.} \approx 4.90625 \text{ in}^2. \end{aligned}$$

The area of the two circles is $2 \cdot 4.90625 \text{ in}^2 = 9.8125 \text{ in}^2$.

Now we find the area of the rectangle. The circumference of the can is

$$\begin{aligned} C &= \pi \cdot d \\ &\approx 3.14 \cdot 2.5 \text{ in.} \approx 7.85 \text{ in.} \end{aligned}$$

The area of a rectangle with length 7.85 in. and width 3.5 in. is

$$\begin{aligned} A &= l \cdot w \\ &= 7.85 \text{ in.} \cdot 3.5 \text{ in.} = 27.475 \text{ in}^2. \end{aligned}$$

Then the surface area of the can is

$$9.8125 \text{ in}^2 + 27.475 \text{ in}^2 = 37.2875 \text{ in}^2.$$

71. Find $3927 \div 1250$ using a calculator.

$$\frac{3927}{1250} = 3.1416 \approx 3.142 \quad \text{Rounding}$$

73. 

Exercise Set 9.3

1. $4 \text{ yd}^2 = 4 \cdot 9 \text{ ft}^2$ Substituting 9 ft^2 for 1 yd^2
$= 36 \text{ ft}^2$

(Had we preferred to use canceling, we could have multiplied 4 yd^2 by $\frac{9 \text{ ft}^2}{1 \text{ yd}^2}$.)

3. $7 \text{ ft}^2 = 7 \cdot 144 \text{ in}^2$ Substituting 144 in^2 for 1 ft^2
$= 1008 \text{ in}^2$

(Had we preferred to use canceling, we could have multiplied 7 ft^2 by $\frac{144 \text{ in}^2}{1 \text{ ft}^2}$.)

5.
$$\begin{aligned} 432 \text{ in}^2 &= 432 \cancel{\text{in}^2} \times \frac{1 \text{ ft}^2}{144 \cancel{\text{in}^2}} \\ &= \frac{432}{144} \times \text{ ft}^2 \\ &= 3 \text{ ft}^2 \end{aligned}$$

7. $22 \text{ yd}^2 = 22 \times 9 \text{ ft}^2$ Substituting 9 ft^2 for 1 yd^2
$= 198 \text{ ft}^2$

9. $44 \text{ yd}^2 = 44 \cdot 9 \text{ ft}^2$ Substituting 9 ft^2 for 1 yd^2
$= 396 \text{ ft}^2$

11. $20 \text{ mi}^2 = 20 \cdot 640 \text{ acres}$ Substituting 640 acres for 1 mi^2

$= 12,800 \text{ acres}$

13. $69 \text{ ft}^2 = 69 \text{ ft}^2 \times \frac{1 \text{ yd}^2}{9 \text{ ft}^2}$

$= \frac{69}{9} \times \text{yd}^2$

$= \frac{23}{3} \text{ yd}^2$, or $7\frac{2}{3} \text{ yd}^2$

15. $720 \text{ in}^2 = 720 \text{ in}^2 \times \frac{1 \text{ ft}^2}{144 \text{ in}^2}$

$= \frac{720}{144} \times \text{ft}^2$

$= 5 \text{ ft}^2$

17. $1 \text{ in}^2 = 1 \text{ in}^2 \times \frac{1 \text{ ft}^2}{144 \text{ in}^2}$

$= \frac{1}{144} \times \text{ft}^2$

$= \frac{1}{144} \text{ ft}^2$

19. $1 \text{ acre} = 1 \text{ acre} \cdot \frac{1 \text{ mi}^2}{640 \text{ acres}}$

$= \frac{1}{640} \cdot \text{mi}^2$

$= \frac{1}{640} \text{ mi}^2$, or 0.0015625 mi^2

21. $17 \text{ km}^2 =$ ______ m^2

Think: A kilometer is 1000 times as big as a meter, so 1 km^2 is 1,000,000 times as big as 1 m^2. We shift the decimal point six places to the right.

17 17.000000.

$17 \text{ km}^2 = 17{,}000{,}000 \text{ m}^2$

23. $6.31 \text{ m}^2 =$ ______ cm^2

Think: A meter is 100 times as big as a centimeter, so 1 m^2 is 10,000 times as big as 1 cm^2. We shift the decimal point 4 places to the right.

6.31 6.3100.

$6.31 \text{ m}^2 = 63{,}100 \text{ cm}^2$

25. $2345.6 \text{ mm}^2 =$ ______ cm^2

Think: To convert from mm to cm, we shift the decimal point one place to the left. To convert from mm^2 to cm^2, we shift the decimal point two places to the left.

2345.6 23.45.6

$2345.6 \text{ mm}^2 = 23.456 \text{ cm}^2$

27. $349 \text{ cm}^2 =$ ______ m^2

Think: To convert from cm to m, we shift the decimal point two places to the left. To convert from cm^2 to m^2, we shift the decimal point four places to the left.

349 0.0349.

$349 \text{ cm}^2 = 0.0349 \text{ m}^2$

29. $250{,}000 \text{ mm}^2 =$ ______ cm^2

Think: To convert from mm to cm, we shift the decimal point one place to the left. To convert from mm^2 to cm^2, we shift the decimal point two places to the left.

250,000 2500.00.

$250{,}000 \text{ mm}^2 = 2500 \text{ cm}^2$

31. $472{,}800 \text{ m}^2 =$ ______ km^2

Think: To convert from m to km, we shift the decimal point three places to the left. To convert from m^2 to km^2, we shift the decimal point six places to the left.

472,800 0.472800.

$472{,}800 \text{ m}^2 = 0.4728 \text{ km}^2$

33. First we convert 3 in. to feet.

$3 \text{ in.} = 3 \text{ in.} \times \frac{1 \text{ ft}}{12 \text{ in.}}$

$= \frac{3}{12} \times \text{ft}$

$= \frac{1}{4} \text{ ft}$

Then we find the area using the formula for the area of a rectangle.

$A = l \cdot w$

$= 8 \text{ ft} \cdot \frac{1}{4} \text{ ft}$

$= \frac{8}{4} \text{ ft}^2$

$= 2 \text{ ft}^2$

35. We convert 4 in. and 7 yd to feet.

$4 \text{ in.} = 4 \text{ in.} \times \frac{1 \text{ ft}}{12 \text{ in.}}$

$= \frac{4}{12} \times \text{ft}$

$= \frac{1}{3} \text{ ft}$

$7 \text{ yd} = 7 \cdot 3 \text{ ft} = 21 \text{ ft}$

Then we find the area using the formula for the area of a trapezoid.

$A = \frac{1}{2} \cdot h \cdot (a + b)$

$= \frac{1}{2} \cdot \frac{1}{3} \text{ ft} \cdot (5 + 21) \text{ ft}$

$= \frac{26}{2} \text{ ft}^2$

$= \frac{13}{3} \text{ ft}^2$, or $4\frac{1}{3} \text{ ft}^2$

37. The area of the large square, before the corners are cut out is

$$A = s \cdot s$$
$$= 2.8 \text{ m} \cdot 2.8 \text{ m}$$
$$= 7.84 \text{ m}^2.$$

The area of each of the squares cut out of the corners of the larger squares is

$$A = s \cdot s$$
$$= 18 \text{ mm} \cdot 18 \text{ mm}$$
$$= 324 \text{ mm}^2.$$

Then the total area of the 4 cut-out squares is $4 \cdot 324 \text{ mm}^2$, or 1296 mm^2.

In order to subtract to find the area of the shaded region, we first convert 1296 mm^2 to m^2. Think: To convert from mm to m, we shift the decimal point 3 places to the left. To convert from mm^2 to m^2, we shift the decimal point 6 places to the left.

1296 0.001296.

$1296 \text{ mm}^2 = 0.001296 \text{ m}^2$

Now we subtract to find the area of the shaded region:

$$7.84 \text{ m}^2 - 0.001296 \text{ m}^2 = 7.838704 \text{ m}^2$$

(This result could also be expressed as $7{,}838{,}704 \text{ mm}^2$.)

39. We first find the interest for 1 year:

$$5\% \times 700 = 0.05 \times 700 = 35$$

Then we multiply that amount by $\frac{1}{2}$:

$$\frac{1}{2} \times 35 = \frac{35}{2} = 17.5$$

The interest for $\frac{1}{2}$ yr is \$17.50.

41. We first find the interest for 1 year:

$$8.9\% \times 1200 = 0.089 \times 1200 = 106.8$$

Then we multiply that amount by $\frac{30}{365}$:

$$\frac{30}{365} \times 106.8 = \frac{3204}{365} \approx 8.78$$

The interest for 30 days is \$8.78.

43. 

45. First we convert 8 in. to ft.

$$8 \text{ in.} = 8 \cancel{\text{in.}} \times \frac{1 \text{ ft}}{12 \cancel{\text{in.}}}$$
$$= \frac{8}{12} \times \text{ft}$$
$$= \frac{2}{3} \text{ ft}$$

Then we find the area of each tile.

$$A = s \cdot s$$
$$= \frac{2}{3} \text{ ft} \cdot \frac{2}{3} \text{ ft}$$
$$= \frac{4}{9} \text{ ft}^2$$

Next we find the area of the dance floor.

$$A = l \cdot w$$
$$= 18 \text{ ft} \cdot 42 \text{ ft}$$
$$= 756 \text{ ft}^2$$

We divide to find the number of tiles needed.

$$756 \div \frac{4}{9} = 756 \cdot \frac{9}{4} = \frac{756 \cdot 9}{4} = \frac{\cancel{4} \cdot 189 \cdot 9}{\cancel{4} \cdot 1} = 1701$$

Thus, 1701 tiles are needed.

Now we find the area of the floor.

$$A = l \cdot w$$
$$= 30 \text{ ft} \cdot 60 \text{ ft}$$
$$= 1800 \text{ ft}^2$$

We use an equation to find the percent of the floor area that is covered by the dance area.

756 is what percent of 1800?

$$756 = n \times 1800$$

To solve the equation we divide on both sides by 1800 and convert to percent notation.

$$756 = n \times 1800$$
$$\frac{756}{1800} = n$$
$$0.42 = n$$
$$42\% = n$$

The dance area is 42% of the area of the floor.

47. $1 \text{ in}^2 \approx (2.54 \text{ cm})^2 \approx 6.4516 \text{ cm}^2$

49.

$$1 \text{ acre} = 43{,}560 \text{ ft}^2$$
$$\approx 43{,}560 \text{ ft}^2 \times \frac{1 \text{ m}}{3.3 \text{ ft}} \times \frac{1 \text{ m}}{3.3 \text{ ft}}$$
$$\approx \frac{43{,}560}{3.3(3.3)} \times \cancel{\text{ft}} \times \cancel{\text{ft}} \times \frac{\text{m}}{\cancel{\text{ft}}} \times \frac{\text{m}}{\cancel{\text{ft}}}$$
$$\approx 4000 \text{ m}^2$$

51. First we convert 2 m to centimeters. Think: Meters are 100 times as large as centimeters (100 cm = 1 m). We move the decimal point two places to the right.

$$2 \text{ m} = 200 \text{ cm}$$

Next we convert 10 in. to centimeters.

$$10 \text{ in.} \approx 10 \times 2.54 \text{ cm} \approx 25.4 \text{ cm}$$

Now we find the area of the scarf.

$$A = l \cdot w$$
$$= 200 \text{ cm} \cdot 25.4 \text{ cm}$$
$$= 5080 \text{ cm}^2$$

The area of the scarf is 5080 cm^2.

Exercise Set 9.4

1. The angle can be named in five different ways:

angle GHI, angle IHG, $\angle\, GHI$, $\angle\, IHG$, or $\angle\, H$.

3. Place the △ of the protractor at the vertex of the angle, and line up one of the sides at 0°. We choose the horizontal side. Since 0° is on the inside scale, we check where the other side of the angle crosses the inside scale. It crosses at 10°. Thus, the measure of the angle is 10°.

5. Place the $\triangle$ of the protractor at the vertex of the angle, point B. Line up one of the sides at $0°$. We choose the side that contains point A. Since $0°$ is on the outside scale, we check where the other side crosses the outside scale. It crosses at $180°$. Thus, the measure of the angle is $180°$.

7. Place the $\triangle$ of the protractor at the vertex of the angle, and line up one of the sides at $0°$. We choose the horizontal side. Since $0°$ is on the inside scale, we check where the other side crosses the inside scale. It crosses at $130°$. Thus, the measure of the angle is $130°$.

9. Using a protractor, we find that the measure of the angle in Exercise 1 is $148°$. Since its measure is greater than $90°$and less than $180°$, it is an obtuse angle.

11. The measure of the angle in Exercise 3 is $10°$. Since its measure is greater than $0°$and less than $90°$, it is an acute angle.

13. The measure of the angle in Exercise 5 is $180°$. It is a straight angle.

15. The measure of the angle in Exercise 7 is $130°$. Since its measure is greater than $90°$and less than $180°$, it is an obtuse angle.

17. The measure of the angle in Margin Exercise 1 is $30°$. Since its measure is greater than $0°$and less than $90°$, it is an acute angle.

19. The measure of the angle in Margin Exercise 3 is $126°$. Since its measure is greater than $90°$and less than $180°$, it is an obtuse angle.

21. Two angles are complementary if the sum of their measures is $90°$.

$$90° - 11° = 79°.$$

The measure of a complement is $79°$.

23. Two angles are complementary if the sum of their measures is $90°$.

$$90° - 67° = 23°.$$

The measure of a complement is $23°$.

25. Two angles are complementary if the sum of their measures is $90°$.

$$90° - 72° = 18°.$$

The measure of a complement is $18°$.

27. Two angles are complementary if the sum of their measures is $90°$.

$$90° - 28° = 62°.$$

The measure of a complement is $62°$.

29. Two angles are supplementary if the sum of their measures is $180°$.

$$180° - 3° = 177°.$$

The measure of a supplement is $177°$.

31. Two angles are supplementary if the sum of their measures is $180°$.

$$180° - 139° = 41°.$$

The measure of a supplement is $41°$.

33. Two angles are supplementary if the sum of their measures is $180°$.

$$180° - 85° = 95°.$$

The measure of a supplement is $95°$.

35. Two angles are supplementary if the sum of their measures is $180°$.

$$180° - 102° = 78°.$$

The measure of a supplement is $78°$.

37. Replace the % symbol with $\times 0.01$: 56.1×0.01

Multiply to move the decimal point two places to the left:

0.56.1

Thus, $56.1\% = 0.561$.

39.

$$\begin{aligned}
3.1x + 4.3 &= x + 9.55 \\
3.1x + 4.3 - x &= x + 9.55 - x \quad \text{Subtracting } x \text{ on both sides} \\
2.1x + 4.3 &= 9.55 \\
2.1x + 4.3 - 4.3 &= 9.55 - 4.3 \quad \text{Subtracting 4.3 on both sides} \\
2.1x &= 5.25 \\
\frac{2.1x}{2.1} &= \frac{5.25}{2.1} \quad \text{Dividing by 2.1 on both sides} \\
x &= 2.5
\end{aligned}$$

The solution is 2.5.

41. $-9.7 + 3.8$

The difference of the absolute values is 5.9. Since -9.7 has the larger absolute value, the answer is negative.

$-9.7 + 3.8 = -5.9$

43. 

45. We find $m\angle 2$:

$$\begin{aligned}
m\angle 6 + m\angle 1 + m\angle 2 &= 180° \\
33.07° + 79.8° + m\angle 2 &= 180° \quad \text{Substituting} \\
112.87° + m\angle 2 &= 180° \\
m\angle 2 &= 180° - 112.87° \\
m\angle 2 &= 67.13°
\end{aligned}$$

The measure of angle 2 is $67.13°$.

We find $m\angle 3$:

$$\begin{aligned}
m\angle 1 + m\angle 2 + m\angle 3 &= 180° \\
79.8° + 67.13° + m\angle 3 &= 180° \\
146.93° + m\angle 3 &= 180° \\
m\angle 3 &= 180° - 146.93° \\
m\angle 3 &= 33.07°
\end{aligned}$$

The measure of angle 3 is $33.07°$.

We find $m\angle 4$:

$$\begin{aligned} m\angle 2 + m\angle 3 + m\angle 4 &= 180^\circ \\ 67.13^\circ + 33.07^\circ + m\angle 4 &= 180^\circ \\ 100.2^\circ + m\angle 4 &= 180^\circ \\ m\angle 4 &= 180^\circ - 100.2^\circ \\ m\angle 4 &= 79.8^\circ \end{aligned}$$

The measure of angle 4 is 79.8°.

To find $m\angle 5$, note that $m\angle 6 + m\angle 1 + m\angle 5 = 180^\circ$. Then to find $m\angle 5$ we follow the same procedure we used to find $m\angle 2$. Thus, the measure of angle 5 is 67.13°.

47. $\angle ACB$ and $\angle ACD$ are complementary angles. Since $m\angle ACD = 40^\circ$and $90^\circ - 40^\circ = 50^\circ$, we have $m\angle ACB = 50^\circ$.

Now consider triangle ABC. We know that the sum of the measures of the angles is 180°. Then

$$\begin{aligned} m\angle ABC + m\angle BCA + m\angle CAB &= 180^\circ \\ 50^\circ + 90^\circ + m\angle CAB &= 180^\circ \\ 140^\circ + m\angle CAB &= 180^\circ \\ m\angle CAB &= 180^\circ - 140^\circ \\ m\angle CAB &= 40^\circ, \end{aligned}$$

so $m\angle CAB = 40^\circ$.

To find $m\angle EBC$ we first find $m\angle CEB$. We note that $\angle DEC$ and $\angle CEB$ are supplementary angles. Since $m\angle DEC = 100^\circ$ and $180^\circ - 100^\circ = 80^\circ$, we have $m\angle CEB = 80^\circ$. Now consider triangle BCE. We know that the sum of the measures of the angles is 180°. Note that $\angle ACB$ can also be named $\angle BCE$. Then

$$\begin{aligned} m\angle BCE + m\angle CEB + m\angle EBC &= 180^\circ \\ 50^\circ + 80^\circ + m\angle EBC &= 180^\circ \\ 130^\circ + m\angle EBC &= 180^\circ \\ m\angle EBC &= 180^\circ - 130^\circ \\ m\angle EBC &= 50^\circ, \end{aligned}$$

so $m\angle EBC = 50^\circ$.

$\angle EBA$ and $\angle EBC$ are complementary angles. Since $m\angle EBC = 50^\circ$ and $90^\circ - 50^\circ = 40^\circ$, we have $m\angle EBA = 40^\circ$.

Now consider triangle ABE. We know that the sum of the measures of the angles is 180°. Then

$$\begin{aligned} m\angle CAB + m\angle EBA + m\angle AEB &= 180^\circ \\ 40^\circ + 40^\circ + m\angle AEB &= 180^\circ \\ 80^\circ + m\angle AEB &= 180^\circ \\ m\angle AEB &= 180^\circ - 80^\circ \\ m\angle AEB &= 100^\circ, \end{aligned}$$

so $m\angle AEB = 100^\circ$.

To find $m\angle ADB$ we first find $m\angle EDC$. Consider triangle CDE. We know that the sum of the measures of the angles is 180°. Then

$$\begin{aligned} m\angle DEC + m\angle ECD + m\angle EDC &= 180^\circ \\ 100^\circ + 40^\circ + m\angle EDC &= 180^\circ \\ 140^\circ + m\angle EDC &= 180^\circ \\ m\angle EDC &= 180^\circ - 140^\circ \\ m\angle EDC &= 40^\circ, \end{aligned}$$

so $m\angle EDC = 40^\circ$. We now note that $\angle ADB$ and $\angle EDC$ are complementary angles. Since $m\angle EDC = 40^\circ$ and $90^\circ - 40^\circ = 50^\circ$, we have $m\angle ADB = 50^\circ$.

Exercise Set 9.5

1. The square roots of 16 are 4 and -4, because $4^2 = 16$ and $(-4)^2 = 16$.

3. The square roots of 121 are 11 and -11, because $11^2 = 121$ and $(-11)^2 = 121$.

5. The square roots of 169 are 13 and -13, because $13^2 = 169$ and $(-13)^2 = 169$.

7. The square roots of 6400 are 80 and -80, because $80^2 = 6400$ and $(-80)^2 = 6400$.

9. $\sqrt{49} = 7$

The square root of 49 is 7, because $7^2 = 49$ and 7 is positive.

11. $\sqrt{81} = 9$

The square root of 81 is 9, because $9^2 = 81$ and 9 is positive.

13. $\sqrt{225} = 15$

The square root of 225 is 15, because $15^2 = 225$ and 15 is positive.

15. $\sqrt{625} = 25$

The square root of 625 is 25 because $25^2 = 625$ and 25 is positive.

17. $\sqrt{400} = 20$

The square root of 400 is 20, because $20^2 = 400$ and 20 is positive.

19. $\sqrt{10,000} = 100$

The square root of 10,000 is 100 because $100^2 = 10,000$ and 100 is positive.

21. $\sqrt{48} \approx 6.928$

23. $\sqrt{8} \approx 2.828$

25. $\sqrt{3} \approx 1.732$

27. $\sqrt{12} \approx 3.464$

29. $\sqrt{19} \approx 4.359$

31. $\sqrt{110} \approx 10.488$

33.

$$\begin{aligned} a^2 + b^2 &= c^2 && \text{Pythagorean equation} \\ 3^2 + 5^2 &= c^2 && \text{Substituting} \\ 9 + 25 &= c^2 \\ 34 &= c^2 \\ \sqrt{34} &= c && \text{Exact answer} \\ 5.831 &\approx c && \text{Approximation} \end{aligned}$$

35.

$$\begin{aligned} a^2 + b^2 &= c^2 && \text{Pythagorean equation} \\ 7^2 + 7^2 &= c^2 && \text{Substituting} \\ 49 + 49 &= c^2 \\ 98 &= c^2 \\ \sqrt{98} &= c && \text{Exact answer} \\ 9.899 &\approx c && \text{Approximation} \end{aligned}$$

37.
$$\begin{aligned} a^2+b^2 &= c^2 \\ a^2+12^2 &= 13^2 \\ a^2+144 &= 169 \\ a^2+144-144 &= 169-144 \\ a^2 &= 169-144 \\ a^2 &= 25 \\ a &= 5 \end{aligned}$$

39.
$$\begin{aligned} a^2+b^2 &= c^2 \\ 6^2+b^2 &= 10^2 \\ 36+b^2 &= 100 \\ 36+b^2-36 &= 100-36 \\ b^2 &= 100-36 \\ b^2 &= 64 \\ b &= 8 \end{aligned}$$

41.
$$\begin{aligned} a^2+b^2 &= c^2 \\ 10^2+24^2 &= c^2 \\ 100+576 &= c^2 \\ 676 &= c^2 \\ 26 &= c \end{aligned}$$

43.
$$\begin{aligned} a^2+b^2 &= c^2 \\ 9^2+b^2 &= 15^2 \\ 81+b^2 &= 225 \\ 81+b^2-81 &= 225-81 \\ b^2 &= 225-81 \\ b^2 &= 144 \\ b &= 12 \end{aligned}$$

45.
$$\begin{aligned} a^2+b^2 &= c^2 \\ 1^2+b^2 &= 32^2 \\ 1+b^2 &= 1024 \\ 1+b^2-1 &= 1024-1 \\ b^2 &= 1024-1 \\ b^2 &= 1023 \\ b &= \sqrt{1023} && \text{Exact answer} \\ b &\approx 31.984 && \text{Approximation} \end{aligned}$$

47.
$$\begin{aligned} a^2+b^2 &= c^2 \\ 3^2+4^2 &= c^2 \\ 9+16 &= c^2 \\ 25 &= c^2 \\ 5 &= c \end{aligned}$$

49. ***Familiarize.*** We first make a drawing. In it we see a right triangle. We let $s =$ the length of the string of lights.

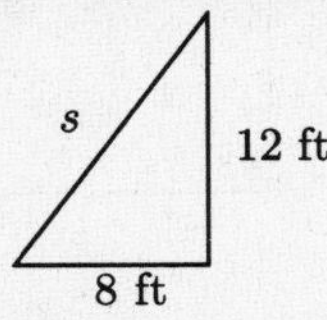

Translate. We substitute 8 for a, 12 for b, and s for c in the Pythagorean equation.

$$\begin{aligned} a^2+b^2 &= c^2 \\ 8^2+12^2 &= s^2 \end{aligned}$$

Solve. We solve the equation for w.

$$\begin{aligned} 64+144 &= s^2 \\ 208 &= s^2 \\ \sqrt{208} &= s && \text{Exact answer} \\ 14.4 &\approx s && \text{Approximation} \end{aligned}$$

Check. $8^2+12^2 = 64+144 = 208 = (\sqrt{208})^2$

State. The length of the string of lights is $\sqrt{208}$ ft, or about 14.4 ft.

51. ***Familiarize.*** We refer to the drawing in the text. We let $d =$ the distance from home plate to second base.

Translate. We substitute 90 for a, 90 for b, and d for c in the Pythagorean equation.

$$\begin{aligned} a^2+b^2 &= c^2 \\ 90^2+90^2 &= d^2 \end{aligned}$$

Solve. We solve the equation for d.

$$\begin{aligned} 8100+8100 &= d^2 \\ 16,200 &= d^2 \\ \sqrt{16,200} &= d \\ 127.3 &\approx d \end{aligned}$$

Check. $90^2+90^2 = 8100+8100 = 16,200 = (\sqrt{16,200})^2$

State. The distance from home plate to second base is $\sqrt{16,200}$ ft, or about 127.3 ft.

53. ***Familiarize.*** We refer to the drawing in the text.

Translate. We substitute in the Pythagorean equation.

$$\begin{aligned} a^2+b^2 &= c^2 \\ 20^2+h^2 &= 30^2 \end{aligned}$$

Solve. We solve the equation for h.

$$\begin{aligned} 400+h^2 &= 900 \\ 400+h^2-400 &= 900-400 \\ h^2 &= 900-400 \\ h^2 &= 500 \\ h &= \sqrt{500} \\ h &\approx 22.4 \end{aligned}$$

Check. $20^2+(\sqrt{500})^2 = 400+500 = 900 = 30^2$

State. The height of the tree is $\sqrt{500}$ ft, or about 22.4 ft.

55. ***Familiarize.*** We refer to the drawing in the text. We let $h =$ the plane's horizontal distance from the airport.

Translate. We substitute 4100 for a, h for b, and 15,100 for c in the Pythagorean equation.

$$\begin{aligned} a^2+b^2 &= c^2 \\ 4100^2+h^2 &= 15,100^2 \end{aligned}$$

Solve. We solve the equation for h.

$$\begin{aligned} 16,810,000+h^2 &= 228,010,000 \\ h^2 &= 228,010,000-16,810,000 \\ h^2 &= 211,200,000 \\ h &= \sqrt{211,200,000} \\ h &\approx 14,532.7 \end{aligned}$$

Check. $4100^2+(\sqrt{211,200,000})^2 = 16,810,000+$ $211,200,000 = 228,010,000 = 15,100^2$

State. The plane's horizontal distance from the airport is $\sqrt{211,200,000}$ ft, or about 14,532.7 ft.

57. a) Replace the percent symbol with $\times 0.01$.

45.6×0.01

b) Multiply to move the decimal point two places to the left.

0.45.6

Thus, $45.6\% = 0.456$.

59. a) Replace the percent symbol with $\times 0.01$.

123×0.01

b) Multiply to move the decimal point two places to the left.

1.23.

Thus, $123\% = 1.23$.

61. a) Replace the percent symbol with $\times 0.01$.

0.41×0.01

b) Multiply to move the decimal point two places to the left.

0.00.41

Thus, $0.41\% = 0.0041$.

63. 

65. We add some labels to the drawing of the polygon.

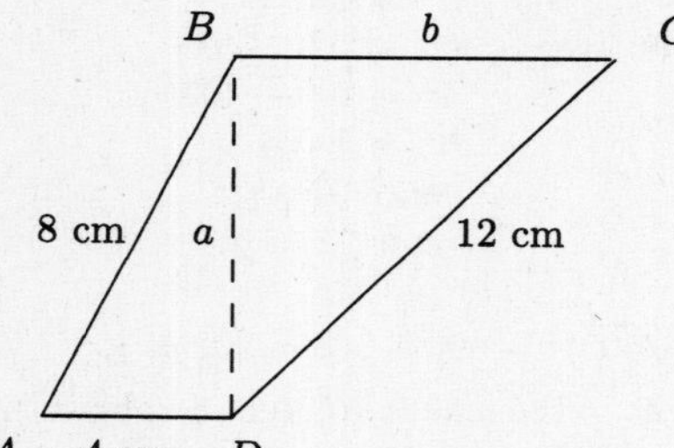

First we consider triangle ABD. We use the Pythagorean equation to find the length a.

$$\begin{aligned} a^2 + b^2 &= c^2 \\ a^2 + 4^2 &= 8^2 \\ a^2 + 16 &= 64 \\ a^2 + 16 - 16 &= 64 - 16 \\ a^2 &= 48 \\ a &= \sqrt{48} \\ a &\approx 6.93 \end{aligned}$$

Now we consider triangle BCD. We use the Pythagorean equation again to find the length b. We will use the exact value of a^2, 48, in this calculation.

$$\begin{aligned} a^2 + b^2 &= c^2 \\ 48 + b^2 &= 12^2 \\ 48 + b^2 &= 144 \\ 48 + b^2 - 48 &= 144 - 48 \\ b^2 &= 96 \\ b &= \sqrt{96} \\ b &\approx 9.80 \end{aligned}$$

Next we find the area of each triangle.

$$\begin{aligned} \text{For triangle } ABD\text{: } A &= \frac{1}{2} \cdot b \cdot h \\ &= \frac{1}{2} \cdot 4 \text{ cm} \cdot 6.93 \text{ cm} \\ &= \frac{27.72}{2} \text{ cm}^2 \\ &= 13.86 \text{ cm}^2 \end{aligned}$$

$$\begin{aligned} \text{For triangle } BCD\text{: } A &= \frac{1}{2} \cdot b \cdot h \\ &= \frac{1}{2} \cdot 9.80 \text{ cm} \cdot 6.93 \text{ cm} \\ &= \frac{67.914}{2} \text{ cm}^2 \\ &= 33.957 \text{ cm}^2 \\ &\approx 33.96 \text{ cm}^2 \end{aligned}$$

We add these two areas to find the area of the polygon:

$$13.86 \text{ cm}^2 + 33.96 \text{ cm}^2 = 47.82 \text{ cm}^2$$

The area is about 47.82 cm^2. (Answers may vary slightly depending on when the rounding was done.)

67. We let $w =$ the width of a 19-in. screen and $l =$ its length. We consider the following similar triangles.

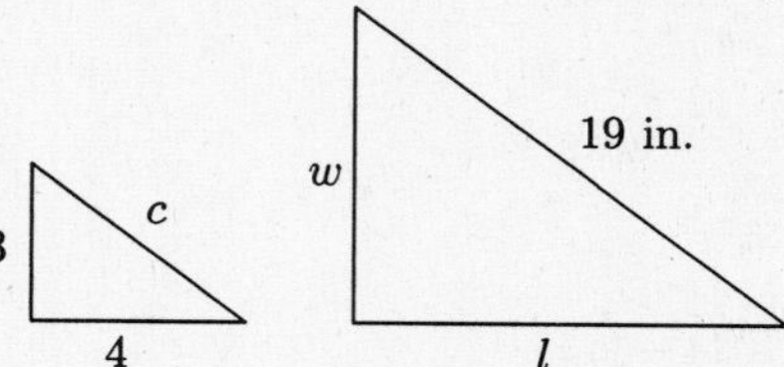

First we use the Pythagorean equation to find the length c.

$$\begin{aligned} a^2 + b^2 &= c^2 \\ 3^2 + 4^2 &= c^2 \\ 9 + 16 &= c^2 \\ 25 &= c^2 \\ 5 &= c \end{aligned}$$

Now we use proportions to find l and w.

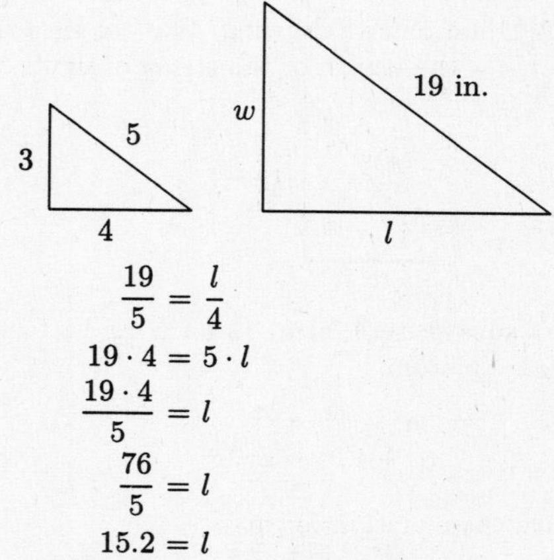

$$\begin{aligned} \frac{19}{5} &= \frac{l}{4} \\ 19 \cdot 4 &= 5 \cdot l \\ \frac{19 \cdot 4}{5} &= l \\ \frac{76}{5} &= l \\ 15.2 &= l \end{aligned}$$

The length of the screen is 15.2 in.

$$\frac{19}{5} = \frac{w}{3}$$
$$19 \cdot 3 = 5 \cdot w$$
$$\frac{19 \cdot 3}{5} = w$$
$$\frac{57}{5} = w$$
$$11.4 = w$$

The width of the screen is 11.4 in.

Exercise Set 9.6

1. $V = l \cdot w \cdot h$
$V = 9 \text{ cm} \cdot 6 \text{ cm} \cdot 6 \text{ cm}$
$V = 54 \cdot 6 \text{ cm}^3$
$V = 324 \text{ cm}^3$

3. $V = l \cdot w \cdot h$
$V = 9 \text{ in.} \cdot 3 \text{ in.} \cdot 5 \text{ in.}$
$V = 27 \cdot 5 \text{ in}^3$
$V = 135 \text{ in}^3$

5. $V = l \cdot w \cdot h$
$V = 10 \text{ m} \cdot 5 \text{ m} \cdot 1.5 \text{ m}$
$V = 50 \cdot 1.5 \text{ m}^3$
$V = 75 \text{ m}^3$

7. $V = l \cdot w \cdot h$
$V = 6\frac{1}{2} \text{ yd} \cdot 5\frac{1}{2} \text{ yd} \cdot 10 \text{ yd}$
$V = \frac{13}{2} \cdot \frac{11}{2} \cdot 10 \text{ yd}^3$
$V = \frac{1430}{4} \text{ yd}^3 = \frac{2 \cdot 715}{2 \cdot 2} \text{ yd}^3$
$V = \frac{715}{2} \text{ yd}^3$
$V = 357\frac{1}{2} \text{ yd}^3$

9. $V = Bh = \pi \cdot r^2 \cdot h$
$\approx 3.14 \times 8 \text{ in.} \times 8 \text{ in.} \times 4 \text{ in.}$
$\approx 803.84 \text{ in}^3$

11. $V = Bh = \pi \cdot r^2 \cdot h$
$\approx 3.14 \times 4 \text{ cm} \times 4 \text{ cm} \times 7.5 \text{ cm}$
$\approx 376.8 \text{ cm}^3$

13. $V = Bh = \pi \cdot r^2 \cdot h$
$\approx \frac{22}{7} \times 210 \text{ yd} \times 210 \text{ yd} \times 300 \text{ yd}$
$\approx 41,580,000 \text{ yd}^3$

15. $V = \frac{4}{3} \cdot \pi \cdot r^3$
$\approx \frac{4}{3} \times 3.14 \times (100 \text{ in.})^3$
$\approx \frac{4 \times 3.14 \times 1,000,000 \text{ in}^3}{3}$
$\approx 4,186,666.\overline{6} \text{ in}^3, \text{ or } 4,186,666\frac{2}{3} \text{ in}^3$

17. $V = \frac{4}{3} \cdot \pi \cdot r^3$
$\approx \frac{4}{3} \times 3.14 \times (3.1 \text{ m})^3$
$\approx \frac{4 \times 3.14 \times 29.791 \text{ m}^3}{3}$
$\approx 124.725 \text{ m}^3$

19. $V = \frac{4}{3} \cdot \pi \cdot r^3$
$\approx \frac{4}{3} \times \frac{22}{7} \times (7 \text{ km})^3$
$\approx \frac{4 \times 22 \times 343 \text{ km}^3}{3 \times 7}$
$\approx 1437\frac{1}{3} \text{ km}^3$

21. $V = \frac{1}{3} \cdot \pi \cdot r^2 \cdot h$
$\approx \frac{1}{3} \times 3.14 \times 12 \text{ ft} \times 12 \text{ ft} \times 20 \text{ ft}$
$\approx 3014.4 \text{ ft}^3$

23. $V = \frac{1}{3} \cdot \pi \cdot r^2 \cdot h$
$\approx \frac{1}{3} \times \frac{22}{7} \times 1.4 \text{ cm} \times 1.4 \text{ cm} \times 12 \text{ cm}$
$\approx 24.64 \text{ cm}^3$

25. $1 \text{ L} = 1000 \text{ mL} = 1000 \text{ cm}^3$

These conversion relations appear in the text on page 551.

27. $53 \text{ L} = 53 \cancel{\text{L}} \times \frac{1000 \text{ mL}}{1 \cancel{\text{L}}}$
$= 53 \times 1000 \text{ mL}$
$= 53,000 \text{ mL}$

29. $29 \text{ mL} = 29 \cancel{\text{mL}} \times \frac{1 \text{ L}}{1000 \cancel{\text{mL}}}$
$= \frac{29}{1000} \text{ L}$
$= 0.029 \text{ L}$

31. $0.401 \text{ mL} = 0.401 \cancel{\text{mL}} \times \frac{1 \text{ L}}{1000 \cancel{\text{mL}}}$
$= \frac{0.401}{1000} \text{ L}$
$= 0.000401 \text{ L}$

33. $9347 \text{ mL} = 9347 \cancel{\text{mL}} \times \frac{1 \text{ L}}{1000 \cancel{\text{mL}}}$
$= \frac{9347}{1000} \text{ L}$
$= 9.347 \text{ L}$

35. $10 \text{ qt} = 10 \cancel{\text{qt}} \times \frac{32 \text{ oz}}{1 \cancel{\text{qt}}}$
$= 10 \cdot 32 \text{ oz}$
$= 320 \text{ oz}$

37. $24 \text{ oz} = 24 \cancel{\text{oz}} \times \frac{1 \text{ cup}}{8 \cancel{\text{oz}}}$
$= \frac{24}{8} \cdot 1 \text{ cup}$
$= 3 \text{ cups}$

39. $8 \text{ gal} = 8 \cancel{\text{gal}} \times \dfrac{4 \text{ qt}}{1 \cancel{\text{gal}}}$

$= 8 \cdot 4 \text{ qt}$

$= 32 \text{ qt}$

41. First we convert 3 gal to quarts:

$$\begin{aligned} 3 \text{ gal} &= 3 \cancel{\text{gal}} \times \frac{4 \text{ qt}}{1 \cancel{\text{gal}}} \\ &= 3 \cdot 4 \text{ qt} \\ &= 12 \text{ qt} \end{aligned}$$

Next we convert 12 qt to pints:

$$\begin{aligned} 12 \text{ qt} &= 12 \cancel{\text{qt}} \times \frac{2 \text{ pt}}{1 \cancel{\text{qt}}} \\ &= 12 \cdot 2 \text{ pt} \\ &= 24 \text{ pt} \end{aligned}$$

Finally we convert 24 pt to cups:

$$\begin{aligned} 24 \text{ pt} &= 24 \cancel{\text{pt}} \times \frac{2 \text{ cups}}{1 \cancel{\text{pt}}} \\ &= 24 \cdot 2 \text{ cups} \\ &= 48 \text{ cups} \end{aligned}$$

43. First we convert 15 pt to quarts:

$$\begin{aligned} 15 \text{ pt} &= 15 \cancel{\text{pt}} \times \frac{1 \text{ qt}}{2 \cancel{\text{pt}}} \\ &= \frac{15}{2} \cdot 1 \text{ qt} \\ &= 7.5 \text{ qt} \end{aligned}$$

Then we convert 7.5 qt to gallons:

$$\begin{aligned} 7.5 \text{ qt} &= 7.5 \cancel{\text{qt}} \times \frac{1 \text{ gal}}{4 \cancel{\text{qt}}} \\ &= \frac{7.5}{4} \cdot 1 \text{ gal} \\ &= 1.875 \text{ gal, or } 1\frac{7}{8} \text{ gal} \end{aligned}$$

45. First we convert 2.5 gal to quarts:

$$\begin{aligned} 2.5 \text{ gal} &= 2.5 \cancel{\text{gal}} \times \frac{4 \text{ qt}}{1 \cancel{\text{gal}}} \\ &= 2.5 \cdot 4 \text{ qt} \\ &= 10 \text{ qt} \end{aligned}$$

Then we convert 10 qt to pints:

$$\begin{aligned} 10 \text{ qt} &= 10 \cancel{\text{qt}} \times \frac{2 \text{ pt}}{1 \cancel{\text{qt}}} \\ &= 10 \cdot 2 \text{ pt} \\ &= 20 \text{ pt} \end{aligned}$$

47. $42.3 \text{ L} = 42.3 \cancel{\text{L}} \times \dfrac{1000 \text{ cm}^3}{1 \cancel{\text{L}}}$

$= 42.3 \times 1000 \text{ cm}^3$

$= 42{,}300 \text{ cm}^3$

49. We convert 0.5 L to milliliters:

$$\begin{aligned} 0.5 \text{ L} &= 0.5 \cancel{\text{L}} \times \frac{1000 \text{ mL}}{1 \cancel{\text{L}}} \\ &= 0.5 \times 1000 \text{ mL} \\ &= 500 \text{ mL} \end{aligned}$$

Dr. Carey ordered 500 mL.

51. First convert 3 L to milliliters:

$$\begin{aligned} 3 \text{ L} &= 3 \cancel{\text{L}} \cdot \frac{1000 \text{ mL}}{1 \cancel{\text{L}}} \\ &= 3 \times 1000 \text{ mL} \\ &= 3000 \text{ mL} \end{aligned}$$

Then translate to an equation.

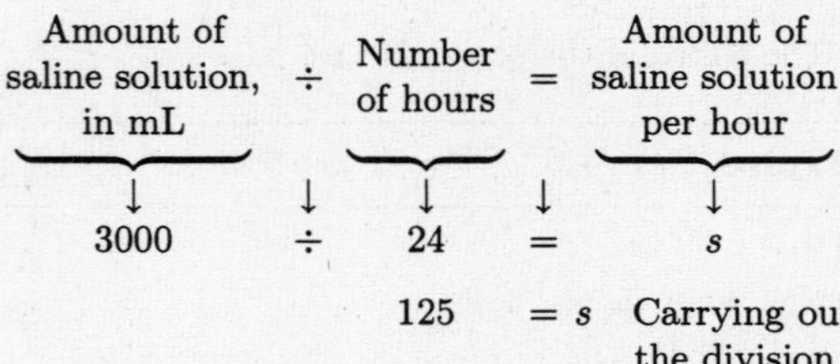

Thus, the patient should receive 125 mL per hour.

53. We must find the radius of the trash can in order to use the formula for the volume of a circular cylinder.

$$\begin{aligned} r &= \frac{d}{2} = \frac{0.7 \text{ yd}}{2} = 0.35 \text{ yd} \\ V &= Bh = \pi \cdot r^2 \cdot h \\ &\approx \frac{22}{7} \times 0.35 \text{ yd} \times 0.35 \text{ yd} \times 1.1\text{yd} \\ &\approx 0.4235 \text{ yd}^3 \end{aligned}$$

55. We must find the radius of the silo in order to use the formula for the volume of a circular cylinder.

$$\begin{aligned} r &= \frac{d}{2} = \frac{6 \text{ m}}{2} = 3 \text{ m} \\ V &= Bh = \pi \cdot r^2 \cdot h \\ &\approx 3.14 \times 3 \text{ m} \times 3 \text{ m} \times 13 \text{ m} \\ &\approx 367.38 \text{ m}^3 \end{aligned}$$

57. First we find the length of the cylindrical section of the submarine. The radius of each end is $\dfrac{d}{2} = \dfrac{8 \text{ m}}{2} = 4 \text{ m}$. Then the length of the cylindrical section is the total length less the radius of the half-sphere at each end, or $10 \text{ m} - 4 \text{ m} - 4 \text{ m}$, or 2 m.

Now we find the volume of the cylindrical portion. It is approximately

$$\begin{aligned} 3.14 \cdot (4 \text{ m})^2 \cdot 2 \text{ m} &= 3.14 \cdot 4 \text{ m} \cdot 4 \text{ m} \cdot 2 \text{ m} \\ &= 100.48 \text{ m}^3. \end{aligned}$$

Next we find the volume of the two ends. Note that together they make a sphere. The volume of the ends is approximately

$$\begin{aligned} \frac{4}{3} \cdot (3.14) \cdot (4 \text{ m})^3 &= \frac{4}{3} \cdot 3.14 \cdot 4 \text{ m} \cdot 4 \text{ m} \cdot 4 \text{ m} \\ &= 267.94\overline{6} \text{ m}^3. \end{aligned}$$

Then the total volume is $100.48 \text{ m}^3 + 267.94\overline{6} \text{ m}^3 = 368.42\overline{6} \text{ m}^3$.

(The result would be 367.7568 m^3 if we had expressed $\frac{4}{3}$ as 1.33 in the computation of the volume of the two ends.)

59. We use the formula for the volume of a sphere.

$$\begin{aligned} V &= \frac{4}{3}\cdot\pi\cdot r^3 \\ &\approx \frac{4}{3}\cdot\frac{22}{7}\cdot(500\text{ km})^3 \\ &\approx 523,809,523.8\text{ km}^3 \end{aligned}$$

61. First convert 32 oz to gallons:

$$\begin{aligned} 32\text{ oz} &= 32\,\cancel{\text{oz}}\cdot\frac{1\,\cancel{\text{pt}}}{16\,\cancel{\text{oz}}}\cdot\frac{1\,\cancel{\text{qt}}}{2\,\cancel{\text{pt}}}\cdot\frac{1\text{ gal}}{4\,\cancel{\text{qt}}} \\ &= \frac{32}{16\cdot 2\cdot 4}\times 1\text{ gal} = 0.25\text{ gal} \end{aligned}$$

Thus 0.25 gal per day are wasted by one person. We multiply to find how many gallons are wasted in a week (7 days) by one person:

$$7\times 0.25\text{ gal} = 1.75\text{ gal}$$

Next we multiply to find how many gallons are wasted in a month (30 days) by one person:

$$30\times 0.25\text{ gal} = 7.5\text{ gal}$$

We multiply again to find how many gallons are wasted in a year (365 days) by one person:

$$365\times 0.25\text{ gal} = 91.25\text{ gal}$$

To find how much water is wasted in this country in a year, we multiply 91.25 gal by 261 million:

$$\begin{aligned} 261,000,000\times 91.25\text{ gal} &= 23,816,250,000\text{ gal} \\ &\approx 24,000,000,000\text{ gal} \end{aligned}$$

63.

$$\begin{aligned} \text{Interest} &= (\text{Interest for 1 year})\times\frac{1}{2} \\ &= (8\%\times\$600)\times\frac{1}{2} \\ &= 0.08\times\$600\times\frac{1}{2} \\ &= \$48\times\frac{1}{2} \\ &= \$24 \end{aligned}$$

The interest is \$24.

65. Let c represent the cost of 12 pens. We translate to a proportion and solve.

$$\begin{array}{l} \text{Pens}\rightarrow \\ \text{Cost}\rightarrow \end{array}\frac{9}{\$8.01} = \frac{12}{c}\begin{array}{l}\leftarrow\text{Pens} \\ \leftarrow\text{Cost}\end{array}$$

$$\begin{aligned} 9\cdot c &= \$8.01\cdot 12 \quad \text{Equating cross-products} \\ c &= \frac{\$8.01\cdot 12}{9} \\ c &= \frac{\$96.12}{9} \\ c &= \$10.68 \end{aligned}$$

12 pens would cost \$10.68.

67.

$$\begin{aligned} -5y+3 &= -12y-4 \\ -5y+3+12y &= -12y-4+12y \quad \text{Adding } 12y \text{ on both sides} \\ 7y+3 &= -4 \\ 7y+3-3 &= -4-3 \quad \text{Subtracting 3 on both sides} \\ 7y &= -7 \\ \frac{7y}{7} &= \frac{-7}{7} \quad \text{Dividing by 7 on both sides} \\ y &= -1 \end{aligned}$$

The solution is -1.

69. 

71. Both cases are rectangular solids. First we find the volume of an audiocassette case:

$$7\text{ cm}\cdot 10.75\text{ cm}\cdot 1.5\text{ cm} = 112.875\text{ cm}^3$$

It holds $\frac{90\text{ min}}{112.875\text{ cm}^3}$, or about $0.797\frac{\text{min}}{\text{cm}^3}$.

Next we find the volume of a compact-disc case:

$$12.4\text{ cm}\cdot 14.1\text{ cm}\cdot 1\text{ cm} = 174.84\text{ cm}^3$$

It holds $\frac{50\text{ min}}{174.84\text{ cm}^3}$, or about $0.286\frac{\text{min}}{\text{cm}^3}$.

An audiocassette case holds more music per cubic centimeter.

73. Let r = the radius of the basketball. Then a side of the cube-shaped box that is just large enough to hold the basketball is the length of the diameter of the ball, or $2r$. The volume of the box is given by

$$\begin{aligned} V &= l\cdot w\cdot h \\ &= 2r\cdot 2r\cdot 2r = 8r^3. \end{aligned}$$

We find r^3 using the formula for the volume of a sphere. Since the volume of the ball is 2304π cm^3 and its radius is r, we have

$$\begin{aligned} V &= \frac{4}{3}\cdot\pi\cdot r^3 \\ 2304\pi\text{ cm}^3 &= \frac{4}{3}\cdot\pi\cdot r^3 \\ \frac{3}{4\pi}\cdot 2304\pi\text{ cm}^3 &= \frac{3}{4\pi}\cdot\frac{4}{3}\cdot\pi\cdot r^3 \\ 1728\text{ cm}^3 &= r^3 \end{aligned}$$

Then the volume of the box is
$8r^3 = 8\cdot 1728\text{ cm}^3 = 13,824\text{ cm}^3$.

Exercise Set 9.7

1. 1 T = 2000 lb

This conversion relation is given in the text on page 559.

3.

$$\begin{aligned} 6000\text{ lb} &= 6000\,\cancel{\text{lb}}\times\frac{1\text{ T}}{2000\,\cancel{\text{lb}}} \quad \text{Writing 1 with tons on the top and pounds on the bottom} \\ &= \frac{6000}{2000}\text{ T} \\ &= 3\text{ T} \end{aligned}$$

5. 5 lb $= 5 \times 1$ lb
$= 5 \times 16$ oz Substituting 16 oz for 1 lb
$= 80$ oz

7. 3.5 T $= 3.5 \times 1$ T
$= 3.5 \times 2000$ lb Substituting 2000 lb for 1 T
$= 7000$ lb

9. $4800 \text{ lb} = 4800 \not{\text{lb}} \times \frac{1 \text{ T}}{2000 \not{\text{lb}}}$ Writing 1 with tons on the top and pounds on the bottom
$= \frac{4800}{2000} \text{ T}$
$= 2.4$ T

11. $72 \text{ oz} = 72 \not{\text{oz}} \times \frac{1 \text{ lb}}{16 \not{\text{oz}}}$
$= \frac{72}{16} \text{ lb}$
$= 4.5$ lb

13. 1 kg = ______ g

Think: A kilogram is 1000 times the mass of a gram. Thus, we move the decimal point 3 places to the right.

1 1.000.

1 kg = 1000 g

15. 1 g = ______ kg

Think: It takes 1000 grams to have 1 kilogram. Thus, we move the decimal point 3 places to the left.

1 0.001.

1 g = 0.001 kg

17. 1 cg = ______ g

Think: It takes 100 centigrams to have 1 gram. Thus, we move the decimal point 2 places to the left.

1 0.01.

1 cg = 0.01 g

19. 1 g = ______ mg

Think: A gram is 1000 times the mass of a milligram. Thus, we move the decimal point 3 places to the right.

1 1.000.

1 g = 1000 mg

21. 1 g = ______ dg

Think: A gram is 10 times the mass of a decigram. Thus, we move the decimal point 1 place to the right.

1 1.0.

1 g = 10 dg

23. Complete: 234 kg = ______ g

Think: A kilogram is 1000 times the mass of a gram. Thus, we move the decimal point 3 places to the right.

234 234.000.

234 kg = 234,000 g

25. Complete: 6345 g = ______ kg

Think: It takes 1000 grams to have 1 kilogram. Thus, we move the decimal point 3 places to the left.

6345 6.345.

6345 g = 6.345 kg

27. 897 mg = ______ kg

Think: It takes 1,000,000 milligrams to have 1 kilogram. Thus, we move the decimal point 6 places to the left.

897 0.000897.

897 mg = 0.000897 kg

29. 7.32 kg = ______ g

Think: A kilogram is 1000 times the mass of a gram. Thus, we move the decimal point 3 places to the right.

7.32 7.320.

7.32 kg = 7320 g

31. Complete: 6780 g = ______ kg

Think: It takes 1000 grams to have 1 kilogram. Thus, we move the decimal point 3 places to the left.

6780 6.780.

6780 g = 6.78 kg

33. Complete: 69 mg = ______ cg

Think: It takes 10 milligrams to have 1 centigram. Thus, we move the decimal point 1 place to the left.

69 6.9.

69 mg = 6.9 cg

35. Complete: 8 kg = ______ cg

Think: A kilogram is 100,000 times the mass of a centigram. Thus, we move the decimal point 5 places to the right.

8 8.00000.

8 kg = 800,000 cg

37. 1 t = 1000 kg

This conversion relation is given in the text on page 560.

39. Complete: 3.4 cg = ______ dag

Think: It takes 1000 centigrams to have 1 dekagram. Thus, we move the decimal point 3 places to the left.

3.4 0.003.4

3.4 cg = 0.0034 dag

41. By laying a straightedge horizontally between the scales on page 561, we see that 178°F ≈ 80°C.

43. By laying a straightedge horizontally between the scales on page 561, we see that 140°F ≈ 60°C.

45. By laying a straightedge horizontally between the scales on page 561, we see that 68°F ≈ 20°C.

47. By laying a straightedge horizontally between the scales on page 561, we see that 10°F ≈ −10°C.

49. By laying a straightedge horizontally between the scales on page 561, we see that 86°C ≈ 190°F.

51. By laying a straightedge horizontally between the scales on page 561, we see that 58°C ≈ 140°F.

53. By laying a straightedge horizontally between the scales on page 561, we see that −10°C ≈ 10°F.

55. By laying a straightedge horizontally between the scales on page 561, we see that 5°C ≈ 40°F.

57.
$$\begin{aligned} F &= \frac{9}{5}\cdot C+32 \\ F &= \frac{9}{5}\cdot 25+32 \\ &= 45+32 \\ &= 77 \end{aligned}$$

Thus, 25°C = 77°F.

59.
$$\begin{aligned} F &= \frac{9}{5}\cdot C+32 \\ F &= \frac{9}{5}\cdot 40+32 \\ &= 72+32 \\ &= 104 \end{aligned}$$

Thus, 40°C = 104°F.

61.
$$\begin{aligned} F &= \frac{9}{5}\cdot C+32 \\ F &= \frac{9}{5}\cdot 3000+32 \\ &= 5400+32 \\ &= 5432 \end{aligned}$$

Thus, 3000°C = 5432°F.

63.
$$\begin{aligned} C &= \frac{5}{9}\cdot (F-32) \\ C &= \frac{5}{9}\cdot (86-32) \\ &= \frac{5}{9}\cdot 54 \\ &= 30 \end{aligned}$$

Thus, 86°F = 30°C.

65.
$$\begin{aligned} C &= \frac{5}{9}\cdot (F-32) \\ C &= \frac{5}{9}\cdot (131-32) \\ &= \frac{5}{9}\cdot 99 \\ &= 55 \end{aligned}$$

Thus, 131°F = 55°C.

67.
$$\begin{aligned} C &= \frac{5}{9}\cdot (F-32) \\ C &= \frac{5}{9}\cdot (98.6-32) \\ &= \frac{5}{9}\cdot 66.6 \\ &= 37 \end{aligned}$$

Thus, 98.6°F = 37°C.

69. 0.0043

a) Multiply by 100 to move the decimal point two places to the right. 0.00.43

b) Write a percent symbol. 0.43%

0.0043 = 0.43%

71. Let c represent the number of cans that can be bought for \$7.45. We translate to a proportion and solve.

$$\begin{array}{l} \text{Cans} \rightarrow \\ \text{Cost} \rightarrow \end{array} \frac{2}{\$1.49} = \frac{c}{\$7.45} \begin{array}{l} \leftarrow \text{Cans} \\ \leftarrow \text{Cost} \end{array}$$

$$\begin{aligned} 2\cdot \$7.45 &= \$1.49\cdot c \quad \text{Equating cross-products} \\ \frac{2\cdot \$7.45}{\$1.49} &= c \\ \frac{\$14.90}{\$1.49} &= c \\ 10 &= c \end{aligned}$$

You can buy 10 cans for \$7.45.

73.
$$\begin{aligned} -8.5x+7.9 &= 28.3-4x \\ -8.5x+7.9+4x &= 28.3-4x+4x \quad \text{Adding } 4x \text{ on both sides} \\ -4.5x+7.9 &= 28.3 \\ -4.5x+7.9-7.9 &= 28.3-7.9 \quad \text{Subtracting 7.9 on both sides} \\ -4.5x &= 20.4 \\ \frac{-4.5x}{-4.5} &= \frac{20.4}{-4.5} \quad \text{Dividing by } -4.5 \text{ on both sides} \\ x &= -4.5\overline{3} \end{aligned}$$

The solution is $-4.5\overline{3}$, or $-\frac{68}{15}$.

75. ◈

77.
$$\begin{aligned} 1\text{ lb} &= 1\,\cancel{\text{lb}}\times\frac{1\text{ kg}}{2.205\,\cancel{\text{lb}}} \\ &= \frac{1}{2.205}\text{ kg} \\ &\approx 0.4535\text{ kg} \end{aligned}$$

79. We substitute C for F in the formula. (We could equivalently substitute F for C.)

$$C = \frac{9}{5}\cdot C + 32$$
$$C - \frac{9}{5}\cdot C = \frac{9}{5}\cdot C + 32 - \frac{9}{5}\cdot C$$
$$\frac{5}{5}\cdot C - \frac{9}{5}\cdot C = 32$$
$$-\frac{4}{5}\cdot C = 32$$
$$-\frac{5}{4}\left(-\frac{4}{5}\cdot C\right) = -\frac{5}{4}\cdot 32$$
$$C = -\frac{5\cdot 32}{4} = -\frac{5\cdot 4\cdot 8}{4}$$
$$C = -40$$

The temperature $-40°$ is the same for both the Fahrenheit and Celsius scales.

81. We solve using a proportion. We let a = the amount of the liquid mixture required.

$$\begin{matrix}\text{Quinidine} \rightarrow \\ \text{Liquid} \rightarrow\end{matrix} \frac{80}{1} = \frac{200}{a} \begin{matrix}\leftarrow \text{Quinidine} \\ \leftarrow \text{Liquid}\end{matrix}$$

Solve: $80\cdot a = 1\cdot 200$ Equating cross-products

$$a = \frac{1\cdot 200}{80}$$
$$a = 2.5$$

Thus, 2.5 mL of the liquid mixture is required.

83. a) First we convert 18 grams to milligrams. (We could equivalently convert 90 milligrams to grams.) Think: A gram is 1000 times the mass of a milligram. Thus, we move the decimal point 3 places to the right.

18 18.000.

$18 \text{ g} = 18{,}000 \text{ mg}$

Now we divide to find the number of actuations in one inhaler:

$$18,000 \div 90 = 200$$

There are 200 actuations in one inhaler.

b) At 4 actuations per day, in one month (30 days) the student will need $4\cdot 30$, or 120 actuations. Then in 4 months the student will need $4\cdot 120$, or 480 actuations.

At 200 actuations per inhaler, the student will need $480 \div 200$, or 2.4 inhalers. Since 2 entire inhalers and 0.4 of a third are needed, the student should take 3 inhalers.

Chapter 10

Polynomials

Exercise Set 10.1

1. $(3x+2)+(-4x+3)=(3-4)x+(2+3)=-x+5$

3. $(-6x+2)+(x^2+x-3)=$
$x^2+(-6+1)x+(2-3)=x^2-5x-1$

5. $(x^2-9)+(x^2+9)=(1+1)x^2+(-9+9)=2x^2$

7. $(3x^2-5x+10)+(2x^2+8x-40)=$
$(3+2)x^2+(-5+8)x+(10-40)=5x^2+3x-30$

9. $(1+4x+6x^2+7x^3)+(5-4x+6x^2-7x^3)=$
$(1+5)+(4-4)x+(6+6)x^2+(7-7)x^3=$
$6+0x+12x^2+0x^3=6+12x^2$, or $12x^2+6$

11. $(9x^8-7x^4+2x^2+5)+(8x^7+4x^4-2x)=$
$9x^8+8x^7+(-7+4)x^4+2x^2-2x+5=$
$9x^8+8x^7-3x^4+2x^2-2x+5$

13. $(9a^3b^2+7a^2b^2+6ab^2)+(3a^3b^2-a^2b^2-5a^2b)=$
$(9+3)a^3b^2+(7-1)a^2b^2-5a^2b+6ab^2=$
$12a^3b^2+6a^2b^2-5a^2b+6ab^2$

15. $(-5x^4y^3+7x^3y^2-4xy^2)+(2x^3y^3-3x^3y^2-5xy)=$
$-5x^4y^3+2x^3y^3+(7-3)x^3y^2-4xy^2-5xy=$
$-5x^4y^3+2x^3y^3+4x^3y^2-4xy^2-5xy$

17. $(17.5abc^3+4.3a^2bc)+(-4.9a^2bc-5.2abc)=$
$17.5abc^3+(4.3-4.9)a^2bc-5.2abc=$
$17.5abc^3-0.6a^2bc-5.2abc$

19. Two equivalent expressions for the additive inverse of $-5x$ are

a) $-(-5x)$ and

b) $5x$. (Changing the sign)

21. Two equivalent expressions for the additive inverse of $-x^2+10x-2$ are

a) $-(-x^2+10x-2)$ and

b) $x^2-10x+2$. (Changing the sign of every term)

23. Two equivalent expressions for the additive inverse of $12x^4-3x^3+3$ are

a) $-(12x^4-3x^3+3)$ and

b) $-12x^4+3x^3-3$. (Changing the sign of every term)

25. We change the sign of every term inside parentheses.
$-(3x-7)=-3x+7$

27. We change the sign of every term inside parentheses.
$-(4x^2-3x+2)=-4x^2+3x-2$

29. We change the sign of every term inside parentheses.
$-\left(-4x^4+6x^2+\frac{3}{4}x-8\right)=4x^4-6x^2-\frac{3}{4}x+8$

31. $(3x+2)-(-4x+3)=3x+2+4x-3$
Changing the sign of every term inside parentheses
$=7x-1$

33. $(-6x+2)-(x^2+x-3)=-6x+2-x^2-x+3$
$=-x^2-7x+5$

35. $(7a^2+5a-9)-(2a^2+7)$
$=7a^2+5a-9-2a^2-7$
$=5a^2+5a-16$

37. $(6x^4+3x^3-1)-(4x^2-3x+3)$
$=6x^4+3x^3-1-4x^2+3x-3$
$=6x^4+3x^3-4x^2+3x-4$

39. $(1.2x^3+4.5x^2-3.8x)-(-3.4x^3-4.7x^2+23)$
$=1.2x^3+4.5x^2-3.8x+3.4x^3+4.7x^2-23$
$=4.6x^3+9.2x^2-3.8x-23$

41. $\left(\frac{5}{8}x^3-\frac{1}{4}x-\frac{1}{3}\right)-\left(-\frac{1}{8}x^3+\frac{1}{4}x-\frac{1}{3}\right)$
$=\frac{5}{8}x^3-\frac{1}{4}x-\frac{1}{3}+\frac{1}{8}x^3-\frac{1}{4}x+\frac{1}{3}$
$=\frac{6}{8}x^3-\frac{2}{4}x$
$=\frac{3}{4}x^3-\frac{1}{2}x$

43. $(5x^3y^3+8x^2y^2+7xy)-(3x^3y^3-2x^2y+3xy)$
$=5x^3y^3+8x^2y^2+7xy-3x^3y^3+2x^2y-3xy$
$=2x^3y^3+8x^2y^2+2x^2y+4xy$

45. $-5x+2=-5\cdot4+2=-20+2=-18$

47. $2x^2-5x+7=2\cdot4^2-5\cdot4+7=2\cdot16-20+7=$
$32-20+7=19$

49. $x^3-5x^2+x=4^3-5\cdot4^2+4=64-5\cdot16+4=$
$64-80+4=-12$

51. $3x+5=3(-1)+5=-3+5=2$

53. $x^2-2x+1=(-1)^2-2(-1)+1=1+2+1=4$

55. $-3x^3+7x^2-3x-2=$
$-3(-1)^3+7(-1)^2-3(-1)-2=$
$-3(-1)+7\cdot1+3-2=3+7+3-2=11$

57. We evaluate the polynomial for $a = 18$:

$$\begin{aligned}0.4a^2 - 40a + 1039 &= 0.4(18)^2 - 40(18) + 1039\\&= 0.4(324) - 40(18) + 1039\\&= 129.6 - 720 + 1039\\&= 448.6\end{aligned}$$

The daily number of accidents involving 18-year-old drivers is 448.6, or about 449.

59. We evaluate the polynomial for $t = 8$:

$$16t^2 = 16(8)^2 = 16 \cdot 64 = 1024$$

The cliff is 1024 ft high.

61. We evaluate the polynomial for $x = 75$:

$$\begin{aligned}280x - 0.4x^2 &= 280(75) - 0.4(75)^2\\&= 280(75) - 0.4(5625)\\&= 21,000 - 2250\\&= 18,750\end{aligned}$$

The total revenue from the sale of 75 stereos is \$18,750.

63. We evaluate the polynomial for $x = 500$:

$$\begin{aligned}5000 + 0.6x^2 &= 5000 + 0.6(500)^2\\&= 5000 + 0.6(250,000)\\&= 5000 + 150,000\\&= 155,000\end{aligned}$$

The total cost of producing 500 stereos is \$155,000.

65. $\dfrac{7 \text{ servings}}{10 \text{ lb}} = \dfrac{7}{10} \dfrac{\text{servings}}{\text{lb}}$, or $0.7 \dfrac{\text{servings}}{\text{lb}}$

67. The sales tax is

$$\underbrace{\text{Sales tax rate}} \times \underbrace{\text{Purchase price}}$$

$$\downarrow \qquad\qquad \downarrow$$

$$6.25\% \quad \times \quad \$2190,$$

or 0.0625×2190, or 136.875. Thus, the tax is \$136.88.

69.

$$\begin{aligned}A &= \pi \cdot r \cdot r\\&\approx 3.14 \cdot 20 \text{ cm} \cdot 20 \text{ cm}\\&\approx 1256 \text{ cm}^2\end{aligned}$$

71. 

73. a) We evaluate the polynomial for $t = 1$:

$$\begin{aligned}&0.5t^4 + 3.45t^3 - 96.65t^2 + 347.7t\\&= 0.5(1)^4 + 3.45(1)^3 - 96.65(1)^2 + 347.7(1)\\&= 0.5 + 3.45 - 96.65 + 347.7\\&= 255\end{aligned}$$

There will be 255 mg of ibuprofen in the bloodstream 1 hr after 400 mg is swallowed.

b) We evaluate the polynomial for $t = 2$:

$$\begin{aligned}&0.5t^4 + 3.45t^3 - 96.65t^2 + 347.7t\\&= 0.5(2)^4 + 3.45(2)^3 - 96.65(2)^2 + 347.7(2)\\&= 0.5(16) + 3.45(8) - 96.65(4) + 347.7(2)\\&= 8 + 27.6 - 386.6 + 695.4\\&= 344.4\end{aligned}$$

There will be 344.4 mg of ibuprofen in the bloodstream 2 hr after 400 mg is swallowed.

c) We evaluate the polynomial for $t = 6$:

$$\begin{aligned}&0.5t^4 + 3.45t^3 - 96.65t^2 + 347.7t\\&= 0.5(6)^4 + 3.45(6)^3 - 96.65(6)^2 + 347.7(6)\\&= 0.5(1296) + 3.45(216) - 96.65(36) + 347.7(6)\\&= 648 + 745.2 - 3479.4 + 2086.2\\&= 0\end{aligned}$$

There will be 0 mg of ibuprofen in the bloodstream 6 hr after 400 mg is swallowed.

75.

$$\begin{aligned}&(7y^2 - 5y + 6) - (3y^2 + 8y - 12) + (8y^2 - 10y + 3)\\&= 7y^2 - 5y + 6 - 3y^2 - 8y + 12 + 8y^2 - 10y + 3\\&= 12y^2 - 23y + 21\end{aligned}$$

77.

$$\begin{aligned}&(-y^4 - 7y^3 + y^2) + (-2y^4 + 5y - 2) - (-6y^3 + y^2)\\&= -y^4 - 7y^3 + y^2 - 2y^4 + 5y - 2 + 6y^3 - y^2\\&= -3y^4 - y^3 + 5y - 2\end{aligned}$$

79. $9x^4 + 3x^4 = 12x^4$, $5x^2 + 0 = 5x^2$, $-7x^3 + 2x^3 = -5x^3$, and $-9 + (-7) = -16$ so we have $9x^4 + \underline{3x^4} + 5x^2 - 7x^3 + \underline{2x^3} - 9 + (-7) = 12x^4 - 5x^3 + 5x^2 - 16$

Exercise Set 10.2

1. $(5a)(9a) = (5 \cdot 9)(a \cdot a)$
$= 45a^2$

3. $(-4x)(15x) = (-4 \cdot 15)(x \cdot x)$
$= -60x^2$

5. $(7x^5)(4x^3) = (7 \cdot 4)(x^5 \cdot x^3)$
$= 28x^8$

7. $(-0.1x^6)(0.2x^4) = (-0.1 \cdot 0.2)(x^6 \cdot x^4)$
$= -0.02x^{10}$

9. $(5x^2y^3)(7x^4y^9) = (5 \cdot 7)(x^2 \cdot x^4)(y^3 \cdot y^9)$
$= 35x^6y^{12}$

11. $(4a^3b^4c^2)(3a^5b^4) = (4 \cdot 3)(a^3 \cdot a^5)(b^4 \cdot b^4)(c^2)$
$= 12a^8b^8c^2$

13. $(3x^2)(-4x^3)(2x^6) = (3)(-4)(2)(x^2 \cdot x^3 \cdot x^6) = -24x^{11}$

15. $3x(-x + 5) = 3x(-x) + 3x \cdot 5$
$= -3x^2 + 15x$

17. $-3x(x - 1) = -3x(x) - 3x(-1)$
$= -3x^2 + 3x$

19. $x^2(x^3 + 1) = x^2 \cdot x^3 + x^2 \cdot 1$
$= x^5 + x^2$

21. $3x(2x^2 - 6x + 1) = 3x \cdot 2x^2 + 3x(-6x) + 3x \cdot 1$
$= 6x^3 - 18x^2 + 3x$

23. $4xy(3x^2 + 2y) = 4xy \cdot 3x^2 + 4xy \cdot 2y$
$= 12x^3y + 8xy^2$

25. $3a^2b(4a^5b^2 - 3a^2b^2) = 3a^2b \cdot 4a^5b^2 - 3a^2b \cdot 3a^2b^2$
$= 12a^7b^3 - 9a^4b^3$

27. $2x + 6 = 2 \cdot x + 2 \cdot 3$
$= 2(x + 3)$

29. $7a - 21 = 7 \cdot a - 7 \cdot 3$
$= 7(a - 3)$

31. $14x + 21y = 7 \cdot 2x + 7 \cdot 3y$
$= 7(2x + 3y)$

33. $9a - 27b + 81 = 9 \cdot a - 9 \cdot 3b + 9 \cdot 9$
$= 9(a - 3b + 9)$

35. $24 - 6m = 6 \cdot 4 - 6 \cdot m$
$= 6(4 - m)$

37. $-16 - 8x + 40y = -8 \cdot 2 - 8 \cdot x - 8(-5y)$
$= -8(2 + x - 5y)$

39. $3x^5 + 3x = 3x \cdot x^4 + 3x \cdot 1$
$= 3x(x^4 + 1)$

41. $a^3 - 8a^2 = a^2 \cdot a - a^2 \cdot 8$
$= a^2(a - 8)$

43. $8x^3 - 6x^2 + 2x = 2x \cdot 4x^2 - 2x \cdot 3x + 2x \cdot 1$
$= 2x(4x^2 - 3x + 1)$

45. $12a^4b^3 + 18a^5b^2 = 6a^4b^2 \cdot 2b + 6a^4b^2 \cdot 3a$
$= 6a^4b^2(2b + 3a)$

47. ***Familiarize***. The tennis court is a rectangle that measures 27 ft by 78 ft. We make a drawing.

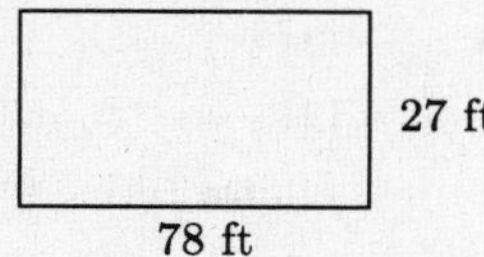

Translate. We substitute in the formula for the area of a rectangle.

$P = 2 \cdot (l + w) = 2 \cdot (78 \text{ ft} + 27 \text{ ft})$

Solve. We carry out the calculation.

$P = 2 \cdot (78 \text{ ft} + 27 \text{ ft})$
$= 2 \cdot 105 \text{ ft}$
$= 210 \text{ ft}$

Check. We repeat the calculation.

State. The perimeter of the tennis court is 210 ft.

49. The sales tax is

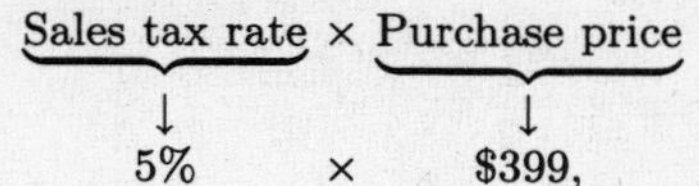

or 0.05 × 399, or 19.95. Thus, the sales tax is $19.95.

The total price is given by the purchase price plus the sales tax:

$399 + $19.95 = $419.85

To check, note that the total price is the purchase price plus 5% of the purchase price, or 105% of the purchase price. Since 1.05 × 399 = 418.95, we have a check. The total price is $418.95.

51. 

53. $391x^{391} + 299x^{299} = 23x^{299} \cdot 17x^{92} + 23x^{299} \cdot 13$
$= 23x^{299}(17x^{92} + 13)$

55. $84a^7b^9c^{11} - 42a^8b^6c^{10} + 49a^9b^7c^8$
$= 7a^7b^6c^8 \cdot 12b^3c^3 - 7a^7b^6c^8 \cdot 6ac^2 + 7a^7b^6c^8 \cdot 7a^2b$
$= 7a^7b^6c^8(12b^3c^3 - 6ac^2 + 7a^2b)$

Exercise Set 10.3

1. $(x + 6)(x + 3) = (x + 6)x + (x + 6)3$
$= x \cdot x + 6 \cdot x + x \cdot 3 + 6 \cdot 3$
$= x^2 + 6x + 3x + 18$
$= x^2 + 9x + 18$

3. $(x + 5)(x - 2) = (x + 5)x + (x + 5)(-2)$
$= x \cdot x + 5 \cdot x + x(-2) + 5(-2)$
$= x^2 + 5x - 2x - 10$
$= x^2 + 3x - 10$

5. $(x - 4)(x - 3) = (x - 4)x + (x - 4)(-3)$
$= x \cdot x - 4 \cdot x + x(-3) - 4(-3)$
$= x^2 - 4x - 3x + 12$
$= x^2 - 7x + 12$

7. $(x + 3)(x - 3) = (x + 3)x + (x + 3)(-3)$
$= x \cdot x + 3 \cdot x + x(-3) + 3(-3)$
$= x^2 + 3x - 3x - 9$
$= x^2 - 9$

9. $(5 - x)(5 - 2x) = (5 - x)5 + (5 - x)(-2x)$
$= 5 \cdot 5 - x \cdot 5 + 5(-2x) - x(-2x)$
$= 25 - 5x - 10x + 2x^2$
$= 25 - 15x + 2x^2$

11. $(2x + 5)(2x + 5) = (2x + 5)2x + (2x + 5)5$
$= 2x \cdot 2x + 5 \cdot 2x + 2x \cdot 5 + 5 \cdot 5$
$= 4x^2 + 10x + 10x + 25$
$= 4x^2 + 20x + 25$

13. $\left(x-\frac{5}{2}\right)\left(x+\frac{2}{5}\right)=\left(x-\frac{5}{2}\right)x+\left(x-\frac{5}{2}\right)\frac{2}{5}$

$$=x\cdot x-\frac{5}{2}\cdot x+x\cdot\frac{2}{5}-\frac{5}{2}\cdot\frac{2}{5}$$
$$=x^2-\frac{5}{2}x+\frac{2}{5}x-1$$
$$=x^2-\frac{25}{10}x+\frac{4}{10}x-1$$
$$=x^2-\frac{21}{10}x-1$$

15. $(x^2+x+1)(x-1)$

$$=(x^2+x+1)x+(x^2+x+1)(-1)$$
$$=x^2\cdot x+x\cdot x+1\cdot x+x^2(-1)+x(-1)+1(-1)$$
$$=x^3+x^2+x-x^2-x-1$$
$$=x^3-1$$

17. $(2x+1)(2x^2+6x+1)$

$$=2x(2x^2+6x+1)+1(2x^2+6x+1)$$
$$=2x\cdot 2x^2+2x\cdot 6x+2x\cdot 1+1\cdot 2x^2+1\cdot 6x+1\cdot 1$$
$$=4x^3+12x^2+2x+2x^2+6x+1$$
$$=4x^3+14x^2+8x+1$$

19. $(y^2-3)(3y^2-6y+2)$

$$=y^2(3y^2-6y+2)-3(3y^2-6y+2)$$
$$=y^2\cdot 3y^2+y^2(-6y)+y^2\cdot 2-3\cdot 3y^2-3(-6y)-3\cdot 2$$
$$=3y^4-6y^3+2y^2-9y^2+18y-6$$
$$=3y^4-6y^3-7y^2+18y-6$$

21. $(x^3+x^2)(x^3+x^2-x)$

$$=x^3(x^3+x^2-x)+x^2(x^3+x^2-x)$$
$$=x^3\cdot x^3+x^3\cdot x^2+x^3(-x)+x^2\cdot x^3+x^2\cdot x^2+x^2(-x)$$
$$=x^6+x^5-x^4+x^5+x^4-x^3$$
$$=x^6+2x^5-x^3$$

23.

$$\begin{array}{rl}
2t^2-\ t-4 & \\
3t^2+2t-1 & \\
\hline
-\ 2t^2+\ t+4 & \text{Multiplying by } -1\\
4t^3-\ 2t^2-8t\quad & \text{Multiplying by } 2t\\
6t^4-3t^3-12t^2\qquad & \text{Multiplying by } 3t^2\\
\hline
6t^4+\ t^3-16t^2-7t+4 &
\end{array}$$

25.

$$\begin{array}{rl}
x\quad -x^3\quad +x^5 & \\
-1\ +x^2\quad +x^4 & \text{Rewriting in ascending order}\\
\hline
x^5-x^7+x^9 & \text{Multiplying by } x^4\\
x^3-x^5+x^7\quad & \text{Multiplying by } x^2\\
-x+\ x^3-x^5\qquad & \text{Multiplying by } -1\\
\hline
-x+2x^3-x^5\qquad +x^9 &
\end{array}$$

27. ***Familiarize.*** We label the width of the sidewalk s.

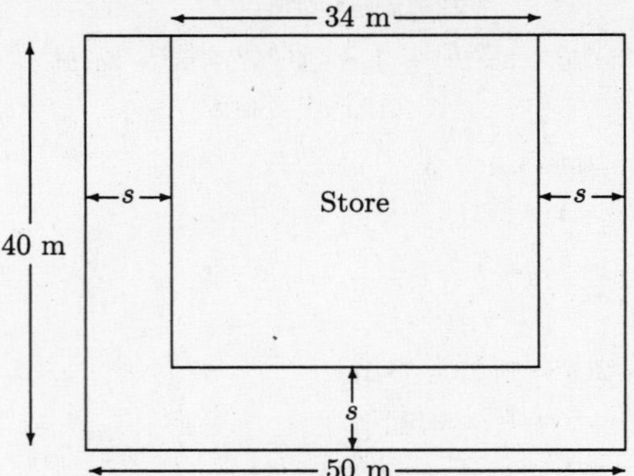

Translate. This is a two-step problem. We first find the width of the sidewalk. Then we find the area of the sidewalk.

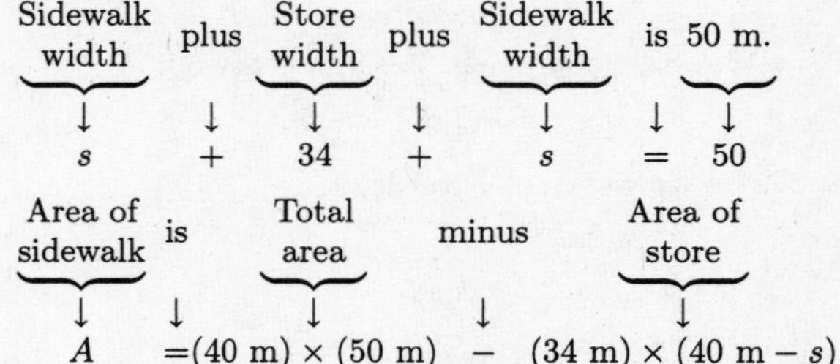

Solve. We solve the first equation.

$s+34+s=50$

$2s+34=50$ Collecting like terms

$2s=16$ Subtracting 34 on both sides

$s=8$ Dividing by 2 on both sides

Thus the width of the sidewalk is 8 m.

Then we solve the second equation.

$A=(40\text{ m})\times(50\text{ m})-(34\text{ m})\times(40\text{ m}-s)$

$A=(40\text{ m})\times(50\text{ m})-(34\text{ m})\times(40\text{ m}-8\text{ m})$

Substituting 8 m for s

$A=(40\text{ m})\times(50\text{ m})-(34\text{ m})\times(32\text{ m})$

$A=(40\times 50\times\text{ m}\times\text{ m})-(34\times 32\times\text{ m}\times\text{ m})$

$A=2000\text{ m}^2-1088\text{ m}^2$

$A=912\text{ m}^2$

Check. We repeat the calculations.

State. The area of the sidewalk is 912 m^2.

29. We rephrase the question and translate.

What percent of 162 is 108?

$n \times 162 = 108$

To solve, we divide on both sides by 162 and convert to percent notation.

$$n \times 162 = 108$$
$$n = \frac{108}{162}$$
$$n = 0.66\overline{6}$$
$$n = 66.\overline{6}\%, \text{ or } 66\frac{2}{3}\%$$

31. 

33. For $x = 5$;

$(x^2 + 2x - 3)(x^2 + 4)$
$= (5^2 + 2 \cdot 5 - 3)(5^2 + 4)$
$= (25 + 10 - 3)(25 + 4)$
$= 32 \cdot 29$
$= 928$

$x^4 + 2x^3 + x^2 + 8x - 12$
$= 5^4 + 2 \cdot 5^3 + 5^2 + 8 \cdot 5 - 12$
$= 625 + 250 + 25 + 40 - 12$
$= 928$

The expressions have the same value for $x = 5$.

For $x = 3.5$;

$(x^2 + 2x - 3)(x^2 + 4)$
$= [(3.5)^2 + 2 \cdot 3.5 - 3][(3.5)^2 + 4]$
$= (12.25 + 7 - 3)(12.25 + 4)$
$= 16.25 \cdot 16.25$
$= 264.0625$

$x^4 + 2x^3 + x^2 + 8x - 12$
$= (3.5)^4 + 2(3.5)^3 + (3.5)^2 + 8(3.5) - 12$
$= 150.0625 + 85.75 + 12.25 + 28 - 12$
$= 264.0625$

The expressions have the same value for $x = 3.5$.

For $x = -1.2$;

$(x^2 + 2x - 3)(x^2 + 4)$
$= [(-1.2)^2 + 2(-1.2) - 3][(-1.2)^2 + 4]$
$= (1.44 - 2.4 - 3)(1.44 + 4)$
$= -3.96(5.44)$
$= -21.5424$

$x^4 + 2x^3 + x^2 + 8x - 12$
$= (-1.2)^4 + 2(-1.2)^3 + (-1.2)^2 + 8(-1.2) - 12$
$= 2.0736 - 3.456 + 1.44 - 9.6 - 12$
$= -21.5424$

The expressions have the same value for $x = -1.2$.

35. 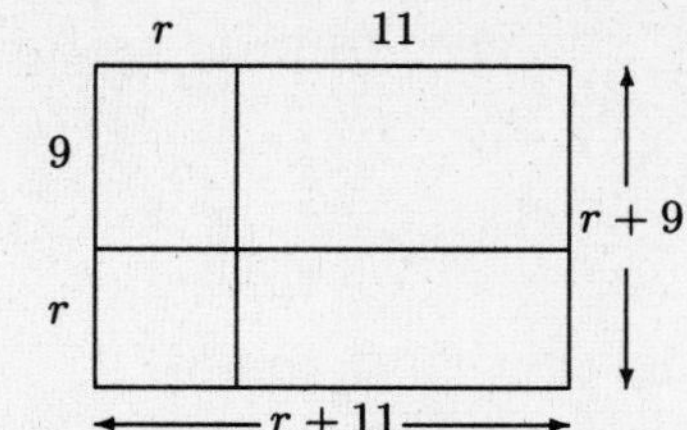

a) The length and width of the figure can be expressed as $r+11$ and $r+9$, respectively. The perimeter of this figure (a rectangle) is given by $P = 2\cdot(l+w) = 2\cdot(r+11+r+9) = 2 \cdot (2r + 20) = 4r + 40$.

b) The area of this figure is given by $A = l \cdot w = (r + 11)(r + 9) = (r+11)r+(r+11)9 = r^2+11r+9r+99 = r^2 + 20r + 99$.

37. The shaded area is the area of the large rectangle less the area of the small rectangle:

$$4t(21t + 8) - 2t(3t - 4) = 84t^2 + 32t - 6t^2 + 8t$$
$$= 78t^2 + 40t$$

Exercise Set 10.4

1. $9^0 = 1$ ($b^0 = 1$, for any nonzero number b.)

3. $3.14^0 = 1$ ($b^0 = 1$, for any nonzero number b.)

5. $(-19.57)^1 = -19.57$

7. $(-5.43)^0 = 1$

9. $x^0 = 1,\ x \neq 0$

11. $(3x - 17)^0 = (3 \cdot 10 - 17)^0$ Substituting
$= (30 - 17)^0$ Multiplying
$= 13^0$
$= 1$

13. $(5x - 3)^1 = (5 \cdot 4 - 3)^1$ Substituting
$= (20 - 3)^1$ Multiplying
$= 17^1$
$= 17$

15. $(4m - 19)^0 = (4 \cdot 3 - 19)^0$
$= (12 - 19)^0$
$= (-7)^0$
$= 1$

17. $3x^0 + 4 = 3(-2)^0 + 4$
$= 3 \cdot 1 + 4$
$= 3 + 4$
$= 7$

19. $(3x)^0 + 4 = [3(-2)]^0 + 4$
$= (-6)^0 + 4$
$= 1 + 4$
$= 5$

21. $(5-3x^0)^1 = (5-3\cdot 19^0)^1$
$= (5-3\cdot 1)^1$
$= (5-3)^1$
$= 2^1$
$= 2$

23. $3^{-2} = \dfrac{1}{3^2} = \dfrac{1}{9}$

25. $10^{-4} = \dfrac{1}{10^4} = \dfrac{1}{10,000}$

27. $a^{-3} = \dfrac{1}{a^3}$

29. $(-5)^{-2} = \dfrac{1}{(-5)^2} = \dfrac{1}{25}$

31. $3x^{-7} = 3\left(\dfrac{1}{x^7}\right) = \dfrac{3}{x^7}$

33. $\dfrac{x}{y^{-4}} = xy^4$ Instead of dividing by y^{-4}, multiply by y^4.

35. $\dfrac{a^3}{b^{-4}} = a^3b^4$ Instead of dividing by b^{-4}, multiply by b^4.

37. $-7a^{-9} = -7\left(\dfrac{1}{a^9}\right) = \dfrac{-7}{a^9}$, or $-\dfrac{7}{a^9}$

39. $\dfrac{m^3n}{p^{-6}} = m^3np^6$

41. $\dfrac{x^3y^{-7}}{z^{-9}} = x^3y^{-7}z^9 = \dfrac{x^3z^9}{y^7}$

43. $\dfrac{1}{4^3} = 4^{-3}$

45. $\dfrac{9}{x^3} = 9x^{-3}$, or $\dfrac{x^{-3}}{9^{-1}}$

47. $x^{-2}\cdot x = x^{-2+1} = x^{-1} = \dfrac{1}{x}$

49. $x^4\cdot x^{-4} = x^{4+(-4)} = x^0 = 1$, assuming $x \neq 0$

51. $x^{-7}\cdot x^{-6} = x^{-7+(-6)} = x^{-13} = \dfrac{1}{x^{13}}$

53. $(3a^2b^{-7})(2ab^9)$
$= 3\cdot 2\cdot a^2\cdot a\cdot b^{-7}\cdot b^9$ Using the commutative and associative laws
$= 6a^{2+1}b^{-7+9}$ Using the product rule
$= 6a^3b^2$

55. $(-2x^{-3}y^8)(3xy^{-2})$
$= -2\cdot 3\cdot x^{-3}\cdot x\cdot y^8\cdot y^{-2}$ Using the commutative and associative laws
$= -6x^{-3+1}y^{8+(-2)}$ Using the product rule
$= -6x^{-2}y^6$
$= -6\left(\dfrac{1}{x^2}\right)y^6$
$= -\dfrac{6y^6}{x^2}$

57. $(3a^{-4}bc^2)(2a^{-2}b^{-5}c)$
$= 3\cdot 2\cdot a^{-4}\cdot a^{-2}\cdot b\cdot b^{-5}\cdot c^2\cdot c$
$= 6a^{-4+(-2)}b^{1+(-5)}c^{2+1}$
$= 6a^{-6}b^{-4}c^3$
$= \dfrac{6c^3}{a^6b^4}$

59. $\dfrac{450 \text{ km}}{9 \text{ hr}} = 50\dfrac{\text{km}}{\text{hr}}$, or 50 km/h

61. *Rephrase:* Sales tax is what percent of purchase price?

Translate: $27.60 = n \times 460$

To solve the equation, we divide by 460 on both sides and convert to percent notation.

$\dfrac{27.60}{460} = n$
$0.06 = n$
$6\% = n$

The sales tax rate is 6%.

63.

65. Use a calculator.
For $x = -4$:
$\dfrac{3^x}{3^{x-1}} = \dfrac{3^{-4}}{3^{-4-1}} = \dfrac{3^{-4}}{3^{-5}} = 3$
For $x = -40$:
$\dfrac{3^x}{3^{x-1}} = \dfrac{3^{-40}}{3^{-40-1}} = \dfrac{3^{-40}}{3^{-41}} = 3$

67.

69. $a^{5k} \div a^{3k} = a^{5k-3k} = a^{2k}$